Quieting the Boom

The Shaped Sonic Boom Demonstrator and the Quest for Quiet Supersonic Flight

D0839865

Lawrence R. Benson

Library of Congress Cataloging-in-Publication Data

Benson, Lawrence R.
  Quieting the boom : the shaped sonic boom demonstrator and the quest
for quiet supersonic flight / Lawrence R. Benson.
     pages cm
  Includes bibliographical references and index.
  1. Sonic boom--Research--United States--History. 2. Noise control--
Research--United States--History. 3. Supersonic planes--Research--United
States--History. 4. High-speed aeronautics--Research--United States--
History. 5. Aerodynamics, Supersonic--Research--United States--History. I.
Title.
  TL574.S55B36 2013
  629.132'304--dc23
                                    2013004829

This publication is available as a free download at
http://www.nasa.gov/ebooks.

ISBN 978-1-62683-004-2

Table of Contents

Preface and Acknowledgments

Quieting the Boom: The Shaped Sonic Boom Demonstrator and the Quest for Quiet Supersonic Flight follows up on a case study I was privileged to write in early 2009, "Softening the Sonic Boom: 50 Years of NASA Research." That relatively short survey was published in volume I of *NASA's Contributions to Aeronautics* (NASA SP-2010-570). Although I was previously familiar with aviation history, initially, I was hesitant to take on what seemed to be such an esoteric and highly technical topic. Thankfully, some informative references on related supersonic programs of the past were already available to help get me started, most notably Erik M. Conway's *High-Speed Dreams: NASA and the Technopolitics of Supersonic Transportation, 1945–1999*, which is cited frequently in "Softening the Sonic Boom" and the first four chapters that follow.

After a 2-year hiatus, I resumed sonic boom research in March 2011 on this new book. I greatly appreciate the opportunity afforded me to write about this fascinating subject by the eminent aviation historian Dr. Richard P. Hallion, editor of *NASA's Contributions to Aeronautics* and the new National Aeronautics and Space Administration (NASA) book series of which this one is a part. While expanding, updating, and, hopefully, improving on my previous account, this book's primary focus is on the breakthrough achieved by the Northrop Grumman Corporation (NGC) and a diverse team of Government and industry partners who proved that aircraft can be designed to significantly lower the strength of sonic booms.

My research into primary sources benefited immeasurably from the help given to me during visits to the Dryden Flight Research Center (DFRC), Edwards, CA, in December 2008 and April 2011 and additional telephone and e-mail communications with DFRC personnel. Librarian Dr. Karl A. Bender introduced me to NASA's superb scientific and technical information resources and, assisted by Freddy Lockarno, helped me collect numerous essential documents. Aviation historian Peter W. Merlin found other sources for me in Dryden's archival collection. Edward A. Haering, Dryden's principal sonic boom investigator, provided valuable source materials, answered questions, and reviewed the chapters covering his projects. Fellow engineer Timothy R. Moes and test pilots James W. Smolka and Dana D. Purifoy helped with additional information and reviewed sections of the draft. Dryden's superb online image gallery provided many of the photographs, and Tony R. Landis provided me with others from his files. Also at Edwards Air Force Base, CA, the long-time Flight Test Center Historian Dr. James O. Young provided me with additional photos and later reviewed and made helpful comments on the first chapter.

Writing a credible history about this subject would have been impossible without extensive help from two of the world's top sonic boom experts—Domenic J. Maglieri of Eagle Aeronautics and Dr. Kenneth J. Plotkin of Wyle Laboratories—both of whose names are scattered throughout the text and notes. In addition to reviewing and commenting on drafts of the chapters, they answered numerous questions and offered valuable suggestions both over the phone and via the Internet. The second and third chapters also benefitted from being reviewed by one of the pioneers of sonic boom theory, professor Albert R. George of Cornell University. Dr. Christine M. Darden and Peter G. Coen, who in turn led NASA's sonic boom research efforts after the mid-1970s, also provided information and reviewed my original study. Peter Coen, who managed the Shaped Supersonic Boom Experiment and has been the principal investigator for NASA's Supersonics Project since 2006, continued to help on this book. His comments, corrections, and guidance were critical to completing chapter 9. Because this final chapter attempts to bring various facets of the as yet unfinished quest for civilian supersonic flight up to date through 2011, its discussion of recent events should be considered provisional pending the availability of more information and the historical perspective that will only come in future years.

For transforming my manuscript into both a printed and electronic book, the author is indebted to the staff of the Communication Support Services Center at Headquarters NASA, especially the careful proofreading and editorial suggestions of Benjamin Weinstein and the attractive design of the final product by Christopher Yates. Because many of the historically significant diagrams, drawings, and other illustrations found in the source materials were of rather poor visual quality, I greatly appreciate the efforts of Chris and his graphics team in trying to make these figures as legible as possible.

The Shaped Sonic Boom Demonstrator culminated four decades of study and research on mitigating the strength of sonic booms. Writing chapters 5 through 8—which cover the origins, design, fabrication, and flight testing of this innovative modification of an F-5E fighter plane—was made possible through the auspices of the Northrop Grumman Corporation. As is evident in the text and notes, the NGC's Joseph W. Pawlowski, David H. Graham, M.L. "Roy" Martin, and Charles W. Boccadoro generously provided informative interviews, detailed documentation, and valuable comments, and they patiently answered numerous questions as I researched, wrote, and coordinated these chapters. I would also like to thank Robert A. "Robbie" Cowart of Gulfstream Aerospace Corporation for his review of the section in chapter 9 on the company's Quiet Spike invention, which subsequently demonstrated another means of mitigating sonic booms.

This book is intended to be a general history of sonic boom research, emphasizing the people and organizations that have contributed, and not a technical study of the science and engineering involved. Any errors in fact or interpretation are those of the author. For more detailed information, interested readers may refer to primary sources referenced in the notes, many of which are available online from the NASA Technical Reports Server; through the American Institute of Aeronautics and Astronautics (AIAA); and in the other professional journals, periodicals, and books cited. I relied on graphs, charts, and drawings in some of these and other original sources for many of the figures presented in this book. Their quality and legibility was often not up to the visual standards desired in current NASA publications, but I believe including them was necessary to illustrate the evolution of knowledge about sonic booms and the related advances in aeronautical design and technology described in the text. In the near future, NASA will also publish what will undoubtedly become the definitive reference work on all aspects of sonic boom science and technology, tentatively titled *Sonic Boom: A Compilation and Review of Six Decades of Research*. Among its coauthors are some of the aforementioned experts who have been so helpful to me.

LAWRENCE R. BENSON
Albuquerque, NM
January 14, 2012

The F-5 Shaped Sonic Boom Demonstrator, piloted by Roy Martin, arriving over Palmdale, California, on July 29, 2003. (Mike Bryan)

A Pelican Flies Cross Country

On a hot and humid July day in 2003, a pair of small supersonic jet airplanes took off together from Cecil Field, a former naval air station on the eastern edge of Jacksonville, FL. Even though the Northrop Corporation had built both planes based on a common design, it was hard at first glance to tell that the two aircraft flying side by side were so closely related. One was a sleek T-38 Talon, a two-seat aircraft that has served as the U.S. Air Force's (USAF's) advanced trainer since the early 1960s. The other was originally an F-5E Tiger II, one of more than 2,000 Northrop F-5s that had equipped air forces around the world with a low-cost, high-performance combat and reconnaissance aircraft. Because of the F-5E's agility and compact size, the U.S. military adopted it as an aggressor aircraft to hone the skills of its own fighter pilots. Both planes attested to the competence of Northrop's design teams. Of all of the many supersonic jets developed for the Air Force and U.S. Navy in the 1950s, the T-38 and F-5 are the only ones still in general use.

Although on loan from the Navy's aggressor training squadron, this par-ticular F-5E no longer looked much like a fighter jet. With what appeared to be a pouch hanging under its chin, the aircraft somewhat resembled an overgrown pelican. In addition to lettering identifying Northrop Grumman Integrated Systems, its white fuselage was decorated with sharply angled blue and red pinstripes along with emblems containing the acronyms "NASA" and "DARPA" while its tail bore an oval logo with the letters "QSP."

After gaining altitude, this odd couple turned west toward their ultimate destination of Palmdale, CA. Roy Martin, the chief test pilot at the Northrop Grumman Corporation's facility in Palmdale, was at the controls of the F-5. Mike Bryan, a Boeing test pilot from Seattle, WA, was flying the T-38. Despite its enlarged nose section, the F-5 no longer had navigational equipment except for a hand-held Global Positioning System (GPS) receiver in the cockpit, so Martin had to stay near the T-38. Their first refueling stop was Huntsville, AL, home of NASA's Marshall Space Flight Center. Next, it was on to the vast Tinker Air Force Base (AFB) in Oklahoma City, OK, where Martin and Bryan

The T-38 and modified F-5E together at Cecil Field, FL, before flying to California. (NGC)

spent the night. The next morning, they stopped to refuel in Roswell, NM, at what had once been Walker Air Force Base, and then they stopped at the former Williams AFB, southeast of Phoenix, AZ, before flying on to California.

At each of these stops, the planes attracted the attention of flight-line personnel and others nearby, most of whom could recognize the strange white jet as some kind of F-5. But many of them still had questions. What's with the big nose? Why is Boeing helping a Northrop Grumman pilot fly across the country? What do those jagged red and blue stripes signify? And why all the various logos?

Unlike a lot of projects sponsored by the Defense Advanced Research Projects Agency (DARPA), the one involving this F-5 was not classified. So the two pilots were happy to explain that the F-5 had been modified for a test to be conducted with the help of NASA called the Shaped Sonic Boom Demonstration (SSBD). It was part of a DARPA program called Quiet Supersonic Platform (QSP). Although Northrop Grumman had won the SSBD contract, Boeing and some other rival companies were also participating and would share in the data collected. The goal of the SSBD was to do something that had never before been accomplished: prove that it was possible to reduce the strength of sonic booms. This experimentation was being undertaken in the hope that civilian airplanes could someday fly at supersonic speeds without disturbing people below. The SSBD team was going to perform this demonstration at Edwards AFB in the very same airspace where supersonic flight had its birth more than 50 years earlier. The pinstripes on the F-5 illustrated the shape of the pressure waves that the team had expected a normal F-5 and the modified F-5 to register on special recording devices.

Since jet aircraft had been making sonic booms for more than half a century, why had this not been done already? Why had the United States, which could land men on the Moon and invent the Internet, never been able to build a supersonic airliner or business jet? With all the advances in science and technology, what is so complicated about the sonic boom that has so far defied solution? Would the SSBD be a significant step toward finding a solution? The rest of this book will attempt to answer these questions.

Bell XS-1 photographed on its way to becoming the first aircraft to exceed Mach 1 in level flight. (USAF)

CHAPTER 1

Making Shock Waves

The Proliferation and Testing of Sonic Booms

Humans have long been familiar with—and often frightened by—natural sonic booms in the form of thunder. Caused by sudden spikes in pressure when strokes of lightning instantaneously heat surrounding columns of air molecules, the sound of thunder varies from low-pitched rumbles to earsplitting bangs, depending on distance. Perhaps the most awesome of sonic booms, heard only rarely, are generated when certain large meteors speed through the atmosphere at just the right trajectories and altitudes. On an infinitesimally smaller scale, the first acoustical shock waves produced by human invention were the modest cracking noises caused by the snapping of a whip. With the perfection of high-powered explosive propellants in the latter half of the 19th century, the muzzle velocity of bullets and artillery shells began to routinely exceed the speed of sound (about 1,125 feet, or 343 meters, per second at sea level), producing noises that firearms specialists call ballistic cracks. These sharp noises result when air molecules cannot be pushed aside fast enough by objects moving at or faster than the speed of sound. The molecules are thereby compressed together into shock waves that surge away from the speeding object at a higher pressure than the atmosphere through which they travel.

Exceeding Mach 1

In the 1870s, an Austrian physicist-philosopher, Ernst Mach, was the first to explain this sonic phenomenon, which he later displayed visually in the 1880s with cleverly made schlieren photographs (from the German word for streaks) showing shadow-like images of the acoustic shock waves formed by high-velocity projectiles. The specific speed of sound, he also determined, depends on the medium through which an object passes. In the gases that make up Earth's atmosphere, sound waves move faster in warm temperatures than cold. In 1929, a Swiss scientist named this variable the "Mach number" in his honor.[1] At 68 degrees Fahrenheit (F) at sea level in dry air, the speed of sound is about 768 miles per hour (mph), or 1,236 kilometers per hour (kph); but at above 40,000 feet at about −70 °F, it is only about 659 mph, or 1,060 kph.[2] The shock waves produced by passing bullets and artillery rounds would be among the cacophony of fearsome sounds heard by millions of soldiers during the two world wars.[3]

On Friday evening, September 8, 1944, a sudden explosion blew out a large crater in Stavely Road, west of London. The first German V-2 ballistic missile aimed at England had announced its arrival. "After the explosion came a double thunderclap caused by the sonic boom catching up with the fallen rocket."[4] For the next 7 months, millions of people would hear this new sound (which became known by the British as a sonic bang) from more than 3,000 V-2s launched at Britain as well as liberated portions of France, Belgium, and the Netherlands. These shock waves would always arrive too late to warn any of those unfortunate enough to be near the missiles' points of impact.[5] After the end of World War II, these strange noises faded into memory until the arrival of supersonic, turbojet-powered fighter planes in the 1950s.

Jet airplanes were preceded in supersonic flight by experimental aircraft powered by rocket engines at Muroc Army Airfield in California's Mojave Desert. Here, a small team of Air Force, National Advisory Committee on Aeronautics (NACA), and contractor personnel were secretly exploring the still largely unknown territory of transonic and supersonic flight. On October 14, 1947, more than 40,000 feet over the desert east of Rogers Dry Lake, Capt. Chuck Yeager broke the fabled sound barrier by flying at Mach 1.06 in a Bell XS-1 (later redesignated the X-1).[6]

Despite hazy memories and legend perpetuated by the best-selling book and hit movie *The Right Stuff*, the shock waves from Yeager's little (31-foot-long) airplane did not reach the ground with a loud boom on that historic day.[7] He flew only 20 seconds at what is considered aerodynamically just a transonic speed (less than Mach 1.15).[8] Yeager's memoir states that NACA personnel in a tracking van heard a sound like distant thunder.[9] This could only have resulted if there had been a strong tailwind and a layer of cooler air near the surface.[10] However, a record of atmospheric soundings from Bakersfield, CA, indicates that a headwind of about 60 knots was more likely.[11] Before long, however, the stronger acoustical signatures generated by faster-flying X-1s and other supersonic aircraft became a familiar sound at and around the isolated air base.

A Swelling Drumbeat of Sonic Booms

In November 1949, the NACA designated its growing detachment at Muroc as the High-Speed Flight Research Station (HSFRS). This came 1 month before the Air Force renamed the installation Edwards Air Force Base after Capt. Glen Edwards, who had perished in the crash of a Northrop YB-49 flying wing the year before.[12] By the early 1950s, the barren dry lakes and jagged mountains around Edwards reverberated with the sonic booms of experimental and prototype aircraft, as did other flight-test locations in the United States, United Kingdom, and Soviet Union. Scientists and engineers were familiar with the ballistic waves of axisymmetric projectiles such as artillery shells (shapes referred to scientifically as "bodies of revolution").[13] This was a reason the fuselage of the XS-1 was shaped like a 50-caliber bullet, which was known to be stable at three times the speed of sound. But these new acoustic phenomena—many of which featured the double-boom sound—hinted that they were more complex than conventional ballistic waves. In late 1952, the editors of the world's oldest aeronautical weekly stated with some hyperbole that "the 'supersonic bang' phenomenon, if only by reason of its sudden incidence and the enormous public interest it has aroused, is probably the most spectacular and puzzling occurrence in the history of aerodynamics."[14]

A perceptive English graduate student, Gerald B. Whitham, accurately analyzed the abrupt rise in air pressure upon arrival of a supersonic object's bow wave, followed by a more gradual but deeper fall in pressure for a fraction of a second, and then a recompression with the passing of the vehicle's tail wave.[15] As shown in a simplified fashion in the upper left corner of figure 1-1, this can be illustrated graphically by an elongated capital N (the solid line) transecting a horizontal axis. The plot of this line represents ambient air pressure during a second or less of elapsed time along a short path, the distance of which depends on the length and altitude of the supersonic body. For Americans, the pressure change (Δp) is usually expressed in pounds per square foot (psf—also abbreviated as lb/ft^2). The shock waves left behind by an aircraft flying faster than Mach 1 on a straight and level course will spread out in a cone-shaped pattern with the sector intersecting the ground being heard as a sonic boom.[16] Even though the shock waves are being left behind by the speeding aircraft (where the pilot and any passengers do not hear their sound), the cone's shock waves are moving forward in the form of acoustic rays, the nature of which would become the subject of future research.

Because a supersonic aircraft is much longer than an artillery shell, the human ear can detect a double boom (or double bang) if the shock

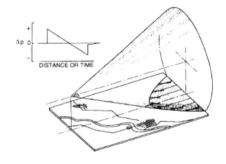

Figure 1-1. Sonic boom signature and shock cone. (NASA)

wave from its tail area arrives a tenth of a second or more after the shock wave from its front portion (sometimes compared to the bow wave of a boat). In some respects, all the sound heard from a subsonic jet airplane as it approaches, flies overhead, and fades away is concentrated in this fraction of a second. Gerald Whitham was first to systematically examine these multiple shock waves, which he called the F-function, generated by the complex nonaxisymmetrical configurations applicable to airplanes.[17] The U.S. Air Force conducted its earliest sonic boom flight test at Edwards AFB in 1956 with an F-100 making in-flight measurements of another F-100 flying at Mach 1.05. Although the instrumentation used was relatively simple, the test found the decay of bow shock pressure and other results to be consistent with Whitham's theory.[18] Later in-flight pressure measurements near supersonic aircraft as well as wind tunnel experiments would reveal a jagged sawtooth pattern that only at much greater distances consolidated into the form of the double-boom-creating N-wave signature. (It would later be determined that the sound waves resulting from the abruptness of the pressure spikes, rather than the overall pressure differential from the ambient level, is what people hear as noise.)

The number of these double booms at Edwards AFB multiplied in the latter half of the 1950s as the Air Force Flight Test Center (AFFTC) at Edwards (assisted by the HSFRS) began putting a new generation of Air Force jet fighters and interceptors of various configurations, known as the Century Series, through their paces. The remarkably rapid advance in aviation technology and priorities of the Cold War arms race is evident in the sequence of their first flights at Edwards (most as prototypes): the YF-100 Super Sabre, May 1953; YF-102 Delta Dagger, October 1953; XF-104 Starfighter, February 1954; F-101 Voodoo, September 1954; YF-105 Thunderchief, October 1955; and F-106 Delta Dart, December 1956.[19]

With the sparse population living in California's Mojave Desert region at the time, disturbances caused by the flight tests of new jet aircraft were not yet an issue, but the Air Force had already become concerned about their future impact. In November 1954, for example, its Aeronautical Research Laboratory at Wright-Patterson AFB, OH, submitted a study to the Air Force Board of top generals on early findings regarding the still-puzzling nature of sonic booms. Although concluding that low-flying aircraft flying at supersonic speeds could cause considerable damage, the report hopefully predicted the possibility of supersonic flight without booms at altitudes over 35,000 feet.[20]

As the latest Air Force and Navy fighters went into full production and began flying from bases throughout the Nation, more of the American public was exposed to jet noise for the first time. This included the thunderclap-like thuds characteristic of sonic booms—often accompanied by rattling windowpanes. Under certain conditions, as the U.S. armed services and British

Six Century Series fighters and interceptors at Edwards AFB. Clockwise from top right: F-100, F-101, F-102, F-104, F-105, F-106. (USAF)

Royal Air Force (RAF) had learned, even maneuvers below Mach 1 (e.g., accelerations, dives, and turns) could generate and focus transonic shock waves in such a manner as to cause localized but powerful sonic booms.[21] Indeed, residents of Southern California began hearing such booms in the late 1940s when North American Aviation was flight testing its new F-86 Sabre. The first civilian claim against the USAF for sonic boom damage was apparently filed at Eglin AFB, FL, in 1951, when only subsonic jet fighters were assigned there.[22]

Much of the rapid progress in supersonic flight was made possible by the famous area rule, discovered in 1951 by the legendary NACA engineer Richard Whitcomb. He subsequently showed how to reduce transonic drag by smoothing out the shock waves that developed along where the wings joined the fuselage of an aircraft approaching Mach 1. The basic solution was to reduce the cross section of the fuselage between the wings so that the combined cross section of the fuselage and wings would gradually increase and decrease in an ideal streamlined shape, allowing jet planes to achieve supersonic speeds much more easily.[23] (Hence the pinched coke-bottle-shaped fuselages of the F-102, F-104, F-105, and F-106 in the photograph.)

Adolf Busemann, a colleague at Langley Research Center in Hampton, VA (the NACA's oldest and largest lab), who had inspired Whitcomb to think of the area rule, also made major contributions to sonic boom theory. For his work as an engineer in Germany before World War II, Busemann is considered the father of supersonic aerodynamics; he is remembered especially for the concept of a swept wing, which he introduced in 1935. By, at the same time, exploring how to eliminate wave drag caused by aircraft volume, he could also be considered as the godfather of sonic boom minimization, even at a time when supersonic flight was only a distant dream. He later contributed more directly to the development of sonic boom theory in a 1955 paper titled "The Relation Between Minimizing Drag and Noise at Supersonic Speeds," which showed the importance of lift effects in creating sonic booms.[24]

Both the area rule and findings about lift during supersonic flight were critical to understanding the effects of wing-body configurations on sonic booms. In 1958, another bright, young English mathematician, Frank Walkden, showed in a series of insightful equations how the lift effect of airplane wings could magnify the strength of sonic booms more than previously estimated.[25] The pioneering work of Whitham and Walkden laid the foundation for the systematic scientific study of sonic booms, especially the formation of N-wave signatures, and provided many of the algorithms and assumptions used in planning future flight tests and wind tunnel experiments.[26]

Sonic boom claims against the U.S. Air Force first became statistically significant in 1957, reflecting the branch's growing inventory of Century fighters and the types of maneuvers they sometimes performed. Such actions could focus the acoustical rays projected by shock waves into what became called super booms. (It was found that these powerful but localized booms had a U-shaped signature with the tail shock as well as that from the nose of the airplane being above ambient air pressure—unlike N-wave signatures, in which the tail shock causes pressure to return only to the ambient level.) Most claims

Convair B-58 Hustler, the first airplane capable of sustained supersonic flight and a major contributor to early sonic boom research. (USAF)

involved broken windows or cracked plaster, but some were truly bizarre, such as the death of pets or the insanity of livestock. In addition to these formal claims, Air Force bases, local police switchboards, and other agencies received an uncounted number of phone calls about booms, ranging from merely inquisitive to seriously irate.[27] Complaints from constituents brought the issue to the attention of the U.S. Congress.[28] Between 1956 and 1968, some 38,831 claims were submitted to the Air Force, which approved 14,006 in whole or in part—65 percent for broken glass, 21 percent for cracked plaster (usually already weakened), 8 percent for fallen objects, and 6 percent for other reasons.[29]

The military's problem with sonic boom complaints peaked in the 1960s. One reason for this peak was the sheer number of fighter-type aircraft stationed around the Nation (more than three times as many as today). Secondly, many of these aircraft had air defense as their mission. This often meant flying at high speed over populated areas to train for defending cities and other key targets from aerial attack, sometimes practicing against Strategic Air Command (SAC) bombers. The North American Air Defense Command (NORAD) conducted the largest such air exercises in history—Skyshield I in 1960, Skyshield II in

1961, and Skyshield III in 1962. The Federal Aviation Agency (FAA) shut down all civilian air traffic while numerous flights of SAC bombers (augmented by some Vulcans from the RAF) attacked from the Arctic and off the coasts. Hundreds of NORAD's interceptors flying thousands of sorties created a sporadic drum beat of sonic booms as F-101, F-102, F-104, and F-106 pilots lit their afterburners in pursuit of the intruders. (About three quarters of the bombers were able to reach their targets, a result kept secret for 35 years.)[30]

Although most fighters and interceptors deployed in the 1960s could readily fly faster than sound, they could only do so for a short distance because of the rapid fuel consumption of jet-engine afterburners. Thus their sonic boom "carpets" (the term used to describe the areas affected on the surface) were relatively short. However, one supersonic American warplane that became operational in 1960 was designed to fly faster than Mach 2 for more than a thousand miles, laying down a continuous sonic boom carpet all the way.

This innovative but troublesome aircraft was SAC's new Convair-built B-58 Hustler medium bomber. On March 5, 1962, the Air Force showed off the long-range speed of the B-58 by flying one from Los Angeles to New York in just over 2 hours at an average pace of 1,215 mph (despite having to slow down for an aerial refueling over Kansas). After another refueling over the Atlantic, the same Hustler "outraced the sun" (i.e., flew faster than Earth's rotation) back to Los Angeles with one more refueling, completing the record-breaking round trip at an average speed of 1,044 mph.[31] The accompanying photo shows one flying over a populated area (presumably at a subsonic speed).

Figure 1-2. Air Force pamphlet for sonic boom claim investigators. (USAF)

Capable of sustained Mach 2+ speeds, the four-engine, delta-winged Hustler (weighing up to 163,000 pounds) helped demonstrate the feasibility of a supersonic civilian transport. But the B-58's performance revealed at least one troubling omen. Almost wherever it flew supersonic over populated areas, the bomber left sonic boom complaints and claims in its wake. Indeed, on its record-shattering flight of March 1962, flown mostly at an altitude of 50,000 feet (except when coming down to 30,000 feet for refueling), "the jet dragged a sonic boom 20 to 40 miles wide back and forth across the country—frightening residents, breaking windows, cracking plaster, and setting dogs to barking."[32] As indicated by figure 1-2, the B-58 (despite its small numbers) became a symbol for sonic boom complaints.

Most Americans, especially during times of increased Cold War tensions, tolerated occasional disruptions that were justified by national defense. But how would they react to constantly repeated sonic booms generated by civilian transports? Could a practical passenger-carrying supersonic airplane be designed to minimize its sonic signature enough to be acceptable to people below? Attempts to resolve these two questions occupy the remainder of this book.

Preparing for an American Supersonic Transport

After its formation in 1958, the National Aeronautics and Space Administration—in keeping with the reason for its creation—began devoting the lion's share of its growing resources to the Nation's new civilian space programs. Yet 1958 also marked the start of a new program in the time-honored aviation mission that the new Agency inherited from the NACA. This new task was to help foster an advanced passenger plane that would fly at rates at least twice the speed of sound, a concept initially named the Supersonic Commercial Air Transport (SCAT).

By the late 1950s, the rapid pace of aeronautical progress—with new turbojet-powered airliners flying twice as fast and high as the propeller-driven transports they were replacing—promised even higher speeds in coming years. At the same time, the perceived challenge to America's technological superiority implied by the Soviet Union's early space triumphs inspired a willingness to pursue ambitious new aerospace ventures. One of these was the Supersonic Commercial Air Transport. This program was further motivated by proposals being made in Britain and France to build a supersonic airliner, a type of airplane that was expected to dominate the future of mid- and long-range commercial aviation.[33]

Because of economic and political factors, developing such an aircraft became more than a purely technological challenge—and thus proved to be in some ways even more problematic than sending astronauts to the Moon. One of the major barriers to producing a supersonic transport involved the still-mysterious phenomenon of how atmospheric shock waves were generated by supersonic flight. Studying sonic booms and learning how to control them became a specialized and enduring field of NASA research for the next five decades.

The recently established Federal Aviation Agency became the major advocate within the U.S. Government for a supersonic transport, with key personnel at three of NACA's former laboratories eager to help in this challenging new program. The Langley Research Center (the NACA's oldest and largest lab), and the Ames Research Center at Moffett Field in Sunnyvale, CA, both had airframe-design expertise and facilities while the Lewis Research Center in Cleveland, OH (later renamed in honor of astronaut and Senator John H. Glenn), specialized in the kind of advanced propulsion technologies needed for supersonic cruise.

The strategy for developing SCAT depended heavily on leveraging technologies being developed for another Air Force bomber—one much larger, faster, and more advanced than the B-58. This would be the revolutionary B-70, designed to cruise several thousand miles at speeds of Mach 3. NACA experts had been helping the Air Force plan this giant intercontinental bomber since the mid-1950s (with aerodynamicist Alfred Eggers of the Ames Laboratory conceiving the innovative design for it to ride partially on compression lift created by its own supersonic shock waves). North American Aviation won the B-70 contract in 1958, but the projected expense of the program and advances in missile technology led President Dwight D. Eisenhower to cancel all but one prototype in 1959. The administration of President John F. Kennedy eventually approved production of two XB-70As. Their main purpose would be to serve as Mach 3 test beds for what was becoming known simply as the SST, for "Supersonic Transport."

NASA continued to refer to specific design concepts for the SST using the older acronym for Supersonic Commercial Air Transport. As shown by the 25 SCAT configurations in figure 1-3, the designers were very creative in exploring a wide variety of shapes for fuselages, wings, tails, engine nacelles, and other surfaces.[34] By early 1963, about 40 concepts had been narrowed down to three Langley designs contributed by well-known Langley aerodynamicists, such as Richard Whitcomb and A. Warner Robins (SCAT-4, SCAT-15, and SCAT-16), and one by a team from Ames (SCAT-17). These became the baselines for subsequent industry studies and proposals.

SCAT-16, with variable sweep wings for improved low-speed handling, and SCAT-17, with a front canard and rear delta wing (based to some extent on the XB-70), were judged as the most promising concepts.[35] But they were still only notional designs. In the judgment of two of the Langley Research Center's supersonic experts, William Alford and Cornelius Driver, "It was obvious that ways would have to be found to obtain further major increases in flight efficiency. It was clear that major attention would have to be paid to the sonic boom, which was shown to have become a dominant factor in aircraft design and operation."[36] Whitcomb later withdrew from working on the SST because of his judgment that it would never be a practical commercial aircraft.[37] Meanwhile, NASA continued research on SCAT concepts 15 through 19.[38]

Even though Department of Defense (DOD) resources—especially the Air Force's—would be important in supporting SST development, the aerospace industry made it clear that direct Federal funding and assistance would be essential. Thus, research and development (R&D) of the SST became a split responsibility between the Federal Aviation Agency and the National Aeronautics and

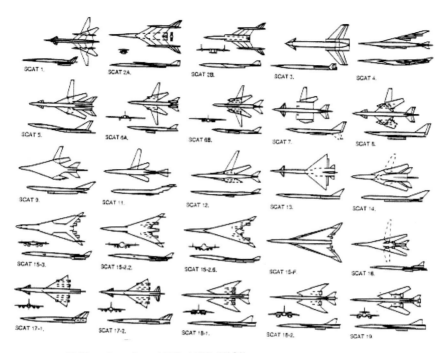

Figure 1-3. SCAT configurations, 1959–1966. (NASA)

Space Administration—with NASA conducting and sponsoring the supersonic research and the FAA overseeing the SST's overall development. The first two leaders of the FAA, retired Lt. Gen. Elwood R. "Pete" Quesada (1958–61) and Najeeb E. Halaby (1961–65), were both staunch proponents of producing an SST, as to a slightly lesser degree was retired Gen. William F. "Bozo" McKee (1965–68). As heads of an independent agency that reported directly to the President, they were at the same level as NASA Administrators T. Keith Glennan (1958–61) and James Webb (1961–68). The FAA and NASA administrators, together with Secretary of Defense Robert McNamara (somewhat of a skeptic on the SST program), provided interagency oversight and composed the Presidential Advisory Committee (PAC) for the SST established in April 1964. This arrangement lasted until 1967, when the Federal Aviation Agency became the Federal Aviation Administration under the new U.S. Department of Transportation (DOT), whose secretary became responsible for the program.[39]

Much of NASA's SST-related research involved advancing the state of the art in such technologies as propulsion, fuels, materials, and aerodynamics. The last item included designing airframe configurations for sustained supersonic cruise at high altitudes, suitable subsonic maneuvering in civilian air-traffic patterns at lower altitudes, safe takeoffs and landings at commercial airports, and acceptable noise levels—to include the still-puzzling matter of sonic booms.

Although the NACA, especially at Langley and Ames, had been doing research on supersonic flight since World War II, none of its technical reports (and only one conference paper) published through 1957 dealt directly with sonic booms.[40] That situation began to change when Langley's long-time manager and advocate of supersonic programs, John P. Stack, formalized the SCAT venture in 1958. During the next year, three Langley employees, whose names would become well-known in the field of sonic boom research, began publishing NASA's first scientific papers on the subject. These were Harry W. Carlson, a versatile supersonic aerodynamicist; Harvey H. Hubbard, chief of the Acoustics and Noise Control Division; and Domenic J. Maglieri, a young engineer who became Hubbard's top sonic boom specialist. Carlson would tend to focus on wind tunnel experiments and sonic boom theory while the two other men specialized in planning and monitoring field tests and recording and analyzing the data collected. Within NASA, the Langley Research Center continued to be the focal point for sonic boom studies throughout the 1960s with the Flight Research Center (FRC) at Edwards AFB increasingly conducting most supersonic tests, often with Air Force support.[41] (The "High Speed" prefix was dropped from the FRC's name in 1959 to indicate a broadening of its experimental activities.)

These research activities began to proliferate under the new pro-SST Kennedy administration in 1961. After the president formally approved development of the supersonic transport in June 1963, sonic boom research really took off. Langley's experts, augmented by NASA contractors and grantees, published 26 papers on sonic booms just 3 years later, with Ames also conducting related research.[42]

Dealing with the sonic boom demanded a multifaceted approach: (1) performing flight tests to better quantify the fluid dynamics and atmospheric physics involved in generating and propagating shock waves as well as their physical effects on structures and people; (2) conducting community surveys to gather public opinion data from sample populations exposed to booms; (3) building and using acoustic simulators to further evaluate human and structural responses in controlled settings; (4) performing field studies of possible effects on animals, both domestic and wild; (5) evaluating shock waves from various aerodynamic configurations in wind tunnel experiments; and (6) analyzing flight-test and wind tunnel data to refine theoretical constructs and create mathematical models for lower-boom aircraft designs. The remainder of this chapter focuses on the first four activities with the final two to be the main subject of the next chapter.

Langley Research Center's first sonic boom testers, Harvey Hubbard and Domenic Maglieri. (NASA)

Early Flight Testing

The systematic sonic boom testing that NASA began in 1958 would expand exponentially the heretofore largely theoretical and anecdotal knowledge about sonic booms with a vast amount of empirical, real-world data. The new information would make possible increasingly sophisticated experiments and provide feedback for checking and refining theories and mathematical models. Because of the priority bestowed on sonic boom research by the SST program and the numerous types of aircraft then available for creating booms (including some faster than any flying today), the data and findings from the tests conducted in the 1960s are still of great value in the 21st century.[43]

The Langley Research Center (often referred to as "NASA Langley") served as the Agency's team leader for supersonic research. Langley's acoustics specialists conducted NASA's initial sonic boom tests in 1958 and 1959 at the Wallops Island Station on Virginia's isolated Delmarva Peninsula. During the first year, they used six sorties by NASA F-100 and F-101 fighters, flying at speeds between Mach 1.1 and Mach 1.4 and altitudes from 25,000 feet to 45,000 feet, to make the first good ground recordings and measurements of sonic booms generated in steady and level flights (the kind of profile a future airliner would mostly fly). Observers judged some of the booms above 1.0 psf to be objectionable, likening them to nearby thunder, and a sample plate glass window was cracked by one plane flying at 25,000 feet. The 1959 test measured shock waves from 26 flights of a Chance Vought F8U-3 (a highly advanced prototype based on the Navy's supersonic Crusader fighter) at speeds up to Mach 2 and altitudes up to 60,000 feet. A much larger B-58 from Edwards AFB also made two supersonic passes at 41,000 feet. Boom intensities from these higher altitudes seemed to be tolerable to observers, with negligible increases in measured overpressures between Mach 1.4 and Mach 2.0 (showing that loudness of sonic booms is based more on aircraft size, altitude, and factors other than extreme speeds). The human response results were, however, very preliminary.[44]

In July 1960, NASA and the Air Force conducted Project Little Boom at a bombing range north of Nellis AFB, NV, to measure the effects on structures and people of extremely powerful sonic booms (which the Air Force thought might have some military value). F-104 and F-105 fighters flew slightly over the speed of sound (Mach 1.09 to Mach 1.2) at altitudes down to 50 feet above ground level. There were more than 50 incidents of sample windows being broken at 20 psf to 100 psf but only a few possible breakages below 20 psf, and there was no physical or psychological harm to volunteers exposed to overpressures as high as 120 psf.[45] At Indian Springs, NV, Air Force fighters flew supersonically over an instrumented C-47 transport from Edwards, both while the aircraft was in the process of landing and while it was on the ground.

Chance Vought's F8U-3 prototype at Wallops Island. (NASA)

Despite 120-psf overpressures, the aircraft was only very slightly damaged when on the ground and there were no problems while it was in flight.[46]

Air Force fighters once again would test powerful sonic booms during 1965 in remote mountain and desert terrain near Tonopah, NV. This was where a special military testing organization from Sandia Base, NM, called Joint Task Force II, was evaluating the low-level penetration capabilities of various fighter aircraft for the Joint Chiefs of Staff. To learn more about possible effects from this kind of low-level training in remote areas, the USAF Aerospace Medical Division's Biomedical Laboratory observed and analyzed the responses of people, structures, and animals to strong sonic booms. As in other tests, the damage to buildings (many in poor condition to begin with) consisted of cracked plaster, items falling from shelves, and broken windows. In some cases, glass fragments were propelled up to 12 feet—a condition not recorded in previous testing. Some campers near the so-called starting gates to the three low-level corridors used for testing also experienced damage, probably from super booms as the fighters maneuvered into the tracks. Cattle and horses did not seem to react much to the noise. Test personnel located in a flat area where the fighters flew at less than 100 feet above ground level and generated shock waves of more than 100 psf felt a jarring sensation against their bodies and were left with temporary ringing or feelings of fullness in their ears, but they experienced no real pain or ill effects. Most, however, could not help involuntarily flinching in anticipation of the booms whenever the speeding jets passed overhead. An Air Force F-4C Phantom II flying Mach 1.26 at 95 feet during this test generated the strongest sonic boom yet recorded: 144 psf.[47] (To put this in perspective, normal air pressure at sea level equates to 14.7 pounds per square inch, or about 2,116 pounds per square foot.)

In late 1960 and early 1961, NASA and AFFTC followed up on Little Boom with Project Big Boom. B-58 bombers made 16 passes flying Mach 1.5 at altitudes of 30,000 feet to 50,000 feet over arrays of sensors, which measured a maximum overpressure of 2.1 psf. Varying the bomber's weight from 82,000 pounds to 120,000 pounds provided the first hard data on how an aircraft's weight and related lift produced higher overpressures than existing theories based on volume alone would indicate.[48]

Throughout the 1960s, Edwards Air Force Base—with its unequalled combination of Air Force and NASA expertise, facilities, instrumentation, airspace, emergency landing space, and types of aircraft—hosted the largest number of sonic boom tests. NASA researchers from Langley's Acoustics Division spent much of their time there working with the Flight Research Center in a wide variety of flight experiments. The Air Force Flight Test Center usually participated as well.

In an early test in 1961, Gareth Jordan of the FRC led an effort to collect measurements from F-104s and B-58s flying at speeds of Mach 1.2 to Mach 2.0 over sensors located along Edward AFB's supersonic corridor and at Air Force Plant 42 in Palmdale, about 20 miles south. Most of the Palmdale measurements were under 1.0 psf, which the vast majority of people surveyed there and in adjacent Lancaster (where overpressures tended to be somewhat higher) considered no worse than distant thunder. But there were some exceptions.[49]

Other experiments at Edwards in 1961 conducted by Langley personnel with support from the FRC and AFFTC contributed a variety of new data. With help from the Goodyear blimp *Mayflower*, hovering at 2,000 feet, they made the first good measurements of atmospheric effects, such as how temperature variations can bend the paths of acoustic rays and how air turbulence in the lower atmosphere near the surface (known as the boundary layer) significantly affected N-wave shape and overpressure.[50]

Testing at Edwards also gathered the first data on booms from very high altitudes. Using an aggressive flight profile, AFFTC's B-58 crew managed to zoom up to 75,000 feet—25,000 feet higher than the bomber's normal cruising altitude and 15,000 feet over its design limit! The overpressures measured from this high altitude proved stronger than predicted (not a promising result for the planned SST). Much lower down, fighter aircraft performed accelerating and turning maneuvers to generate the kind of acoustical rays that amplified shock waves and produced multiple booms and super booms. The various experiments showed that a combination of atmospheric conditions, altitude, speed, flight path, aircraft configuration, and sensor location determined the shape and strength of the pressure signatures.[51]

Of major significance for future boom minimization efforts, NASA also began taking in-flight shock wave measurements. The first of these, at Edwards in 1960, had used an F-100 with a sensor probe to measure supersonic shock waves from the sides of an F-100, F-104, and B-58 as well as from F-100s speeding past with only 100 feet of separation. The data confirmed Whitham's overall theory with some discrepancies. In early 1963, an F-106 equipped with a sophisticated new sensor probe designed at Langley flew seven sorties both above and below a B-58 at speeds of Mach 1.42 to Mach 1.69 and altitudes of approximately 40,000 feet to 50,000 feet. The data gathered confirmed Walkden's theory about how lift as well as volume increase peak shock wave pressures. As indicated by figure 1-4, analysis of the readings also found that the bow and tail shock waves spread farther apart as they flowed from the B-58. Perhaps most significant, the probing measurements revealed how the multiple, or saw tooth, shock waves (sudden increases in pressure) and expansions (regions of decreasing pressure) produced by the rest of an airplane's structure (canopy, wings, engine nacelles, weapons pod, etc.) merged with the stronger bow and tail waves until—at a distance of between 50 body lengths and 90 body lengths—they began to coalesce into the classic N-shaped signature.[52] This historic flight test, which hinted at how shock waves might be modified to reduce peak overpressures, marked a major milestone in sonic boom research.

One of the most publicized and extended flight-test programs at Edwards had begun in 1959 with the first launch from a B-52 of the fastest piloted aircraft ever flown: the rocket-propelled X-15. Three of these legendary aerospace vehicles expanded the envelope and gathered data on supersonic and hypersonic flight for the next 8 years. Although the X-15 was not specifically dedicated to sonic boom tests, the Flight Research Center did begin placing microphones and tape recorders under the X-15s' flight tracks in the fall of 1961 to gather boom data. Much later, FRC researchers reported on the measurements of these sonic booms, which were made at speeds of Mach 3.5 and Mach 4.8.[53]

For the first few years, NASA's sonic boom tests occurred in

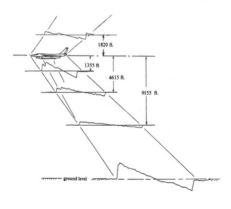

Figure 1-4. Shock wave signature of a B-58 at Mach 1.6. (NASA)

relative isolation within military airspace in the desert Southwest or over Virginia's rural Eastern Shore and adjacent waters. A future SST, however, would have to fly over heavily populated areas. Thus, from July 1961 through January 1962, NASA, the FAA, and the Air Force carried out the Community and Structural Response Program at St. Louis, Missouri. In an operation nicknamed "Bongo," the Air Force sent B-58 bombers on 76 supersonic training flights over the city at altitudes from 31,000 to 41,000 feet, announcing them as routine SAC radar bomb-scoring mis-

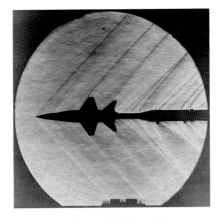

Shock waves from an X-15 model in Langley's 4-by-4-foot Supersonic Pressure Tunnel. (NASA)

sions. F-106 interceptors flew 11 additional flights at 41,000 feet. Langley personnel installed sensors on the ground, which measured overpressures up to 3.1 psf. Investigators from Scott AFB, Illinois, or for a short time, a NASA-contracted engineering firm, responded to damage claims, finding some possibly legitimate minor damage in about 20 percent of the cases. Repeated interviews with more than 1,000 residents found 90 percent were at least somewhat affected by the booms and about 35 percent were annoyed. Scott AFB (a long distance phone call from St. Louis) received about 3,000 complaints during the test and another 2,000 in response to 74 sonic booms in the following three months. The Air Force eventually approved 825 claims for $58,648. These results served as a warning that repeated sonic booms could indeed pose an issue for SST operations.[54]

To obtain more definitive data on structural damage, NASA in December 1962 resumed tests at Wallops Island using various sample buildings. Air Force F-104s and B-58s and Navy F-4H Phantom IIs flew at altitudes from 32,000 feet to 62,000 feet, creating overpressures up to 3 psf. Sonic booms triggered cracks to plaster, tile, and other brittle materials in spots where the materials were already under stress (a finding that would be repeated in later, more comprehensive tests).[55]

In February 1963, NASA, the FAA, and the USAF conducted Project Littleman at Edwards AFB to measure the results of subjecting two specially instrumented light aircraft to sonic booms. F-104s made 23 supersonic passes as close as 560 feet from a small Piper Colt and a two-engine Beech C-45, creating overpressures up to 16 psf. Their responses were "so small as to be insignificant," dismissing one possible concern about SST operations.[56]

The St. Louis survey had left many unanswered questions about public opinion. To learn more, the FAA's Supersonic Transport Development Office with support from NASA Langley and the USAF (including Tinker AFB) conducted the Oklahoma City Public Reaction Study from February through July 1964. This was a much more intensive and systematic test. In an operation named Bongo II, B-58s, F-101s, F-104s, and F-106s were called upon to deliver sonic booms eight times per day, 7 days a week for 26 weeks, with another 13 weeks of followup activities. The aircraft flew a total of 1,253 supersonic flights at Mach 1.2 to Mach 2.0 and altitudes between 21,000 feet and 50,000 feet.

The FAA (which had the resources of a major field organization available in Oklahoma City) instrumented nine control houses scattered throughout the metropolitan area with various sensors to measure structural effects while experts from Langley instrumented three houses and set up additional sensors throughout the area to record overpressures, wave patterns, and meteorological conditions. The National Opinion Research Center at the University of Chicago interviewed a sample of 3,000 adults three times during the study.[57] By the end of the test, 73 percent of those surveyed felt that they could live with the number and strength of the booms experienced, but 40 percent believed they caused some structural damage (even though the control houses showed no significant effects), and 27 percent would not accept indefinite booms at the level tested. Analysis of the shock wave patterns by NASA Langley showed that a small number of overpressure measurements were significantly higher than expected, indicating probable atmospheric influences, including heat rising from urban landscapes.[58] Sometimes, the effects of even moderate turbulence near the surface could be dramatic, as shown in figure 1-5 by the rapid change in pressure measurements from an F-104 flying Mach 1.4 at 28,000 feet recorded by an array of closely spaced microphones.[59]

The Oklahoma City study added to the growing knowledge of sonic booms and their acceptance or nonacceptance by the public at the cost of $1,039,657, seven lawsuits, and some negative publicity for the FAA. In view of the public and political reactions to the St. Louis and Oklahoma City tests, plans for another extended sonic boom test over a different city, including flights at night, never materialized.[60]

The FAA and Air Force conducted the next series of tests from November 1964 into February

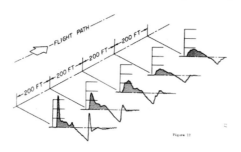

Figure 1-5. Effect of turbulence in just 800 feet of an F-104's sonic boom carpet. (NASA)

1965 in a much less populated place: the remote Oscura camp in the Army's vast White Sands Missile Range, NM. Here, 21 structures of various types and ages with a variety of plaster, windows, and furnishings were studied for possible damage. F-104s from nearby Holloman AFB and B-58s from Edwards AFB generated 1,494 booms, producing overpressures from 1.6 psf to 19 psf. The 680 sonic booms of up to 5.0 psf caused no real problems, but those above 7.9 psf caused varying degrees of damage to glass, plaster, tile, and stucco that were already in vulnerable condition. A parallel study of several thousand incubated chicken eggs showed no reduction in hatchability, and audiology tests on 20 personnel subjected daily to the booms showed no hearing impairment.[61]

Before the White Sands tests ended, NASA Langley personnel began collecting boom data from a highly urbanized setting in winter weather. During February 1965 and March 1965, they recorded data at five ground stations as B-58 bombers flew 22 training missions in a corridor over downtown Chicago at speeds from Mach 1.2 to Mach 1.66 and altitudes from 38,000 feet to 48,000 feet. The results demonstrated further that amplitude and wave shape varied widely depending upon atmospheric conditions. These 22 flights and 27 others resulted in the Air Force approving 1,442 of 2,964 damage claims for a total of $114,763. Figure 1-6 shows how a gusty day in the "Windy City" greatly increased the strength of sonic booms (N-wave signatures, shown on the right) over those created by a B-58 flying at the same speed and altitude on a more tranquil day (left) as measured by microphones placed at 100-foot intervals in a cruciform pattern.[62] The planned SST would, of course, encounter similar enhanced boom conditions.

Also in March 1965, the FAA and NASA, in cooperation with the U.S. Forest Service, studied the effects of Air Force fighters creating boom overpressures up to 5.0 psf over hazardous mountain snow packs in the Colorado Rockies. Because of the stable snow conditions, these booms did not created any avalanches. Interestingly enough, in the early 1960s, the National Park Service tried to use newly deployed F-106s at Geiger Field, WA, to create controlled avalanches in Glacier National Park (known as Project Safe Slide), but, presumably, it found traditional methods such as artillery fire more suitable.[63]

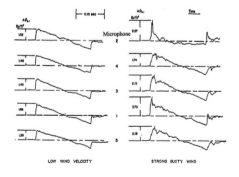

Figure 1-6. Effect of winds on B-58 sonic boom signatures. (NASA)

XB-70 Valkyrie, the largest of the sonic boom test aircraft. (USAF)

Enter the Valkyrie and the Blackbird

From the beginning of the SST program, the aircraft most desired for experiments was, of course, the North American XB-70 Valkyrie. The first of the giant test beds (XB-70-1) arrived at Edwards AFB in September 1964, and the better-performing, better-instrumented second aircraft (XB-70-2) arrived in July 1965. With a length of 186 feet, a wingspan of 105 feet, and a gross weight of about 500,000 pounds, the six-engine would-be bomber was considerably heavier but less than two thirds as long as some of the later SST concepts, but it was the best real-life surrogate available.[64]

Even during the initial flight-envelope expansion by contractor and AFFTC test pilots, the Flight Research Center began gathering sonic boom data, including direct comparisons of its shock waves with those of a B-58 flying only 800 feet behind.[65] Using an array of microphones and recording equipment at several ground stations, NASA researchers built a database of boom signatures from 39 flights made by the XB-70s (10 with B-58 chase planes) from March 1965 through May 1966.[66] Because "the XB-70 is capable of duplicating the SST flight profiles and environment in almost every respect," the FRC was looking forward to beginning its own experimental research program using the second Valkyrie on June 15, 1966, with sonic boom testing listed as the first priority.[67]

On June 8, however, the XB-70-2 crashed on its 47th flight as the result of a midair collision during an infamous publicity flight for General Electric (GE) to advertise its jet engines. Despite this tragic setback to the overall test program, the less capable XB-70-1 (which underwent modifications until November) eventually proved useful for many purposes. After 6 months of joint AFFTC-FRC operations (with a total of 60 flights, including the boom testing described below), the Air Force turned the plane over full time to NASA in April 1967. The FRC, with a more limited budget, then used the Valkyrie for 23 more test missions until February 1969, when the unique aircraft was retired to the USAF Museum in Dayton, OH.[68] All told, NASA acquired sonic boom measurements from 51 of the 129 total flights made by the XB-70s using two ground stations on Edwards AFB, one at nearby Boron, CA, and two in Nevada.[69] These data would be of great value in the future.

The loss of one XB-70 and retirement of the other from supersonic testing was made somewhat less painful by the availability of two smaller (107 feet long) but even faster products of advanced aviation technology: the Lockheed YF-12 and its cousin, the SR-71—both nicknamed Blackbirds. On May 1, 1965, shortly after arriving at Edwards, a YF-12A set nine new world records, including a closed-course speed of 2,070 mph (Mach 3.14) and a sustained altitude of 80,257 feet. Four of that day's five flights also yielded sonic boom measurements. At speeds of Mach 2.6 to Mach 3.1 and altitudes of 60,000 feet to 76,500 feet above ground level, overpressures varied from 1.2 psf to 1.7 psf depending on distance from the flight path. During another series of flight tests at slower speeds and lower altitudes, overpressures up to 5.0 psf were measured during accelerations after having slowed down to refuel. These early results proved consistent with previous B-58 data.[70] Data gathered from ground arrays measuring the sonic signatures from YF-12s, XB-70s, B-58s, and smaller aircraft flying at various altitudes also showed that the lateral spread of a boom carpet (without the influence of atmospheric variables) could be roughly equated to 1 mile for every 1,000 feet of altitude with the N-signatures becoming more rounded with distance until degenerating into the approximate shape of a sine wave.[71] In all cases, however, acoustic rays reflected off the ground along with those that propagated above the aircraft could be refracted or bent by the conditions in the thermosphere and intersect the ground as a much weaker over-the-top or secondary boom carpet.

Although grateful to benefit from the flights of the AFFTC's Blackbirds, the FRC wanted its own YF-12 or SR-71 for supersonic research. It finally gained the use of two YF-12s through a NASA-USAF Memorandum of Understanding signed in June 1969, paying for operations with funding left over from the termination of the X-15 and XB-70 programs.[72]

The National Sonic Boom Evaluation

In the fall of 1965, with public acceptance of sonic booms becoming a significant public and political issue, the White House Office of Science and Technology established the National Sonic Boom Evaluation Office (NSBEO) under the interagency Coordinating Committee on Sonic Boom Studies. The new organization, which was attached to Air Force Headquarters for administrative purposes, planned a comprehensive series of tests known as the National Sonic Boom Evaluation Program, which was to be conducted primarily at Edwards AFB. NASA (in particular, the Flight Research Center and Langley Research Center) would be responsible for test operations and data collection with the Stanford Research Institute (SRI) hired to help analyze the findings.[73]

After careful preparations (including specially built structures and extensive sensor and recording arrays), the National Sonic Boom Evaluation began in June 1966. Its main objectives were to address the many issues left unresolved from previous tests. Unfortunately, the loss of the XB-70-2 on June 8 forced a 4-month break in the test schedule, and the limited events completed in June became designated as Phase I. The second phase began in November 1966, when the XB-70-1 returned to flight status, and lasted into January 1967. A total of 367 supersonic missions were flown by XB-70s, B-58s, YF-12s, SR-71s, F-104s, and F-106s during the two phases. These were supplemented by 256 subsonic flights by KC-135s, WC-135Bs, C-131Bs, and Cessna 150s. In addition, the Goodyear blimp *Mayflower* was used in the June phase to measure sonic booms at 2,000 feet.[74]

By the end of testing, the National Sonic Boom Evaluation had obtained new and highly detailed acoustic and seismic signatures from all the different supersonic aircraft in various flight profiles during a variety of atmospheric conditions. The data from 20 XB-70 flights at speeds from Mach 1.38 to Mach 2.94 were to be of particular long-term interest. For example, Langley's sophisticated nose probe used for the pioneering in-flight flow-field measurements of the B-58 in 1963 was installed on one of the FRC's F-104s to do the same for the XB-70. A comparison of data between blimp and ground sensors and variations between the summer and winter tests confirmed the significant influence that atmospheric conditions, such as turbulence and convective heating near the surface, have on boom propagation.[75] (In general, rising temperatures near the surface bend the path followed by the acoustic rays accompanying the shock waves upward, sometimes away from the ground altogether while cooler air near the surface, as with a temperature inversion, does just the opposite, bending the rays downward so that they bounce off the ground.[76]) Also, the evaluation provided an opportunity to gather data

An Air Force YF-12, which provided valuable sonic boom data for NASA, taking off at Edwards AFB. (USAF)

on more than 1,500 sonic boom signatures created during 35 flights by the recently available SR-71s and YF-12s at speeds up to Mach 3.0 and altitudes up to 80,000 feet.[77]

Some of the findings portended serious problems for planned SST operations. The program obtained responses from several hundred participating volunteers, both outdoors and inside houses, to sonic booms of different intensities produced by each of the supersonic aircraft. The time between the peak overpressure of the bow and tail shocks for aircraft at high altitudes ranged from about one-tenth of a second for the F-104, two-tenths of a second for the B-58, and three-tenths of a second for the XB-70. (See figure 1-7.) The respondents also compared sonic booms to the jet-engine noise of subsonic aircraft. Although data varied for each of the criteria measured, significant minorities tended to find the booms either just acceptable or unacceptable and the sharper N-wave signature from the lower flying F-104 more annoying outdoors than the more rounded signatures from the larger aircraft, which had to fly at higher altitudes to create the same overpressure. Other factors included the frequency, time of day or night, and type of boom signature. Correlating how the subjects responded to jet noise (measured in decibels) and sonic booms (normally measured in psf), the SRI researchers used a criterion called the

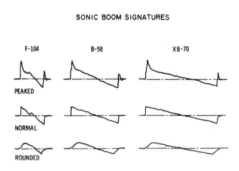

SONIC BOOM SIGNATURES

Figure 1-7. Variations in N-waves caused by aircraft size and atmospheric conditions. (NASA)

perceived noise decibel (PNdB) level to assess how loud booms seemed to human ears.[78]

Employing sophisticated sensors, civil engineers measured the physical effects on houses and a building with a large interior space (the base's bowling alley) from varying degrees of booms created by the F-104s, B-58s, and XB-70. Of special concern for the SST's acceptability, the engineers found the XB-70's elongated N-wave (although less bothersome to observers outdoors) created more of the ultralow frequencies that cause indoor vibrations, such as rattling windows, which many of the respondents considered objectionable. And although no significant harm was detected to the instrumented structures, 57 complaints of damage were received from residents in the surrounding area, and three windows were broken on the base. Finally, monitoring by the U.S. Department of Agriculture detected no ill effects on farm animals in the area, although avian species (chickens, turkeys, etc.) reacted more than livestock.[79] The National Sonic Boom Evaluation remains the most comprehensive test program of its kind ever conducted.[80]

Last of the Flight Tests

Even with the advantages offered by sophisticated simulators, researchers continued to look for ways to obtain human-response data from live sonic booms. In the spring of 1967, the opportunity for collecting additional survey data presented itself when the FAA and NASA learned that SAC was starting an extensive training program for its growing fleet of SR-71s. TRACOR, Inc., of Austin, TX, which was already under contract to NASA doing surveys on airport noise, had its contract's scope expanded in May 1967 to include public responses to the SR-71s' sonic booms in Dallas, Los Angeles, Denver, Atlanta, Chicago, and Minneapolis. Between July 3 and October 2, Air Force SR-71s made 220 high-altitude supersonic flights over these cities, ranging from 5 over Atlanta to 60 over Dallas. Those sonic booms that were measured were almost all N-waves with overpressures ranging from slightly less than 1.0 psf to 2.0 psf.

Although the data from this impromptu test program were less than definitive, its overall findings (based on 6,375 interviews) were fairly consistent with the previous human-response surveys. For example, after an initial dropoff, the level of annoyance with the booms tended to increase over time, and almost all those who complained were worried about damage. Among 15 different adjectives supplied to describe the booms (e.g., disturbing, annoying, irritating), the word "startling" was chosen much more frequently than any other.[81] The tendency of people to be startled by the suddenness of sonic booms was becoming recognized as their most problematic attribute in gaining public acceptance.

Although the FRC and AFFTC continued their missions of supersonic flight testing and experimentation at Edwards, what might be called the heroic era of sonic boom testing was drawing to a close. The FAA and the Environmental Science Services Administration (a precursor of the Environmental Protection Agency) did some sophisticated testing of meteorological effects at Pendleton, OR, from September 1968 until May 1970, using a dense grid of recently invented, unattended transient data recorders to measure random booms from SR-71s. On the other side of the continent, NASA and the Navy studied sonic booms during Apollo missions in 1970 and 1971.[82]

The most significant NASA testing in 1970 took place from August to October at the Atomic Energy Commission's Jackass Flats test site in Nevada. In conjunction with the FAA and the National Oceanic and Atmospheric Administration (NOAA), NASA took advantage of the 1,527-foot tall Bare Reactor Experiment Nevada (BREN) Tower, which had been named for its original nuclear radiation tests in 1962. The researchers installed a vertical array of 15 microphones as well as meteorological sensors at various levels

Interagency team at the base of the BREN Tower. NASA personnel include Herbert Henderson, second from left; Domenic Maglieri, third from left; and David Hilton, far right. (Maglieri)

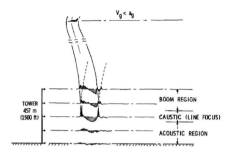

Figure 1-8. BREN Tower measurements of Mach cutoff signatures. (NASA)

along the tower. (Until then, a 250-foot tower at Wallops Island had been the highest used in sonic boom testing.[83])

During the summer and fall of 1970, the FRC's F-104s made 121 boom-generating flights from Edwards AFB to provide measurements of several still inadequately understood aspects of the sonic boom, especially the conditions known as caustics, in which acoustical rays can converge and focus in a nonlinear manner.[84] Frequently caused by aircraft at transonic speeds or during acceleration, they can result in normal N-wave signatures being distorted as they pass through caustic regions into U-shaped signatures, sometimes with bow and tail wave overpressures strong enough to create super booms. Such signatures, however, are also sensitive to turbulence and prone to refracting before reaching the surface (rather than reflecting off the ground as with N-waves). The BREN Tower allowed such measurements to be made in the vertical dimension for the first time. This testing resulted in definitive data on the formation and nature of caustics as well as the Mach cutoff—information that would be valuable in planning boomless transonic flights and helping pilots to avoid making focused booms.[85] Figure 1-8 illustrates the combined results of 3 days of testing by F-104s flying at about Mach 1.1 at 30,000 feet (with the solid lines representing shock waves and the dotted lines their reflection).[86]

For all intents and purposes, the results of earlier testing and human-response surveys had already helped seal the fate of the SST before the reports on this latest test began coming in. Even so, the test results garnered from 1958 through 1970 during the SCAT and SST programs contributed tremendously to the international aeronautical and scientific communities' understanding of one of the most baffling and complicated aspects of supersonic flight. As Harry W. Carlson told the Nation's top sonic-boom experts on the same day the last F-104 flew over Jackass Flats: "The importance of flight-test programs cannot be overemphasized. These tests have provided an impressive amount of high-quality data."[87] Unfortunately, however, learning about the nature of sonic booms did not yet translate into learning how to control them.

As will be described in the next chapter, the American SST program proved to be too ambitious for the technology of its time despite a concerted effort by many of the best minds in aeronautical science and engineering. Yet for all its disappointments and controversies, the program's proliferation of data and scientific knowledge about supersonic flight, including sonic booms, would be indispensable for progress in the future.

Endnotes

1. Paul Pojman, "Ernst Mach," in *The Stanford Encyclopedia of Philosophy* (Summer 2011 Edition), ed. Edward N. Zalta, accessed ca. June 1, 2011, *http://plato.stanford.edu/archives/sum2011/entries/ernst-mach/*.

2. Because the effects of pressure and density on the speed of sound offset each other in most gases, temperature is the major variable. As a result, the speed of sound in Earth's lower atmosphere (the troposphere) typically decreases as air cools with altitude until reaching the tropopause, which is a boundary layer of stable (isothermal) temperatures between the troposphere and the upper stratosphere. Although the altitude of the tropopause varies depending on season and latitude, the speed of sound usually travels at its uniformly slowest speed between about 36,000 feet and 85,000 feet. See "Layers of the Earth's Atmosphere," Windows to the Universe, accessed June 9, 2011, *http://www.windows2universe.org/earth/Atmosphere/layers.html*; Steven A. Brandt et al., *Introduction to Aeronautics: A Design Perspective* (Reston, VA: American Institute of Aeronautics and Astronautics [AIAA], 2004), Appendix B, 449–451.

3. J.W.M. Dumond et al., "A Determination of the Wave Forms and Laws of Propagation and Dissipation of Ballistic Shock Waves," *Journal of the Acoustical Society of America (JASA)* 18, no. 1 (January 1946): 97–118. By the end of World War II, ballistic waves were well understood.

4. David Darling, *The Complete Book of Spaceflight: From Apollo 1 to Zero Gravity* (Hoboken, NJ: John Wiley and Sons, 2003), 457. See also "Airpower: Missiles and Rockets in Warfare," accessed December 21, 2009, *http://www.centennialofflight.gov/essay/Air_Power/Missiles/AP29.htm*; and Bob Ward, *Dr. Space: The Life of Wernher von Braun* (Annapolis, MD: Naval Institute, 2005), 43.

5. The definitive biography, Michael J. Neufeld's *Von Braun: Dreamer of Space, Engineer of War* (New York: Alfred A. Knopf, 2007), 133–36, leaves open the question of whether the Germans at Peenemünde heard the first humanmade sonic booms in 1942 when their A-4 test rockets exceeded Mach 1 about 25 seconds after launch.

6. For its development and testing, see Richard P. Hallion, *Supersonic Flight: Breaking the Sound Barrier and Beyond—The Story of the Bell X-1 and Douglas D-558* (New York: The Macmillan Co. in association with the Smithsonian Institution National Air and Space Museum, 1972). For a detailed analysis of the XS-1's first supersonic

flight by a former NASA engineer, see Robert W. Kempel, *The Conquest of the Sound Barrier* (Beirut, Lebanon: HPM Publications, 2007), 31–49.

7. Tom Wolfe, *The Right Stuff* (New York: Bantam Books, 1980), 47. Some personnel stationed at Muroc when Yeager broke the sound barrier would later recall hearing a full sonic boom, but these may have been memories of subsequent flights at higher speeds.

8. One of NASA's top sonic boom experts, using a computer program called PCBoom4, has calculated that at Mach 1.06 and 43,000 feet in a standard atmosphere, refraction and absorption of the shock waves would almost certainly have dissipated the XS-1's sonic boom before it could reach the surface; Edward A. Haering, Dryden Flight Research Center (DFRC), to Lawrence R. Benson, e-mail message, April 8, 2009.

9. Chuck Yeager and Leo Janis, *Yeager* (New York: Bantam Books, 1985), 165.

10. Domenic J. Maglieri to Christine M. Darden, NASA Dryden, January 13, 1993, with "A Note on the Existence of Sonic Booms from Aircraft Flying at Speeds Less Than Mach 1," Eagle Engineering (January 1993). This analysis shows that it is physically possible that Bell test pilot Chalmers ("Slick") Goodlin may have created a localized sonic boom when making an 8-g pull-up from a Mach 0.82 dive during an XS-1 envelope-expansion flight many months earlier.

11. Domenic Maglieri to Lawrence Benson, e-mail message, September 22, 2011.

12. For the authoritative history of the NACA-NASA mission at Edwards AFB, see Richard P. Hallion and Michael H. Gorn, *On the Frontier: Experimental Flight at NASA Dryden* (Washington, DC: Smithsonian Institution Press, 2003).

13. Kenneth J. Plotkin and Domenic J. Maglieri, "Sonic Boom Research: History and Future," AIAA paper no. 2003-3575 (June 2003), 2. This publication is recommended for any reader who would like a succinct introduction to the subject. Maglieri and Percy J. Bobbitt also compiled a highly detailed, 372-page reference, "History of Sonic Boom Technology Including Minimization" (Hampton, VA: Eagle Aeronautics, November 1, 2001), a copy of which was kindly provided to the author in early 2009 for his reference.

14. Introduction to "The Battle of the Bangs," *Flight and Aircraft Engineer* 61, no. 2289 (December 5, 1952): 696, accessed

ca. January 15, 2009, *http://www.flightglobal.com/pdfarchive/ view/1952/1952%20-%203468.html.*

15. G.B. Whitham, "The Flow Pattern of a Supersonic Projectile," *Communications on Pure and Applied Mathematics* 5, no. 3 (1952): 301–348, accessed ca. January 30, 2009, *http://www3.interscience. wiley.com/journal/113395160/issue.* Whitham received his Ph.D. from the University of Manchester in 1953.

16. Figure 1-1 copied from Peter Coen and Roy Martin, "Fixing the Sound Barrier: Three Generations of U.S. Research into Sonic Boom Reduction … and What it Means to the Future," PowerPoint presentation presented at the Experimental Aircraft Association AirVenture Oshkosh, Oshkosh, WI, July 2004, slide no. 3, "Sonic Boom Basics."

17. G.B. Whitham, "On the Propagation of Weak Shock Waves," *Journal of Fluid Dynamics* 1, no. 3 (September 1956): 290–318, accessed ca. January 30, 2009, *http://journals.cambridge.org/action/ displayJournal?jid=JFM.* Both papers are described in Larry J. Runyan et al., "Sonic Boom Literature Survey. Volume 2. Capsule Summaries," Boeing Commercial Airplane Co. for the FAA (September 1973), Defense Technical Information Center (DTIC) no. AD 771274, 6-8 and 59-60. Whitham later taught at both the Massachusetts and California Institutes of Technology.

18. Marshall E. Mullens, "A Flight Test Investigation of the Sonic Boom," AFFTC Technical Note (TN) no. 56-20 (May 1956), as summarized in Runyan, "Sonic Boom Capsule Summaries," 46–47.

19. Air Force Flight Test Center History Office, *Ad Inexplorata: The Evolution of Flight Testing at Edwards Air Force Base* (Edwards, CA: AFFTC, 1996), appendix B, 55. Photo provided courtesy of this office. Most aircraft names are assigned in later stages of development or production.

20. John G. Norris, "AF Says 'Sonic Boom' Can Peril Civilians," *Washington Post* and *Times Herald*, November 9, 1954, 1, 12.

21. One of the first studies on focused booms was G.M. Lilley et al., "Some Aspects of Noise from Supersonic Aircraft," *Journal of the Royal Aeronautical Society* 57 (June 1953): 396–414, as described in Runyan, "Sonic Boom Capsule Summaries," 54.

22. History of the 3201 Air Base Group, Eglin AFB, July–September 1951, abstract from Information Retrieval and Indexing System (IRIS), no. 438908, Air Force Historical Research Center, Maxwell AFB, AL.

23. F. Edward McLean, *Supersonic Cruise Technology*, Special Publication (SP)-472 (Washington, DC: NASA, 1985), 31–32; Richard P. Hallion, "Richard Whitcomb's Triple Play," *Air Force Magazine* 93, no. 2 (February 2010): 70–71. The following abbreviations are used for NASA publications cited in the notes: Conference Publication (CP), Contractor Report (CR), Reference Publication (RP), Special Publication (SP), Technical Memorandum (TM), formerly classified Tech Memo (TM-X), Technical Note (TN), Technical Paper (TP), and Technical Report (TR). Bibliographic information and, often, full-text copies can be accessed through the NASA Technical Reports Server (NTRS), *http://ntrs.nasa.gov/search.jsp*. The NTRS is a publicly accessible database maintained by NASA's Center for AeroSpace Information (CASI).

24. Richard Seebass, "Sonic Boom Minimization," paper presented at the North Atlantic Treaty Organization (NATO) Research and Technology Organization Applied Vehicle Technology course "Fluid Dynamics Research on Supersonic Aircraft," Rhode Saint-Genèse, Belgium, May 25–29, 1998; Robert T. Jones, "Adolf Busemann, 1901–1986," in *Memorial Tributes: National Academy of Engineering* 3 (Washington, DC: The National Academies Press, 1989), 62–67, accessed June 7, 2011, *http://www.nap.edu/openbook. php?record_id=1384*.

25. F. Walkden, "The Shock Pattern of a Wing-Body Combination Far from the Flight Path," *Aeronautical Quarterly* 9, pt. 2 (May 1958): 164–94; described in Runyan, "Sonic Boom Capsule Summaries," 8–9. Both Walkden and Whitman did their influential studies at the University of Manchester.

26. Plotkin and Maglieri, "Sonic Boom Research," 2.

27. Fred Keefe and Grover Amen, "Boom," *New Yorker*, May 16, 1962, 33–34.

28. Albion B. Hailey, "AF Expert Dodges Efforts to Detail 'Sonic Boom' Loss," *Washington Post*, August 25, 1960, A15.

29. J.P. and E.G.R. Taylor, "A Brief Legal History of the Sonic Boom in America," in *Aircraft Engine Noise and Sonic Boom*, Conference Proceedings (CP) no. 42, presented at the NATO Advisory Group for Aerospace Research and Development (AGARD), Neuilly sur Seine, France, 1969, 2-1–2-11.

30. Roger A. Mola, "This Is Only a Test," *Air & Space Magazine* 21, no. 2 (March–April 2006), accessed ca. March 1, 2011, *http://www. airspacemag.com/history-of-flight/this-is-only-a-test.html*. For contemporary accounts, see "Warplanes Fill Skies Over U.S. and Canada,"

Los Angeles Times, September 10, 1960, 4; Albion B. Halley and Warren Kornberg, "U.S. Tests Air Defenses in 3000-Plane 'Battle,' " *Washington Post*, October 15, 1961, A1, B1; Richard Witkin, "Civilian Planes Halted 12 Hours in Defense Test," *New York Times*, October 15, 1961, 1, 46.

31. Marcelle S. Knaack, *Post–World War II Bombers, 1945–1973* 2 of *Encyclopedia of U.S. Air Force Aircraft and Missile Systems* (Washington, DC: USAF, 1988), 394–395.

32. "Jet Breaks 3 Records—and Many Windows," *Los Angeles Times*, March 6, 1962, 1. In reality, most of the damage was done while accelerating after the refuelings. The Air Force pamphlet shown in figure 1-2 is in the Dryden Flight Research Center's archival collection.

33. For the political and economic aspects of the SST, see Mel Horwitch, *Clipped Wings: The American SST Conflict* (Cambridge: Massachusetts Institute of Technology [MIT] Press, 1982) and Erik M. Conway's definitive account *High-Speed Dreams: NASA and the Technopolitics of Supersonic Transportation, 1945–1999* (Baltimore, MD: Johns Hopkins University Press, 2005), which also covers subsequent programs and includes many technical details. For an informative earlier study by an insider, see the previously cited McLean, *Supersonic Cruise Technology*.

34. M. Leroy Spearman, "The Evolution of the High-Speed Civil Transport," NASA TM no. 109089 (February 1994), figure 1-3 extracted from 26.

35. McLean, *Supersonic Cruise Technology*, 35–46; Joseph R, Chambers, *Innovation in Flight: Research of the NASA Langley Research Center on Revolutionary Concepts for Aeronautics*, SP-2005-4539 (Washington, DC: NASA, 2005), 25–28.

36. William J. Alford and Cornelius Driver, "Recent Supersonic Transport Research," *Astronautics & Aeronautics* 2, no. 9 (September 1964): 26.

37. Conway, *High-Speed Dreams*, 55.

38. Spearman, "Evolution of the HSCT," 7.

39. *FAA Historical Chronology, 1926–1996*, accessed February 15, 2009, *http://www.faa.gov/about/media/b-chron.pdf*. For Quesada's role, see Stuart I. Rochester, *Takeoff at Mid-Century: Federal Civil Aviation Policy in the Eisenhower Years, 1953–1961* (Washington, DC: FAA, 1976). For the activism of Halaby and the demise of the SST after his departure, see Richard J. Kent, Jr., *Safe, Separated, and Soaring: A*

History of Civil Aviation Policy, 1961–1972 (Washington, DC: FAA, 1980).

40. Based on author's review of Section 7.4, "Noise, Aircraft," in volumes of the *Index of NACA Technical Publications* (Washington DC: NACA Division of Research Information, n.d.) covering the years 1915–1957.

41. For an overall summary of Langley's supersonic activities, see Chambers, *Innovations in Flight*, chapter 1, "Supersonic Civil Aircraft: The Need for Speed," 7–70; Domenic Maglieri by Lawrence Benson, telephone interview, February 6, 2009.

42. A.B. Fryer et al., "Publications in Acoustics and Noise Control from the NASA Langley Research Center during 1940–1976," NASA TM-X-74042 (July 1977).

43. For a chronological summary of selected projects during the first decade of sonic boom research, see Johnny M. Sands, "Sonic Boom Research (1958–1968)," FAA, DTIC no. AD 684806, November 1968.

44. Domenic J. Maglieri, Harvey H. Hubbard, and Donald L. Lansing, "Ground Measurements of the Shock-Wave Noise from Airplanes in Level Flight at Mach Numbers to 1.4 and Altitudes to 45,000 Feet," NASA TN D-48 (September 1959); Lindsay J. Lina and Domenic J. Maglieri, "Ground Measurements of Airplane Shock-Wave Noise at Mach Numbers to 2.0 and at Altitudes to 60,000 Feet," NASA TN D-235 (March 1960).

45. Domenic J. Maglieri, Vera Huckel, and Tony L. Parrott, "Ground Measurements of Shock-Wave Pressure for Fighter Airplanes Flying at Very Low Altitudes and Comments on Associated Response Phenomena," NASA TN D-3443 (July 1966) (which superseded classified TM-X-611 [1961]).

46. Gareth H. Jordan, "Flight Measurements of Sonic Booms and Effects of Shock Waves on Aircraft," in *Society of Experimental Test Pilots Quarterly Review* 5, no. 1 (1961): 117–131, presented at the Society of Experimental Test Pilots (SETP) Supersonic Symposium, September 29, 1961.

47. C.W. Nixon et al., "Sonic Booms Resulting from Extremely Low-Altitude Supersonic Flight: Measurements and Observations on Houses, Livestock, and People," Aerospace Medical Research Laboratories (AMRL) Technical Report (TR) 68-52 (October 1968), DTIC AD 680800; USAF Fact Sheet, "Sonic Boom," February 23, 2011, *http://www.af.mil/information/factsheets/factsheet.*

asp?id=184; Domenic Maglieri by Lawrence Benson, telephone interview, March 19, 2009.

48. Domenic J. Maglieri and Harvey H. Hubbard, "Ground Measurements of the Shock-Wave Noise from Supersonic Bomber Airplanes in the Altitude Range from 30,000 to 50,000 Feet," NASA TN D-880 (July 1961).

49. Jordan, "Flight Measurements of Sonic Booms."

50. Domenic J. Maglieri and Donald L. Lansing, "Sonic Booms from Aircraft in Maneuvers," NASA TN D-2370 (July 1964) (based on the tests in 1961); Domenic Maglieri, *Sonic Boom Research: Some Effects of Airplane Operations and the Atmosphere on Sonic Boom Signature*, NASA SP-147 (1967), 25–48. D.J. Maglieri, J.O. Powers, and J.M. Sands, "Survey of United States Sonic Boom Overflight Experimentation," NASA TM-X-66339 (May 30, 1969), 15–17.

51. Harvey H. Hubbard et al., "Ground Measurements of Sonic-Boom Pressures for the Altitude Range of 10,000 to 75,000 Feet," NASA TR R-198 (July 1964) (this report was based on testing in 1961).

52. Harriet J. Smith, "Experimental and Calculated Flow Fields Produced by Airplanes Flying at Supersonic Speeds," NASA TN D-621 (November 1960); D.J. Maglieri and V.S. Richie, "In-Flight Shock-Wave Measurements Above and Below a Bomber Airplane at Mach Numbers from 1.42 to 1.69," NASA TN D-1968 (October 1963). Figure 1-4 extracted with permission from Domenic Maglieri and Percy Bobbitt from "History of Sonic Boom Technology Including Minimization," Eagle Aeronautics (November 1, 2001), 145.

53. NASA Flight Research Center, "X-15 Program" (monthly report), September 1961, Dryden archive, file LI-6-10A-13. (Peter Merlin assisted the author in finding this and other archival documents.); Karen S. Green and Terrill W. Putnam, "Measurements of Sonic Booms Generated by an Airplane Flying at Mach 3.5 and 4.8," NASA TM-X-3126 (October 1974). (Since hypersonic speeds were not directly relevant for the SST, a formal report was delayed until NASA began planning re-entry flights by the Space Shuttle.) For the X-15 program, see Hallion and Gorn, *On the Frontier*, 101–125. For a discussion of its shock wave propagation, see Wendell H. Stillwell, *X-15 Research Results*, SP-60 (Washington, DC: NASA, 1965), 46–66.

54. Charles W. Nixon and Harvey H. Hubbard, "Results of the USAF-NASA-FAA Flight Program to Study Community Response to Sonic Booms in the Greater St. Louis Area," NASA TN no. D-2705 (May

1965); Clark et al., "Studies of Sonic Boom Damage," NASA CR 227 (May 1965).

55. Sands, "Sonic Boom Research (1958–1968)," 3.

56. Domenic J. Maglieri and Garland J. Morris, "Measurement of Response of Two Light Airplanes to Sonic Booms," NASA TN D-1941 (August 1963).

57. Paul M. Borsky, "Community Reactions to Sonic Booms in the Oklahoma City Area—Volume II: Data on Community Reactions and Interpretations," USAF Aerospace Medical Research Laboratory, Wright-Patterson AFB (August 1965), accessed ca. February 15, 2009, *http://www3.norc.org/NR/rdonlyres/255A2AA2-B953-4305-9AD0-B8ABCC824FA9/0/NORCRpt_101B.pdf*.

58. D.A. Hilton, D.J. Maglieri, and R. Steiner, "Sonic-Boom Exposures during FAA Community Response Studies over a 6-Month Period in the Oklahoma City Area," NASA TN D-2539 (December 1964).

59. Source for figure 1-5: Ibid., 68.

60. D.J. Maglieri (NASA), D.J. Powers, and J.M. Sands (FAA), "Survey of United States Sonic Boom Overflight Experimentation," NASA TM-X-66339 (May 30, 1969), 37–39; Conway, *High-Speed Dreams*, 121–122.

61. Thomas H. Higgins, "Sonic Boom Research and Design Considerations in the Development of a Commercial Supersonic Transport," *JASA* 39, no. 5, pt. 2 (November 1966): 526–531.

62. David. A. Hilton, Vera Huckel, and Domenic J. Maglieri, "Sonic Boom Measurements During Bomber Training Operations in the Chicago Area," NASA TN D-3655 (October 1966), figure 1-6 on N-wave signatures copied from 7.

63. Histories of the 4700 Air Defense Wing, January–March and April–June 1960, IRIS abstracts; History of the 84th Fighter Group, January–December 1961, IRIS abstract; Benson, Maglieri interview, March 19, 2009.

64. For a definitive history of this remarkable aircraft, see Dennis R. Jenkins and Tony R. Landis, *Valkyrie: North America's Mach 3 Superbomber* (North Branch, MN: Specialty Press, 2004).

65. William H. Andrews, "Summary of Preliminary Data Derived from the XB-70 Airplanes," NASA TM-X-1240 (June 1966), 11–12. Despite being more than three times heavier than the B-58, the XB-70's bow wave proved to be only slightly stronger, reflecting its more tailored aerodynamic design and the benefits of its large size.

66. Domenic J. Maglieri et al., "A Summary of XB-70 Sonic Boom Signature Data, Final Report," NASA CR 189630 (April 1992).

Until this report, the 1965–1966 findings were filed away unpublished. The original oscillographs were also scanned and digitized at this time for use in the High-Speed Research (HSR) program.

67. FRC, "NASA XB-70 Flight Research Program," April 1966, Dryden archive, File L2-4-4D-3, 10 quoted. See also C.M. Plattner, "XB-70A Flight Research: Phase 2 to Emphasize Operational Data," *Aviation Week* (June 13, 1966): 60–62.

68. NASA Dryden Fact Sheet, "XB-70," accessed May 10, 2011, *http://www.nasa.gov/centers/dryden/news/FactSheets/FS-084-DFRC.html*; Hallion and Gorn, *On the Frontier*, 176–85, 421.

69. Maglieri, "Summary of XB-70 Sonic Boom," 4–5. Photo of BREN Tower courtesy of Mr. Maglieri.

70. R.T. Klinger, "YF-12A Flight Test Sonic Boom Measurements," Lockheed Advanced Development Projects Report SP-815 (June 1, 1965), Dryden archive, File LI-4-10A-1.

71. John O. Powers, Johnny M. Sands, and Domenic J. Maglieri, "Survey of United States Sonic Boom Overflight Experimentation," NASA TM-X-66339 (May 1969), 9, 12–13.

72. Peter W. Merlin, *From Archangel to Senior Crown: Design and Development of the Blackbird* (Reston, VA: AIAA, 2008), 106–107, 116–118, 179; Hallion and Gorn, *On the Frontier*, 187.

73. NSBEO, "Sonic Boom Experiments at Edwards Air Force Base; Interim Report" (July 28, 1967), 1–2 (hereinafter cited as the Stanford Research Institute [SRI], "Edwards AFB Report" as it was prepared under contract by the SRI). For background on the NSBEO, see Conway, *High-Speed Dreams*, 122–123.

74. SRI, "Edwards AFB Report," 9.

75. D.A. Hilton, D.J. Maglieri, and N.J. McLeod, "Summary of Variations of Sonic Boom Signatures Resulting from Atmospheric Effects," NASA TM-X-59633 (February 1967), and D.J. Maglieri, "Preliminary Results of XB-70 Sonic Boom Field Tests During National Sonic Boom Evaluation Program," March 1967, annexes C-1 and C-2, in SRI, "Edwards AFB Report;" H.H. Hubbard and D.J. Maglieri, "Sonic Boom Signature Data from Cruciform Microphone Array Experiments during the 1966–67 EAFB National Sonic Boom Evaluation Program," NASA TN D-6823 (May 1972).

76. Maglieri and Bobbitt, "History of Sonic Boom Technology," 60–63.

77. SRI, "Edwards AFB Report," 17–20, annexes C–F; Domenic J. Maglieri et al., "Sonic Boom Measurements for SR-71 Aircraft Operating at Mach Numbers to 3.0 and Altitudes to 24834 Meters," NASA TN D-6823 (September 1972).

78. SRI, "Edwards AFB Report," 11–16, annex B; K.D. Kryter, "Psychological Experiments on Sonic Booms Conducted at Edwards Air Force Base, Final Report," SRI (1968), summarized by Richard M. Roberds, "Sonic Boom and the Supersonic Transport," *Air University Review* 22, no. 7 (July–August 1971): 25–33.

79. SRI, "Edwards AFB Report," 20–23, annexes G and H; David Hoffman, "Sonic Boom Tests Fail to Win Any Boosters," *Washington Post*, August 3, 1967, A3; A.J. Bloom, G. Kost, J. Prouix, and R.L. Sharpe, "Response of Structures to Sonic Booms Produced by XB-70, B-58, and F-104 Aircraft: Based on Sonic Boom Experiments at Edwards Air Force Base, Final Report," NSBEO 2-67, (October 1967); D.S. Findley et al., "Vibration Responses of Test Structure No. 1 During the Edwards Air Force Base Phase of the National Sonic Boom Program," NASA TM-X-72706 (June 1975), and "Vibration Responses of Test Structure No. 2 During the Edwards Air Force Base Phase of the National Sonic Boom Program," NASA TM-X-72704 (June 1975).

80. Source for figure 1-4: D.J. Maglieri (NASA), J.O. Powers, and J.M. Sands (FAA), "Survey of United States Sonic Boom Overflight Experimentation," NASA TM-X-66339 (May 30, 1969), 4.

81. TRACOR, Inc., "Public Reactions to Sonic Booms," NASA CR 1665 (September 1970).

82. David A. Hilton and Herbert R. Henderson documented the sonic boom measurements from the Apollo 15, 16, and 17 missions in NASA TNs D-6950 (1972), D-7606 (1974), and D-7806 (1974).

83. Maglieri and Bobbitt, "History of Sonic Boom Technology," 69–72.

84. For the governing equations for the wave field near a caustic, see J.P. Guiraud, "Acoustique Géométrique Bruit Ballistique des Avions Supersoniques et Focalisation," *Journal Mécanique*, 4 (1965): 215–267, cited by Plotkin and Maglieri, "Sonic Boom Research," 5.

85. George T. Haglund and Edward J. Kane, "Flight Test Measurements and Analysis of Sonic Boom Phenomena Near the Shock Wave Extremity," NASA CR 2167 (February 1973); Benson, Maglieri, interview, March 19, 2009.

86. Domenic Maglieri et al.," Measurement of Sonic Boom Signatures from Flights at Cutoff Mach Number," in *Third Conference on Sonic Boom Research, October 29–30, 1970*, ed. Ira R. Schwartz, SP-255 (Washington, DC: NASA, 1971), 243–254. Figure 1-7 extracted from 252.

87. Harry W. Carlson, "Some Notes on the Present Status of Sonic Boom Prediction and Minimization Research," in Schwartz, *Third Conference on Sonic Boom Research*, 395.

Langley's Unitary Plan Wind Tunnel, shown here upon completion in 1955, had two 4-by-4-by-7-foot test sections and could generate speeds up to Mach 4.63. (NASA)

CHAPTER 2

The SST's Sonic Boom Legacy

The rapid progress made in understanding the nature and significance of sonic booms during the 1960s stemmed from the synergy among flight testing, wind tunnel experiments, psychoacoustical studies, theoretical refinements, and powerful new computing capabilities. Vital to this process was the largely free exchange of information by NASA, the FAA, the USAF, the airplane manufacturers, academia, and professional organizations such as the American Institute of Aeronautics and Astronautics (AIAA) and the Acoustical Society of America (ASA). The sharing of much of this information even extended to counterparts in Europe, where the rival Anglo-French Concorde supersonic airliner got off to a head start on the more ambitious American program.

Designing commercial aircraft has long required a variety of tradeoffs involving cruising, landing, and takeoff speeds; range; passenger or cargo capacity; weight, with and without payload; durability; comfort; safety; and, of course, costs—both for manufacturing and operations. Balancing such factors was especially challenging with an aircraft as revolutionary as the SST, which was expected to cruise at about Mach 3 while still being able to take off and land at existing airports. Unlike with previous supersonic military aircraft, NASA's scientists and engineers and their partners in industry increasingly had to also consider the environmental impacts of their designs, including engine noise around airports, the effects of high-altitude exhaust on the upper atmosphere—especially the little understood ozone layer—and, of course, the inevitable sonic boom.[1]

As the program progressed, the FAA set a desired goal for the SST's sonic boom level of 2.0 psf when accelerating and 1.5 psf during cruise in hopes that this would be acceptable to the average person exposed to the booms on the ground. At NASA's aeronautical centers, especially Langley, aerodynamicists tried to incorporate the growing knowledge about the physics of sonic booms into their equations, models, and wind tunnel experiments to meet or exceed this goal—even as the research described in the previous chapter revealed more about the psychoacoustics of human response.

Wind Tunnel Experimentation

The National Advisory Committee for Aeronautics, in making many if not most of its contributions to aviation technology, had relied heavily on ever more powerful and inventive wind tunnels. By nurturing the development and improvement of both military and civilian aircraft, the NACA's wind tunnels became true national treasures. "They were logical and flexible instruments, useful for theoretical explorations as well as highly applied studies. More crucially, tunnels allowed researchers to shift back and forth from mathematical models to flight [data], thus increasing the reliability of models while also serving to predict aircraft performance."[2] These wind tunnels remained just as essential during the early years of NASA, especially in helping design supersonic and hypersonic vehicles.

The aerodynamic and structural characteristics traditionally examined in a wind tunnel—including lift, drag, stability, control, angle of attack (AOA)—were determined largely by measuring forces applied to the model itself, measuring air pressure with special sensors, and using schlieren photography and other techniques to show shock waves or airflow close to its surface. Air is considered a fluid when moving past objects (hence the inseparable relationship between aerodynamics and fluid dynamics). The effects of air flowing over the surfaces of a wind tunnel model, whether laminar (smooth) or turbulent, can be correlated directly with those of full-scale airframes by a scaling parameter known as the Reynolds number.[3]

Using models in supersonic wind tunnels to examine sonic booms posed new and difficult challenges. Ever since Whitham's analyses, sonic boom researchers have known the area within a few body lengths of an airframe—where it generates multiple shock waves at transonic and supersonic speeds—as the near field. Wind tunnel models could provide fairly accurate results at this distance but not too far beyond. To examine how these shock waves propagate and begin to coalesce, researchers needed to measure them many more body lengths away in the midfield. And to determine the final outcome—specifically, how they evolve into the typical N-shaped signature of a sonic boom—results were needed at an even greater distance, known as the far field. As time went on, researchers developed clever innovations and techniques to work around these inherent limitations as best as they could.

In 1959, Harry W. Carlson of Langley conducted what may have been the first wind tunnel experiment on sonic boom generation. As reported that December, he had tested seven models of various geometrical and airplane-like shapes at differing angles of attack in Langley's original 4-by-4-foot supersonic wind tunnel (not the one in the photo) at a speed of Mach 2.01. The tunnel's relatively limited interior space mandated building extremely small models to obtain useful shock wave signatures: about 2 inches in length for measuring them at 8-body-lengths distance and only three-quarters of an inch for trying to measure them at 32 body lengths (as close as possible to the far field). Compatible with Whitham's theory, many of Carlson's models consisted of the better understood and easier to measure "equivalent bodies of revolution." (This was the accepted technique for translating the complex shape of airframes with their wings and other surfaces using the area rule into standard aerodynamic principles governing simpler projectiles with rounder cross sections). Carlson determined these models to be suitable substitutes for more realistic, nonaxisymmetrical airplane-shaped models in obtaining theoretical estimates of far-field bow shock pressures. Although his more realistic, airplane-shaped model could not reach far-field conditions, the overall results correlated with existing theory, such as Whitham's formulas on volume-induced overpressures and Walkden's on those caused by lift.[4] Carlson's attempt to design one of the models to alleviate the strength of the bow shock was unsuccessful, but this can be considered NASA's first experimental attempt at boom minimization.

In April 1959, before the results of either Carlson's wind tunnel or those of the first flight tests at Wallops Island were published, he and Domenic Maglieri advised about sonic boom implications early in the Supersonic Commercial Air Transport program. Based on existing theory, some USAF and British reports, and preliminary findings in their own experiments, they concluded "that for the proposed supersonic transport airplanes of the future, booms on the ground will most probably be experienced during the major portion of the flight plan. The boom pressures will be most severe during the climb and descent phases of the flight plan."[5] Although they warned that sonic booms during cruise would extend laterally for many miles, it was hoped that special operating procedures and high altitudes could help alleviate both problems to some extent.

The extreme precision demanded in making the tiny models needed for early sonic boom experiments, the disruptive effects of the sting assemblies needed to mount them (which inevitably distorted tail shocks), the vibration by the models, the extra sensitivity required of pressure-sensing devices, and the interactions with a tunnel's walls all limited a wind tunnel's ability to measure the type of shock waves that would reach the ground from a full-sized aircraft, especially one as large as the planned SST. Even so, substantial progress continued, and the data served as useful cross-checks on flight-test data and mathematical formulas.[6] For example, in 1962 Harry Carlson used a 1-inch model of a B-58 to make the first direct correlation of recent flight-test data (described in the previous chapter) with wind tunnel results and sonic boom theory. His findings proved that wind tunnel readings, with appropriate analysis, could be used with some confidence to estimate sonic boom signatures.[7] Several months later, he concluded that locating the major portion of an SST's

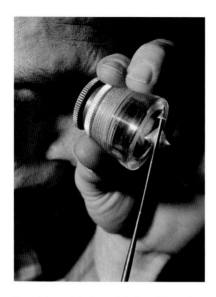

Examining a 1-inch model of the XB-70 in 1961. (NASA)

lift-generating surface aft of the maximum cross-sectional area could lower sonic boom overpressure, a principle thereafter considered in the design of most planned SST configurations.[8]

Exactly 5 years after publishing results of his first wind tunnel sonic boom experiment, Harry Carlson was able to report, "In recent years, intensive research efforts treating all phases of the problem have served to provide a basic understanding of this phenomenon. The theoretical studies [of Whitham and Walkden] have resulted in correlations with the wind tunnel data…and with the flight data."[9] As for the prospect of minimizing the strength of sonic booms, wind tunnel tests of SCAT models had revealed that some configurations (e.g., the arrow wing) produced lower overpressures.[10] The challenge was to find configurations that would reduce sonic booms without significantly sacrificing other needed attributes.

In 1967, Ames researchers Raymond Hicks and Joel Mendoza greatly improved the ability of wind tunnels to predict sonic boom characteristics. Experimenting with a 12-inch model of the XB-70 in the Ames 7-by-9-foot supersonic wind tunnel at Mach 1.8, they applied Whitham's near-field F-function theory to compare pressure readings at one body length in the wind tunnel with actual flight-test data from 4.5 body lengths and 290 body lengths from a real XB-70. This resulted in a new, more reliable method for extrapolating near-field F-function measurements to the far field, allowing the use of much larger and therefore more accurate models for that purpose.[11]

Mobilizing Brainpower To Minimize the Boom

Motivated by the SST and Concorde programs, scientists and engineers rapidly expanded the knowledge and understanding of sonic boom theory during the 1960s. Much of their efforts focused on ways to predict the sonic booms that would be produced by various aircraft configurations and how to modify them to lower the impact of the shock waves that reached the surface—a goal that became known as minimization or mitigation.

At the very start of the decade, when the British Aircraft Corporation (BAC) was exploring options for a supersonic airliner, L.B. Jones, an aerodynamicist at English Electric Aviation (a BAC subsidiary), added to the fundamental understanding of sonic booms (or bangs) pioneered by Whitham and Walkden.[12] Noting with some understatement that "the sonic bangs caused by supersonic aircraft can be a nuisance [with] a level of noise...on the borderline of acceptable value," Jones introduced his theory by observing that "it seems important to examine ways of reducing them at the aircraft design stage." He presented equations for ways of lowering shock waves in the far field caused by lift, volume, and lift plus volume.[13] Although these hypothetical designs were too blunt to be practical, his work marked the first significant theory on how supersonic aircraft might be designed to reduce boom intensity. Such possibilities were soon being explored by NASA aerodynamicists and a growing number of NASA partners in the American aerospace industry and university engineering departments.

In addition to publishing results of their tests and experiments in technical reports and academic journals, researchers began presenting their findings at special conferences and professional symposia dealing with supersonic flight. One of the earliest such gatherings took place from September 17 to September 19, 1963, when NASA Headquarters sponsored an SST feasibility studies review at the Langley Research Center—attended by Government, contractor, and airline personnel—that examined every aspect of the planned SST. In a session on noise, Harry Carlson warned that "sonic boom considerations alone may dictate allowable minimum altitudes along most of the flight path and have indicated that in many cases the airframe sizing and engine selection depend directly on sonic boom."[14] On top of that, Harvey Hubbard and Domenic Maglieri discussed how atmospheric effects and community response to building vibrations might pose problems with the current SST sonic boom objectives (2 psf during its acceleration and 1.5 psf while cruising).[15]

The conferees discussed various other technological challenges for the planned American SST, some indirectly related to the sonic boom issue. For example, because of frictional heating, an airframe covered largely with stainless steel (such as the XB-70) or with titanium (such as the still-top-secret A-12/YF-12) would cruise at Mach 2.7+ and over 60,000 feet, an altitude which many still hoped would allow the sonic boom to weaken by the time it reached the surface. Manufacturing such a plane, however, would be much more expensive than manufacturing a Mach 2.2 SST with aluminum skin, such as the design being planned for the British-French Concorde, which the United Kingdom and France had formally approved for joint development on November 29, 1962. (Interestingly, this agreement had no provision for either side to back out.)[16]

Despite serious and potentially incurable problems raised at the NASA conference concerning cost and feasibility, the FAA, spurred on by the Concorde agreement, had already released the SST Request for Proposals (RFP) on August 15, 1963. Thereafter, as explained by Langley's long-time supersonic expert, F. Edward McLean, "NASA's role changed from one of having its own concepts evaluated by the airplane industry to one of evaluating the SST concepts of the airplane industry."[17] By January 1964, Boeing, Lockheed, North American, and their jet-engine partners had submitted initial proposals, with Boeing drawing upon NASA's swing-wing SCAT-16 concept and Lockheed's proposal resembling the SCAT-17 with its canard and delta-wing configuration. North American's design, which relied heavily on its XB-70 but did not benefit from NASA's concepts, was soon eliminated from the competition.[18] In retrospect, the manufacturers and Government advocates of the SST were obviously hoping that technology would catch up with requirements before it went into production. The SST program schedule was too compressed, however, for many of the emerging concepts on controlling sonic booms to be incorporated or retrofitted into the contractors' designs.

With the SST program now well under way, a growing awareness of the public response to booms became one factor among those that tri-agency (FAA-NASA-DOD) groups in the mid-1960s, including the PAC chaired by Robert McNamara, considered in evaluating the proposed SST designs. The sonic boom issue also became the focus of a rather skeptical committee of the National Academy of Sciences between 1964 and 1965 and attracted growing attention from the academic and scientific community at large, much of it increasingly negative.[19]

By 1965, NASA specialists at Langley had been studying possible ways to address the sonic boom problem for the past 5 years. In June, Ed McLean pointed out that, contrary to current asymptotic far-field theory, the near-field shock waves from a transonically accelerating SST do not necessarily have to evolve into the final form of an N-wave. This opened the prospect for a properly designed, large supersonic aircraft flying at the right altitude to avoid projecting a full sonic boom to the surface.[20]

Of major significance at the time and even more potentially for the future, improved data-reduction methods and numerical evaluations of sonic boom theory were being adapted for processing with new codes in the latest International Business Machines (IBM) computers. Langley used this capability for the first application of high-speed computers on the aerodynamic design of supersonic aircraft—which was considered a "quantum leap in engineering analysis capability."[21] Meanwhile, Boeing developed one of the most widely known of the early sonic boom computer programs to help in designing its SST candidate.[22] Automated data processing allowed faster and more precise

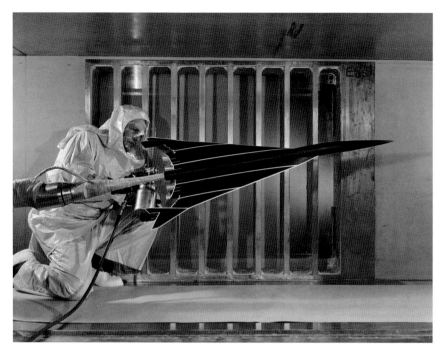

Wind tunnel test of SCAT-15F model. (NASA)

correlations between wind tunnel and flight-test data, leading to continued refinements in sonic boom theory (although still mainly applicable to bow shocks during steady and level flight in a standard atmosphere).[23] Applying these new capabilities, Carlson, McLean, A. Warner Robins, and their colleagues at Langley designed the SCAT-15F, an improved SST concept with a highly swept "arrow wing" optimized for highly efficient cruise (and, to some extent, a lower sonic boom).[24] Solving resultant problems with stability and control at low speeds was more difficult and came too late for Boeing to adapt this design for its SST in the late 1960s, but the lessons learned from the SCAT-15F would be of value in future supersonic transport studies.[25]

The Acoustical Society of America, made up of professionals from all fields involving sound (ranging from music to audiology, and from noise to vibration), sponsored its first Sonic Boom Symposium on November 3, 1965, as part of its 70th meeting in—appropriately enough—St. Louis. McLean, Hubbard, Carlson, Maglieri, and other Langley experts presented papers on the background and techniques of sonic boom research as well as their latest findings.[26] The paper by McLean and Barrett L. Shrout included details on the potential breakthrough in using near-field shock waves to evaluate wind tunnel models for boom minimization—in this case, a reduction in maximum overpressure in a climb profile from 2.2 psf to 1.1 psf. This technique also allowed for the use of 4-inch models, which were easier to fabricate to the close tolerances required for accurate shock wave measurements.[27] Harry Carlson described how the Langley Research Center's latest high-speed computer programs for analyzing the lift and drag of aerodynamic configurations were being used by both NASA and the manufacturers to calculate F-function results and theoretical pressure signatures at various distances.[28]

In addition to the scientists and engineers employed by the aircraft manufactures, many eminent researchers in academia took on the challenge of discovering ways to minimize the sonic boom, usually with NASA's sponsorship and support. These included the influential team of Albert R. George and A. Richard Seebass of Cornell University, which had one of the Nation's premier aeronautical laboratories. Seebass, already prominent in the field of aerospace engineering at 29 years old, edited the proceedings of NASA's first sonic boom research conference, held on April 12, 1967. The meeting was chaired by Wallace D. Hayes of Princeton University, who was now devoting much of his attention to sonic boom mitigation. Hayes was well known for his groundbreaking work in supersonic and hypersonic aerodynamics, which began with his 1947 dissertation, "Linearized Supersonic Flow," written while at the California Institute of Technology (which, in mathematical terms, foreshadowed Whitcomb's area rule).[29] The conference was attended by more than 60 other Government, industry, and university experts in aeronautics and related fields. In reviewing the area rule as it applied to supersonic flight, Hayes cautioned "that the total equivalent source strength connected with the sonic boom cannot ever be zero. Thus the sonic boom below the aircraft is truly inescapable. The best we can hope for is that the boom is a minimum for given values of this parameter, with limits on the magnitude of the drag."[30]

Boeing had been selected over Lockheed as the SST prime contractor less than 4 months earlier, but public acceptance of even a somewhat reduced sonic boom was becoming recognized far and wide as a possibly fatal flaw for its future production or at least for allowing it to fly supersonically over land.[31] The two most obvious theoretical ways to minimize sonic booms during supersonic cruise—flying much higher with no increase in weight or building an airframe 50-percent longer at half the weight—were not considered realistic.[32] Furthermore, as was made apparent from a presentation by Domenic Maglieri on flight-test findings, such an airplane would still have to deal with the problem of the as-yet somewhat unpredictable, stronger booms caused by maneuvering, accelerating, and atmospheric conditions.[33]

The stated purpose of this conference was "to determine whether or not all possible aerodynamic means of reducing sonic boom overpressure were being explored."[34] In that regard, Harry Carlson showed how the various computer programs being used at Langley (mentioned above) were complementing improved wind tunnel experiments for examining boom minimization concepts. As shown by figure 2-1, even minor changes in an aircraft's configuration could result in a significant alteration of its shock waves. In this case, a wind tunnel comparison of 4-inch

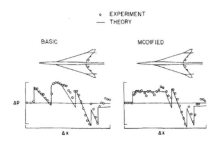

Figure 2-1. Wind tunnel verification of a flat-top signature. (NASA)

models at Mach 1.4 confirmed that the shape on the right rendered a quieter flattop signature for shock waves, albeit at a distance of only five body lengths.[35]

Boeing's sonic boom expert, Edward J. Kane, presented recent research on dealing with the tricky problems of atmospheric effects, while Albert George from Cornell explored the potential for reducing sonic boom overpressures reaching the surface by designing airframes that could disperse some portion of shock waves caused by volume off to the sides of the flightpath.[36]

Notable aeronautics pioneer Adolf Busemann (see chapter 1), now at the University of Colorado, expressed both frustration with the current situation and guarded hope for a solution. His outlook probably reflected the feelings of many other SST proponents in both Government and industry.

> Since people are not satisfied with the sonic boom reduction which the reasonable altitude for supersonic flights provides naturally, further means for reductions must either be found or proved to be impossible. However, to call something impossible is dangerous. Our time is full of innovations in physics and technology, and although we have certain laws of conservation which we accept as being invariably valid, many scientists who declared that desirable effects were impossible have been proved wrong.[37]

After all the papers were read and discussed, many of the attendees agreed that additional avenues of research were promising enough to be explored, but they were still concerned whether low-enough sonic booms were possible using existing technologies. Accordingly, NASA's Office of Advanced Research and Technology, which hosted the conference, established specialized research programs on seven aspects of sonic boom theory and mitigation at five American universities—Columbia, Colorado, Cornell, Princeton, and New York University (NYU)—and the Aeronautical Research Institute of Sweden.[38] This mobilization of aeronautical brainpower almost immediately began to pay dividends.

Seebass and Hayes cochaired NASA's second sonic boom conference from May 9 to May 10, 1968. This came just a few weeks after Boeing replaced its variable-sweep 2707-200 SST design, which was found to be too heavy, with the 2707-300 (figure 2-2), a delta-wing configuration similar to the losing Lockheed proposal. The conference included 19 papers on the latest boom-related testing, research, experimentation, and theory by specialists from NASA and participating universities. The advances made in 1 year were impressive. In the area of theory, for example, the fairly straightforward linear techniques for predicting the propagation of sonic booms from slender airplanes such as the SST had proven reliable, even for calculating some nonlinear (mathematically more complex and unpredictable) aspects of their signatures. Additional field testing had improved understanding of the geometrical acoustics caused by atmospheric conditions. Many of the papers—including those from Seebass, George, Hayes, McLean, and Carlson—presented promising aerodynamic techniques for reducing the strength of sonic booms.[39]

One of the most celebrated aerodynamicists recruited by NASA to work on the sonic boom problem was Antonio Ferri of New York University. An Italian air force officer and pioneer of supersonic research in prewar Italy, he had joined the anti-Nazi resistance movement after the collapse of the Mussolini regime and became a partisan leader before escaping in 1944 to the United States. There, he continued advancing high-speed research for the NACA at Langley for several years before entering the academic world. At the conference, he reported several innovative ideas on how to design a 300-foot SST airframe with reduced sonic booms by spreading lift along almost its entire length by means of suitable volume adjustments. Following these principles, he predicted, could yield maximum overpressures of about 1 psf while cruising at 60,000 feet (as compared to the 2 psf expected with existing SST designs) without seriously hurting their lift-to-drag ratio.[40]

In retrospect, two of the papers that would prove most significant to future progress dealt with new computer processing capabilities. Representing Aeronautical Research Associates of Princeton (ARAP), Wallace Hayes reported on what became known as the ARAP Program. Using the then-ubiquitous FORTRAN computer language, it consisted of a master program called SONIC with 19 subroutines. NASA had sponsored this project to clarify the confusion that existed among the various complex numerical techniques being used for calculating the propagation of sonic boom signatures and comparisons with flight-test measurements. Based on linear geometric acoustics, the ARAP Program used F-function effects from a supersonic airframe at various Mach numbers and lift coefficients combined with acoustic-ray tracing and an age variable to define (if not yet solve) the nonlinear effects that help shape sonic boom signatures. It was the first computer program with an algorithm comprehensive enough to accurately calculate a full range of overpressure signatures in a standard, horizontally stratified atmosphere with winds—a major advance.[41] Using this, Hayes extended McLean's 1965 hypothesis on the persistence of near-field pressure signatures by showing that effects in the real atmosphere would tend to "freeze" the signature from supersonic aircraft at cruise altitudes before it reached the surface.[42]

Harvard Lomax of Ames offered a sneak preview of a more distant digital future that would eventually be possible through the marriage of computational fluid dynamics (CFD), which was being pioneered at Ames, with computer graphics. He reported preliminary results on using a cathode-ray tube monitor directly connected to the core processor of a mainframe computer to show in real time the results of three-dimensional, nonlinear flow-field analyses of a dozen diverse aircraft configurations in a search for lower boom signatures. After describing the mathematical principles involved, Lomax presciently predicted that "the ability to compute flow fields for airplanes traveling at supersonic speeds with the aid of an immediate visual display of the calculations as they proceed opens the possibility of devising new, or revising parts of old, numerical techniques."[43] As will be shown in later chapters, the full realization of this capability with the development of super computers and massively parallel processors would eventually prove to be the key to the successful design of low-boom airplane configurations.

Despite these signs of considerable progress made by 1968, several important theoretical problems remained unresolved, such as the prediction of sonic boom signatures near a caustic (a major objective of the 1970 Jackass Flats testing described in the previous chapter), the diffraction of shock waves into "shadow zones" (areas normally skipped over between primary and secondary sonic boom carpets), nonlinear shock wave behavior near an aircraft, and the still somewhat mystifying effects of turbulence. Ira R. Schwartz of NASA's Office of Advanced Research and Technology summed up the state of sonic boom minimization as follows: "It is yet too early to predict whether any of these design techniques will lead the way to development of a domestic SST that will be allowed to fly supersonic over land as well as over water."[44]

The challenge to the SST posed by the sonic boom became even more serious shortly after the conference. In July 1968, President Lyndon Johnson signed into law a bill requiring the Federal Aviation Administrator to "prescribe and amend such rules and regulations as he may find necessary to provide for the control and abatement of aircraft noise and sonic boom."[45] Many expected this authority would effectively prohibit supersonic flight over land as several other nations were already considering.

Rather than conduct another meeting the following year, NASA deferred to a conference hosted by the North Atlantic Treaty Organization's (NATO's)

Advisory Group for Aerospace Research & Development (AGARD) on aircraft engine noise and sonic boom, which was held in Paris in May 1969. Experts from the United States and five other nations—including the two facing similar issues with the Concorde—attended this forum, which consisted of seven sessions. Three sessions and a roundtable dealt with the status of boom research and the challenges ahead.[46]

As reflected in these conferences, the three-way partnership between NASA, Boeing, and the academic aeronautical community during the late 1960s continued to yield new knowledge about sonic booms as well as scientific and technological advances in exploring ways to deal with them. In addition to the flight-test and wind tunnel data described in the previous chapter, some of this progress came from new experimental techniques, some of them quite ingenious.

Laboratory Devices and Experiments

NASA and its contractors developed several types of simulators, both large and small, that proved useful in studying the physical and psychoacoustic effects of sonic booms. The smallest (and least expensive) was a spark-discharge system. Langley and other laboratories used these bench-type devices for basic research into the physics of pressure waves. Langley's system created miniature sonic booms by using parabolic, or two-dimensional mirrors to focus the shock waves caused by discharging high-voltage bolts of electricity between tungsten electrodes toward precisely placed microphones. Such experiments were used to verify laws of geometrical acoustics. The system's ability to produce shock waves that spread out spherically proved useful for investigating how the cone-shaped waves generated by aircraft will interact with buildings.[47]

For studying the effects of temperature gradients on boom propagation, Langley used a ballistic range consisting of a helium-gas launcher that shot miniature projectiles at constant Mach numbers through a partially enclosed chamber. The atmosphere inside could be heated to ensure a stable atmosphere for accuracy in boom measurements.

Innovative NASA-sponsored simulators included Ling-Temco-Vought's shock-expansion tube—basically a mobile, 13-foot-diameter conical horn mounted on a trailer—and General American Research Division's explosive, gas-filled envelopes suspended above sensors at Langley's sonic boom simulation range.[48] Other simulators were devised to handle both human and structural response to sonic booms. (The need to better understand effects on people was called for in a report released in June 1968 by the National Academy of Sciences.)[49] Unlike the previously described studies using actual sonic booms created by aircraft, these devices had the advantages of a controlled laboratory environment. They allowed researchers

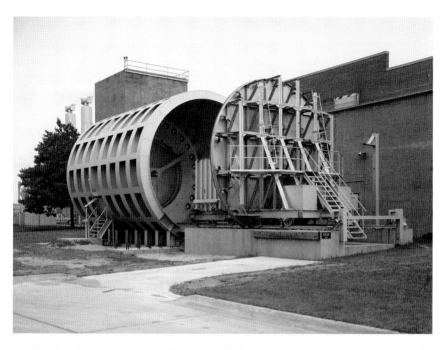

Langley's Low-Frequency Noise Facility, built originally for testing the extremely loud sounds of Apollo booster rockets. (NASA)

to produce multiple boom signatures of varying shapes, pressures, and durations as often as needed at a relatively low cost.[50]

Langley's Low-Frequency Noise Facility—built earlier in the 1960s to generate the intense, chest-pounding, eardrum-splitting sounds of giant Saturn boosters during Apollo launches—also performed informative sonic boom simulation experiments. As indicated by the photograph, it was a large, cylindrical test chamber 24 feet in diameter and 21 feet long that could accommodate people, small structures, and materials for testing. The facility's electrohydraulically operated 14-foot piston was capable of producing low-frequency sound waves from 1 hertz (Hz) to 50 Hz (sort of a super subwoofer) and sonic boom N-waves from 0.5 psf to 20 psf at durations from 100 milliseconds to 500 milliseconds.[51]

To provide an even more versatile system designed specifically for sonic boom research, NASA contracted with General Applied Sciences Laboratories (GASL) of Long Island, NY, to develop an ideal simulator using a quick-action valve and shock-tube design. (Antonio Ferri was the president of GASL, which he had cofounded with the illustrious Hungarian-born scientist and airpower visionary Theodore von Kármán in 1956). Completed in 1969, this new simulator consisted of a high-speed flow valve that sent pressure-wave bursts through a heavily reinforced, 100-foot long conical duct that expanded into an 8-by-8-foot test section with an instrumentation and model room. It could generate overpressures up to 10 psf with durations from 50 milliseconds to 500 milliseconds. Able to operate at less than a 1-minute interval between bursts, its sonic boom signatures proved very accurate and easy to control.[52] In the opinion of Ira Schwartz, "the GASL/NASA facility represents the most advanced state of the art in sonic boom simulation."[53]

Losing the Battle Against the Boom

While NASA and its partners were learning more and more about the nature of sonic booms, the SST was becoming mired ever deeper in controversy. Many in the public, the press, and the political arena were concerned about the noise SSTs would create—both around airports with its powerful engines and elsewhere with its sonic boom carpet—with a growing number expressing hostility to the entire SST program. As one of the more reputable critics wrote in 1966, with a map showing a dense network of future boom carpets crossing the United States, "the introduction of supersonic flight, as it is at present conceived, would mean that hundreds of millions of people would not only be seriously disturbed by the sonic booms … they would also have to pay out of their own pockets (through subsidies) to keep the noise-creating activity alive."[54]

Opposition to the SST grew rapidly in the late 1960s, becoming a cause célèbre for the emerging environmental movement as well as a target for small-Government conservatives opposed to federal subsidies.[55] Typical of the growing trend among opinion makers, the *New York Times* published its first strongly anti–sonic boom editorial in June 1968, linking the SST's potential sounds with an embarrassing incident the week before when an F-105 flyover shattered 200 windows at the Air Force Academy, injuring a dozen people.[56] The next 2 years brought a swelling crescendo of complaints about the supersonic transport, both for its expense and the problems it could cause—even as research on controlling sonic booms began to bear some fruit.

By the time 150 scientists and engineers gathered in Washington, DC, for NASA's third sonic boom research conference on October 29 and 30, 1970, the American supersonic transport program had less than 6 months left to live. Thus, the 29 papers presented at this conference and other papers at the ASA's second sonic boom symposium in Houston, TX, the following month might be considered, in their entirety, a final status report on sonic boom research during the SST decade. The reports in Washington, many of which followed up on presentations at the 1968 conference, covered the full range of topics related to nature, measurement, and mitigation of sonic booms. The otherwise more limited agenda in Houston also included papers on the issues of human and animal response.[57]

Of future if not near-term significance, NASA and its partners were making considerable progress in understanding how to design airplanes that could fly faster than sound while leaving behind a gentler sonic footprint. As summarized by Ira Schwartz:

> In the area of boom minimization, the NASA program has utilized the combined talents of Messrs. E. McLean, H.L. Runyan, and H.R. Henderson at NASA Langley Research Center, Dr. W.D. Hayes at Princeton University, Drs. R. Seebass and A.R. George at Cornell University, and Dr. A. Ferri at New York University to determine the optimum equivalent bodies of rotation that minimize the overpressure, shock pressure rise, and impulse [i.e., the total amount of pressure variation] for given aircraft weight, length, Mach number, and altitude of operation. Simultaneously, research efforts of NASA and those of Dr. A. Ferri at New York University have provided indications of how real aircraft can be designed to provide values approaching these optimums.... This research must be continued or even expanded if practical supersonic transports with minimum and acceptable sonic boom characteristics are to be built.[58]

Any consensus among the attendees about the progress they were making on the sonic boom issue was tempered by their awareness of the financial problems now plaguing the Boeing Company and the political difficulties facing the administration of President Richard M. Nixon in continuing to subsidize the American SST. Many attendees also seemed resigned to the reality that Boeing's final 2707-300 design (figure 2-2), with its 306-foot length and 64,000-foot cruising altitude, would never come close to passing the overland sonic boom criteria for civil aircraft being proposed by the FAA.[59]

Although the noise level ultimately deemed acceptable by the public was still uncertain, the consensus was that the N-wave signature of an acceptable SST must be reduced to at least 1 psf to allow cruising at supersonic speeds over the United States. As Antonio Ferri lamented, "programs for the first

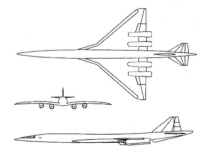

Figure 2-2. Boeing 2707-300 SST, final design. (NASA)

generation of these airplanes have been initiated without a complete under-standing of the effects of the sonic boom on the population and their reaction against it."[60] For designing the next generation of large supersonic transports, he offered several concepts developed since the last conference (including an imaginative biplane configuration) as well as operational techniques and flight profiles to reduce overpressures to a suitable level.[61]

The shock waves from cruising at a low supersonic speed (up to about 840 mph) and the right altitude, a condition known as Mach cutoff (see figure 1-8), would under the proper conditions either refract away from the surface or not coalesce enough to cause a full sonic boom. In view of this, there was some hope that carefully planned operations while over land might still make the SST economically practical. This, however, would require solu-tions to issues associated with such variables as meteorological conditions, typographic effects, building vibrations, caustics, and super booms—some of which were being clarified by the BREN Tower tests described at the end of the previous chapter.[62]

On related issues, Albert George and one of his graduate students, Kenneth Plotkin, substantiated much about the complex relationship between turbu-lence and shock wave scattering as well as N-wave distortions,[63] phenomena that were further refined by an examination of multiple scattering of shock waves in a turbulent atmosphere by two researchers at Columbia University, W.J. Cole and M.B. Freidman.[64] George had reported earlier on how ways of lowering tail shock as well as bow shock and other factors could reduce the lower bounds for sonic booms well below the accepted levels calculated by L.B. Jones and at shorter distances.[65] For his part, Jones continued to extend his earlier results on lower bounds for bow-pressure shocks to the midfield and far field, although only in a homogenous atmosphere.[66]

Thanks to new computer capabilities, Richard Seebass reported on progress in using linear equations to study the nonlinear characteristics of shock waves at a caustic by means of automated numerical analysis and graphical represen-tation.[67] Figure 2-3, representing acoustic rays interacting to create a caustic, shows this early computer-generated graphing capability (which can be com-pared with the increasingly detailed and sophisticated CFD images illustrated in later chapters).[68] Much more study and testing would be needed, however, to make the necessary quantitative predictions of sonic boom intensities needed for even transonic civilian flight.

In view of such unresolved technical issues as well as overriding political and economic factors, Seebass generally echoed Ferri's opinion, noting, "We should adopt the view that the first few generations of supersonic transport (SST) aircraft, if they are built at all, will be limited to supersonic flight over oceanic and polar regions."[69] In view of such concerns, some of the attendees were even looking toward a more distant future when hypersonic aerospace vehicles might be able to cruise high enough to leave only an acceptable boom carpet down at the surface.

As for the ongoing and future technological challenges of quieter supersonic flight, Lynn Hunton of the Ames Research Center warned that "with regard to experimental problems in sonic boom research, it is essential that the techniques and assumptions used be continuously questioned as a requisite for assuring the maximum in reliability."[70] Harry Carlson probably expressed the general opinion of NASA's aero-

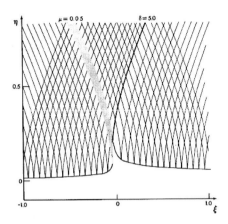

Figure 2-3. An early computer-generated depiction of a simple shock wave (shaded area) among acoustic rays at a caustic, by Richard Seebass, 1970. (NASA)

dynamicists when he cautioned that "the problem of sonic boom minimization through airplane shaping is inseparable from the problems of optimization of aerodynamic efficiency, propulsion efficiency, and structural weight.... In fact, if great care is not taken in the application of sonic boom design principles, the whole purpose can be defeated by performance degradation, weight penalties, and a myriad of other practical considerations."[71]

In view of the SST's other technical, operational, political, and economic hurdles, lowering its sonic boom in time for the final design of the Boeing 2707-300 would probably not have been enough to save the program. In any case, after both the U.S. House of Representatives and U.S. Senate voted in March 1971 to discontinue SST funding, a joint conference committee confirmed its termination in May.[72]

A Decade of Progress in Understanding Sonic Booms

The cancellation of the SST and related cuts in supersonic research inevitably slowed the momentum for dealing with sonic booms. Even so, the ill-fated SST program, which invested approximately $1 billion in supersonic research, left behind a wealth of data and discoveries about sonic booms—including measurements of more than 100,000 booms. As documented evidence of part of this effort, the Langley Research Center alone produced or sponsored more than 200 technical publications on the subject over a span of 19 years, most related to the SST program. (Many of those published in the early 1970s were

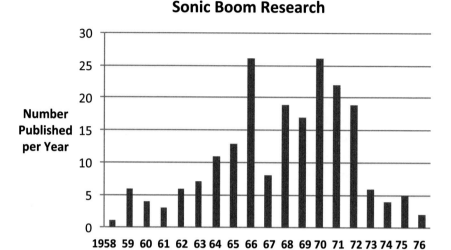

Figure 2-4. Reports produced or sponsored by NASA Langley, 1958–1976. (Author)

based on previous SST-related research and testing.) This literature, graphically depicted in figure 2-4, would be of enduring value in the future.[73]

Thanks mostly to the SST program, great progress had been made in the understanding and application of sonic boom theory at speeds up to Mach 3.0. Using the latest algorithms and geometric techniques, experts could now predict the evolution of a sonic boom signature from its shock pattern near an aircraft configuration to the surface either by extrapolating measurements from several body lengths away in a supersonic wind tunnel model or by using Whitham's F-function as calculated from the volume and lift distribution of the aircraft. By using acoustic-ray tracing and other techniques, sonic boom theory could also account for the effects of variations of temperature, humidity, winds, and turbulence on sonic boom strength and behavior, and it could even predict the approximate location of focused super booms.[74]

As for the need to mitigate sonic booms, the research of the SST era pointed toward the avenues to be followed in the future. Flying careful profiles at speeds of about Mach 1.15 could avoid creating sonic booms on the surface,[75] but this probably would not provide enough of a speed advantage over conventional airliners, unless most of a supersonic airplane's route could be flown at much faster speeds. (Some would later consider such low-Mach speeds as appropriate on overland routes for small passenger airplanes.)

For cruising at higher Mach numbers, it had become apparent that lowering sonic booms to acceptable levels would require either a reduction in a bow shock's overpressure or an increase in its rise time (i.e., lessening the steepness of the initial spike in pressure). The latter solution was deemed impractical since extending and smoothing out the N-wave to lengthen the rise time and make the initial shock wave less shocking would require an extremely long airplane. Also, as indicated by the XB-70 tests and other human-response data, this proposed solution would not necessarily help to reduce annoying indoor vibrations caused by sonic booms, which some experts (such as George and Seebass) proposed were the result of a sonic boom's impulse. That impulse was a measure of the total momentum that a sonic boom signature could impart, for example, on a building, which tends to vibrate at low frequencies. (The rapid pressure changes heard during an N-wave's double boom will produce sound waves at a rate of 30 Hz to 300 Hz, but the relatively gradual drop in pressure between the bow shock and rear shock will produce vibrations between 3 Hz to 8 Hz, which is far below the range of human hearing.[76]) The impulse of a simplified signature is depicted along with an N-wave's initial rise time in figure 2-5.[77]

To help alleviate sonic booms, one futuristic idea that received some attention at the time was the projection of a heat or force field to create a long phantom body in the front and rear of the fuselage for eliminating troublesome shock waves. Although not quite impos-

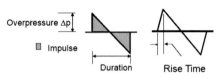

Figure 2-5. Simplified illustration of impulse and rise time. (NASA)

sible (assuming new inventions and ideal conditions), aerodynamic issues as well as enormous power requirements (not to mention the additional weight) made this proposal exceedingly unrealistic.[78]

Reducing overpressure—which could be predicted based on an aircraft's Mach number, length, weight, altitude, and equivalent area distribution—therefore seemed to be the most feasible solution. As pointed out in numerous studies, the two basic ways of doing so were to lighten aircraft weight, thereby decreasing its need for boom-producing lift, and specially shaping an airframe to modify its shock waves, such as by creating a flattop signature.[79] The challenges lay in knowing exactly how to design such an airframe and knowing what level of overpressure would be acceptable to the general public.

In May 1971, the same month that the House-Senate conference committee put the last nails in the coffin of an American SST, Albert George and Richard Seebass completed a short but extremely influential treatise on sonic boom minimization theory that culminated their past several years of collaborative NASA-sponsored research. Published in the *AIAA Journal* that October under the descriptive title "Sonic Boom Minimization Including Both Front and Rear Shocks," its compact presentation of equations and graphs analyzed the parametric relations of shock waves to lift and area distribution as they affect the full sonic boom signature. Their findings offered the prospect of design- ing aircraft not only for controlling abrupt pressure rises to achieve what they referred to as "a bangless boom" but also for possibly reducing the vibrations that annoy people indoors.[80] Future psychoacoustical studies would show that this outcome indeed would seem significantly quieter than the normal N-wave signature. Efforts during the coming decades to design supersonic aircraft that could reshape the sonic boom would cite the George and Seebass minimization theory as a cornerstone.[81]

Endnotes

1. The previously cited McLean, *Supersonic Cruise Technology*, provides an overview of these challenges from the perspective of a designer and program manager.
2. Robert G. Ferguson, "Evolution of Aeronautics Research at NASA," in *NASA's First 50 Years: Historical Perspectives*, ed. Steven J. Dick, SP-2010-4704 (Washington, DC: NASA, 2010), 208–209. For an earlier summary, see Donald D. Baals and William R. Corliss, *Wind Tunnels of NASA*, SP-440 (Washington, DC: NASA, 1981).
3. Named after Osborne Reynolds of the University of Manchester in the 1880s. See Jeremy Kinney, "NASA and the Evolution of the Wind Tunnel," in, *NASA's Contributions to Aeronautics* 2, ed. Richard P. Hallion (Washington, DC: NASA, 2011), 313.
4. Harry W. Carlson, "An Investigation of Some Aspects of the Sonic Boom by Means of Wind-Tunnel Measurements of Pressures About Several Bodies at a Mach Number of 2.01," NASA TN D-161 (December 1959). Carlson used Langley's 4-by-4-foot Supersonic Pressure Tunnel, completed in 1948, for most of his experiments.
5. Domenic J. Maglieri and Harry W. Carlson, "The Shock-Wave Noise Problem of Supersonic Aircraft in Steady Flight," NASA Memo 3-4-59L (April 1959), 9.
6. For numerous examples of these wind tunnel experiments, see Runyan, "Sonic Boom Capsule Summaries" as well as the NTRS bibliographical database.
7. Harry W. Carlson, "Wind Tunnel Measurements of the Sonic-Boom Characteristics of a Supersonic Bomber Model and a Correlation with Flight-Test Ground Measurements," NASA TM-X-700 (July 1962).
8. Harry W. Carlson, "The Lower Bound of Attainable Sonic-Boom Overpressure and Design Methods of Approaching This Limit," NASA TN D-1494 (October 1962).
9. Harry W. Carlson, "Correlation of Sonic-Boom Theory with Wind Tunnel and Flight Measurements," NASA TR R-213 (December 1964), 1.
10. Evert Clark, "Reduced Sonic Boom Foreseen for New High-Speed Airliner," *New York Times*, January 14, 1965, 7, 12 (based on a visit to NASA Langley).
11. Raymond M. Hicks and Joel P. Mendoza, "Prediction of Sonic Boom Characteristics from Experimental Near Field Results," NASA TM-X-1477 (November 1967).

12. As research progressed, some experts began to define the sharp noise accompanying the sudden spikes in pressure as a sonic bang and the overall sound and vibrations of an extended N-wave as a sonic boom.

13. L.B. Jones, "Lower Bounds for Sonic Bangs," *Journal of the Royal Aeronautical Society* 65, no. 606 (June 1961): 433–436; sentences quoted are on page 433.

14. Harry W. Carlson, "Configuration Effects on Sonic Boom," in *Proceedings of NASA Conference on Supersonic-Transport Feasibility Studies and Supporting Research, September 17–19, 1963, Hampton, Virginia*, NASA TM-X-905 (December 1963), 381.

15. Harvey H. Hubbard and Domenic J. Maglieri, "Factors Affecting Community Acceptance of the Sonic Boom," in *Proceedings of NASA Conference on Supersonic-Transport Feasibility Studies and Supporting Research*, 399–412.

16. "Anglo-French Agreement," London, UK, November 29, 1962, accessed ca. February 1, 2009, *http://www.concordesst.com/history/docs/agreement.html*.

17. McLean, *Supersonic Transport Technology*, 46.

18. M. Leroy Spearman, "The Evolution of the High-Speed Civil Transport," NASA TM no. 109089 (February 2004), 6–7.

19. Conway, *High-Speed Dreams*, 122–124.

20. F. Edward McLean, "Some Nonasymptotic Effects of the Sonic Boom of Large Airplanes," NASA TN D-2877 (June 1965).

21. Chambers, *Innovations in Flight*, 32.

22. M.P. Friedman, E.J. Kane, and A. Sigalla, "Effects of Atmosphere and Aircraft Motion on the Location and Intensity of a Sonic Boom," *AIAA Journal* 1, no. 6 (June 1963): 1327–1335.

23. Ibid.; Carlson, "Correlation of Sonic-Boom Theory," 2–23.

24. For a status report on supersonic work at Langley and some at Ames, see William J. Alford and Cornelius Driver, "Recent Supersonic Transport Research," *Astronautics and Aeronautics* 2, no. 9 (September 1964): 26–37.

25. Chambers, *Innovation in Flight*, 32–35.

26. *JASA* 39, no. 5, pt. 2 (November 1966).

27. F. Edward McLean and Barrett L. Shrout, "Design Methods for Minimization of Sonic Boom Pressure-Field Disturbances," *JASA* 39, no. 5, pt. 2 (November 1966): 519–525. See also Harry W. Carlson, F. Edward McLean, and Barrett L. Shrout, "A Wind Tunnel Study of Sonic-Boom Characteristics for Basic and Modified Models

of a Supersonic Transport Configuration," NASA TM-X-1236 (May 1966).

28. Harry W. Carlson, Robert J. Mack, and Odell A. Morris, "Sonic Boom Pressure-Field Estimation Techniques," *JASA* 39, no. 5, pt. 2 (November 1966): 510–518.

29. Lane E. Wallace, "The Whitcomb Area Rule: NACA Aerodynamics Research and Innovation," chapter 5 in *From Engineering Science to Big Science*, ed. Pamela E. Mack, NASA SP-4219 (Washington, DC: NASA, 1998), 8, available online at *http://history.nasa.gov/SP-4219/Chapter5.html*.

30. Wallace Hayes, "Brief Review of the Basic Theory," *JASA* 39, no. 5, pt. 2 (Nov. 1966): 6.

31. Evert Clark, "Sonic Boom to Limit Speed of Superjets Across U.S.," *New York Times*, October 31, 1966, 1, 71; George Gardner, "Overland Flights by SST Still in Doubt," *Washington Post*, July 10, 1967, A7.

32. A.R. Seebass, ed., "Preface," in *Sonic Boom Research: Proceedings of a Conference Held at the National Aeronautics and Space Administration, Washington, DC, April 12, 1967*, SP-147 (Washington, DC: NASA, 1967), iv.

33. Domenic J. Maglieri, "Sonic Boom Flight Research—Some Effects of Airplane Operations and the Atmosphere on Sonic Boom Signatures," in *Sonic Boom Research: Proceedings of a Conference*, 25–48.

34. Seebass, "Preface," in *Sonic Boom Research: Proceedings of a Conference*, iii.

35. Harry Carlson, "Experimental and Analytic Research on Sonic Boom Generation at NASA," in *Sonic Boom Research: Proceedings of a Conference*, 9–23. Figure 2-1 is extracted from page 21.

36. Harry W. Carlson, "Experimental and Analytical Research on Sonic Boom Generation at NASA," in *Sonic Boom Research: Proceedings of a Conference*, 9–23; Edward J. Kane, "Some Effects of the Atmosphere on Sonic Boom," in *Sonic Boom Research: Proceedings of a Conference*, 49–64; A.R. George, "The Possibilities for Reducing Sonic Booms by Lateral Redistribution," in *Sonic Boom Research: Proceedings of a Conference*, 83–93.

37. Adolf Busemann, "Sonic Boom Reduction," in *Sonic Boom Research: Proceedings of a Conference*, 79. Among Busemann's later accomplishments was suggesting the use of ceramic tiles for the Space Shuttle.

38. Ira R. Schwartz, ed., "Preface," *Second Conference on Sonic Boom Research: Proceedings of a Conference Held at the National Aeronautics*

   and Space Administration, Washington, DC, May 9–10, 1968, SP-180 (Washington, DC: NASA, 1968), iv–v.

39. Ibid., vi.

40. Antonio Ferri and Ahmed Ismail, "Analysis of Configurations," in *Second Conference on Sonic Boom Research*, 73–88. See also Percy J. Bobbitt and Domenic Maglieri, "Dr. Antonio Ferri's Contribution to Supersonic Transport Sonic-Boom Technology," *Journal of Spacecraft and Rockets* 40, no. 4 (July–August 2003): 459–466.

41. Wallace D. Hayes and Rudolph C. Haefeli, "The ARAP Sonic Boom Computer Program," in *Second Conference on Sonic Boom Research*, 151–158. For a more complete description, see Wallace D. Hayes, "Sonic Boom Propagation in a Stratified Atmosphere with Computer Program," NASA CR 1299 (April 1969).

42. Wallace Hayes, "State of the Art of Sonic Boom Theory," in *Second Conference on Sonic Boom Research*, 182.

43. Harvard Lomax, "Preliminary Investigation of Flow Field Analysis on Digital Computers with Graphic Display," in *Second Conference on Sonic Boom Research*, 72.

44. Schwartz, "Preface," *Second Conference on Sonic Boom Research*, vii.

45. Ibid., iii.

46. *Aircraft Engine Noise and Sonic Boom*, Conference Proceedings (CP) no. 42, Paris, France, May 1969 (Neuilly Sur Seine, France: NATO AGARD, 1969).

47. W.D. Beasley, J.D. Brooks, and R.L. Barger, "A Laboratory Investigation of N-Wave Focusing," NASA TN D-5306 (July 1969); R.L. Barger, W.D. Beasley, and J.D. Brooks, "Laboratory Investigation of Diffraction and Reflection of Sonic Booms by Buildings," NASA TN D-5830 (June 1970).

48. Phillip M. Edge and Harvey H. Hubbard, "Review of Sonic Boom Simulation Devices and Techniques," *JASA* 51, no. 2, pt. 3 (February 1972): 724–728; Hugo E. Dahlke et al., "The Shock-Expansion Tube and its Application as a Sonic Boom Simulator," NASA CR-1055 (June 1968); R.T. Sturgielski et al., "The Development of a Sonic Boom Simulator with Detonable Gases," NASA CR 1844 (November 1971).

49. David Hoffman, "Report Sees Need for Study on Sonic Boom Tolerance," *Washington Post*, June 26, 1968, A3.

50. Ira R. Schwartz, "Sonic Boom Simulation Facilities," AGARD, *Aircraft Engine Noise and Sonic Boom*, 29-1.

51. Philip M. Edge and William H. Mayes, "Description of Langley Low-Frequency Noise Facility and Study of Human Response to

Noise Frequencies below 50 cps," NASA TN D-3204 (January 1966).

52. Roger Tomboulian, "Research and Development of a Sonic Boom Simulation Device," NASA CR-1378 (July 1969); Stacy V. Jones, "Sonic Boom Researchers Use Simulator," *New York Times*, May 10, 1969, 37, 41.

53. Ira R. Schwartz, "Sonic Boom Simulation Facilities," 29-6.

54. B.K.O. Lundberg, "Aviation Safety and the SST," *Astronautics and Aeronautics* 3, no. 1 (January 1966), 28. Lundberg, a distinguished Swedish engineer, was a very effective critic of SSTs.

55. See Conway, *High-Speed Dreams*, 118–156.

56. "The Shattering Boom," *New York Times*, June 8, 1968, 30.

57. Ira R. Schwartz, ed., *Third Conference on Sonic Boom Research: Proceedings of a Conference Held at the National Aeronautics and Space Administration, Washington, DC, Oct. 29–30, 1970*, SP-255 (Washington, DC: NASA, 1971). The papers from the ASA's Houston symposium were published in *JASA* 51, no. 2, pt. 2 (February 1972).

58. Ira Schwartz, "Preface," *Third Conference on Sonic Boom Research*, iv.

59. "Civil Aircraft Sonic Boom," *Federal Register* 35, no. 4 (April 16, 1970): 6189–6190. Source for figure 2-3: Spearman, "Evolution of the HSCT," 36.

60. Antonio Ferri, "Airplane Configurations for Low Sonic Boom," *Third Conference on Sonic Boom Research*, 255.

61. Ibid., 255–256.

62. Domenic Maglieri et al., "Measurements of Sonic Boom Signatures from Flights at Cutoff Mach Number," *Third Conference on Sonic Boom Research*, 243–254.

63. Albert George, "The Effects of Atmospheric Inhomogeneities on Sonic Boom," and Kenneth J. Plotkin, "Perturbations Behind Thickened Shock Waves," both in *Third Conference on Sonic Boom Research*, 33–66.

64. M.J. Cole and M.B. Freeman, "Analysis of Multiple Scattering of Shock Waves by a Turbulent Atmosphere," in *Third Conference on Sonic Boom Research*, 67–74.

65. A.R. George, "Lower Bounds for Sonic Booms in the Midfield," *AIAA Journal* 7, no. 8 (August 1969): 1542–1545. Plotkin recalled George receiving the draft of this article back from the publisher with a comment by an anonymous reviewer: "This paper must be published!" They deduced that Wallace Hayes must have been the reviewer.

66. L.B. Jones, "Lower Bounds for Sonic Bang in the Far Field," *Aeronautical Quarterly* 18, pt. 1 (February 1967): 1–21; L.B. Jones, "Lower Bounds for the Pressure Jump of the Bow Shock of a Supersonic Transport," *Aeronautical Quarterly*, 21 (February 1970): 1–17.

67. A.R. Seebass, "Nonlinear Acoustic Behavior at a Caustic," in *Third Conference on Sonic Boom Research*, 87–122, with figure 2-3 extracted from page 100.

68. Although sonic boom shock waves expand behind a moving super-sonic body in a three-dimensional, cone-shaped band (figure 1-1), their energy is disseminated outward and thereby forward as acoustic rays. Aircraft maneuvers, atmospheric conditions, and other factors determine the paths they follow, which can take the form of discrete ray tubes.

69. A.R. Seebass, "Comments on Sonic Boom Research," in *Third Conference on Sonic Boom Research*, 411.

70. Lynn W. Hunton, "Comments on Low Sonic Boom Configuration Research, in *Third Conference on Sonic Boom Research*, 417.

71. Harry W. Carlson, "Some Notes on the Status of Sonic Boom Prediction and Minimization Research," in *Third Conference on Sonic Boom Research*, 397.

72. For a detailed postmortem, see Edward Wenk, "SST—Implications of a Political Decision," *Astronautics and Aeronautics* 9, no. 10 (October 1971): 40–49.

73. Compiled by screening B.A. Fryer, "Publications in Acoustics and Noise Control from the NASA Langley Research Center During 1940–1976," NASA TM-X-7402 (July 1977). Five reports for 1967 that Domenic Maglieri (in reviewing the draft of this chapter) found to be missing from Fryer's compilation have been added to that column.

74. Comments by Herbert Ribner in *The Proceedings of the Second Sonic Boom Symposium of the Acoustical Society of America*, Houston, TX, November 1970, cited by Christine M. Darden, "Affordable/Acceptable Supersonic Flight: Is It Near?" 40th Aircraft Symposium, Japan Society for Aeronautical and Space Sciences, Yokohama, Japan, October 9–11, 1973, 2.

75. This cutoff speed is based on a standard atmosphere without winds.

76. John Morgenstern, "Fixing the Sonic Boom," a presentation at the FAA Workshop on Advanced Technologies and Supersonics, Palm Springs, CA, March 1, 2009, slide no. 3.

77. Figure 2-5 extracted from Peter G. Coen and Roy Martin, "Fixing the Sound Barrier: The DARPA/NASA/Northrop-Grumman Shaped Sonic Boom Flight Demonstration," PowerPoint presentation, July 2004, slide no. 3. Later analysis by civil engineers determined that impulse is only an approximate parameter for assessing structural response; e-mail, Kenneth Plotkin, Wyle Laboratories, to Lawrence Benson, June 5, 2011.

78. David S. Miller and Harry W. Carlson, "A Study of the Application of Heat or Force Fields to the Sonic-Boom-Minimization Problem," NASA TN D-5582 (December 1969).

79. Domenic J. Maglieri, Harry L. Carlson, and Norman J. McLeod, "Status of Studies of Sonic Boom," in *NASA Aircraft Safety and Operating Problems* (Washington, DC: GPO, 1971), 439–456.

80. A.R. George and R. Seebass, "Sonic Boom Minimization Including Both Front and Rear Shocks," *AIAA Journal* 9, no. 10 (October 1971): 2091–2093. They published a more detailed description, "Sonic-Boom Minimization," in *JASA* 51, no. 2, pt. 3 (February 1972): 686–694. See also next chapter.

81. Plotkin and Maglieri, *Sonic Boom Research*, 5.

Vincent R. Mascitti, F. Edward McLean, and Cornelius Driver in the mid-1970s with a Lockheed AST model. (NASA)

CHAPTER 3

Continuing the Quest
Supersonic Cruise Research

"The number one technological tragedy of our time."[1] That was how President Nixon characterized the votes by Congress to stop funding an American supersonic transport. Despite the SST's cancellation, the White House, the Department of Transportation and its Federal Aviation Administration, and NASA—with help from some members of Congress—did not allow the SST's progress in supersonic technologies to completely dissipate. During 1971 and 1972, the DOT and NASA allocated funds for completing some of the research and experiments that were under way when the program was terminated. The administration then added line-item funding to NASA's fiscal year (FY) 1973 budget for scaled-down supersonic research, especially as related to environmental issues raised during the SST program. In response, NASA established the Advanced Supersonic Technology (AST) program in July 1972. Thus resumed what became a half-century pattern of on-again, off-again efforts to solve the problems of faster-than-sound civilian flight, with the sonic boom remaining one of the most difficult challenges of all.

Changing Acronyms:
An Overview of the AST/SCAR/SCR Program

To indicate more clearly the exploratory nature of this effort and allay fears that it might be a potential follow-on to the SST, the overall AST program was renamed Supersonic Cruise Aircraft Research (SCAR) in 1974. When the term "aircraft" in the program's title continued to raise suspicion in some quarters that the goal might be some sort of prototype, NASA shortened the program's name to Supersonic Cruise Research (SCR) in 1979, not long before the program's demise.[2] For the sake of simplicity, the latter name is often applied to all 9 years of the program's existence.

To NASA, the principal purpose of the AST/SCAR/SCR program was to conduct and support focused research into the problems of supersonic flight while advancing related technologies. As with the SST (albeit more modestly), NASA's aeronautical centers, most of the major airframe manufactures, and many research organizations and universities participated.[3] From Washington, NASA's Office of Aeronautics and Space Technology (OAST) provided overall supervision but delegated day-to-day management to the Langley Research Center, which established an AST Project Office in its Directorate of Aeronautics (soon placed under a new Aeronautical Systems Division). The AST program was organized into several major elements: propulsion; structure and materials; stability and control; aerodynamic performance; and airframe-propulsion integration. NASA spun off propulsion work on a variable cycle engine (VCE) as a separate program in 1976. (A variable cycle engine is similar to a conventional mixed-flow turbofan except that it has an additional secondary outer duct to increase the overall bypass ratio and, thus, the airflow handling capability desirable at very high speeds.) Sonic boom research, which fell under aerodynamic performance, was but one of 16 AST subelements.[4]

At Langley's Aeronautical Systems Division, Cornelius "Neil" Driver, who headed the Vehicle Integration Branch, and F. Edward McLean, as chief of the AST Project Office, were key officials in planning and managing the AST/SCAR effort. After McLean retired in 1978, the AST Project Office passed on to a fellow aerodynamicist, Vincent R. Mascitti, while Driver took over the Aeronautical Systems Division. (All three are shown in the accompanying photo.) One year later, Domenic Maglieri replaced Mascitti in the AST Project Office.[5] Despite Maglieri's sonic boom expertise, the goal of minimizing the AST's sonic boom for overland cruise had by then long since ceased being an SCR objective. As later explained by McLean: "The basic approach of the SCR program … was to search for the solution of supersonic problems through disciplinary research. Most of these problems were well known, but no satisfactory solution had been found. When the new SCR research suggested a potential solution … the applicability of the suggested solution was assessed by determining if it could be integrated into a practical commercial supersonic airplane and mission…. If the potential solution could not be integrated, it was discarded."[6]

To meet the practicality standard for integration into a supersonic airplane, the scientists and engineers trying to solve the sonic boom problem had to clear a new and almost insurmountable hurdle less than a year into the AST effort. In April 1973, responding to concerns raised since the SST program, the FAA announced a new rule, effective on September 30, banning commercial or civil aircraft from supersonic flight over the landmass or territorial waters of the United States if measurable overpressure would reach the surface.[7] One of the initial objectives of AST's sonic boom research had been to establish a metric for public acceptability of sonic boom signatures for use in the aerodynamic design process. The FAA's stringent new regulation seemed to rule out any such flexibility.

As a result, when Congress cut FY 1974 funding for the AST program from $40 million to about $10 million, the subelement for sonic boom research went on NASA's chopping block. The design criteria for the SCAR program became a 300-foot-long, 270-passenger airplane that could fly as effectively as possible over land at subsonic (or possibly low-transonic) speeds yet still cruise efficiently at 60,000 feet and Mach 2.2 over water. To meet these less ambitious criteria, Langley aerodynamicists modified their SCAT-15F design from the late 1960s into a notional concept with better low-speed performance (but higher sonic boom potential) called the ATF-100. This served as a baseline for three industry teams in coming up with their own designs.[8]

Learning More About Sonic Booms

Back when the AST program began, however, prospects for a significant quieting of its sonic footprint for operations over land still appeared possible. Sonic boom theory had advanced significantly during the 1960s, and some promising ideas for reducing boom signatures had begun to emerge. As indicated in figure 2-4, these endeavors continued to bear fruit into the early 1970s.

As far back as 1965, NASA's Ed McLean had predicted that the sonic boom signature from a very long supersonic aircraft flying at the proper altitude could be nonasymptotic (i.e., not reach the ground in the form of an N-wave).[9] The most radical ideas for lengthening an aircraft by projecting a phantom body as with a heat shield continued to be set aside (but another idea—projecting a long extension from its nose to slow the rise of the bow shock wave—would eventually prove more realistic).[10] Most of the ongoing research on sonic boom minimization, however, tended to follow the general course set by Albert George and Richard Seebass with their landmark treatise in 1971.

George and Seebass went into more detail about their sonic boom minimization concepts during the next few years.[11] This culminated in "Design and Operation of Aircraft to Minimize their Sonic Boom," an article published in its final form in October 1974.[12] Although by then not especially relevant to the AST, since overland supersonic operations were no longer part of the program, it marked another milestone in the development of sonic boom theory. Because there was as yet no accepted standard on "what is to be reduced or minimized in order to make the sonic boom more acceptable,"[13] they considered remedies for treating each of its symptoms individually. These included shock strength (both front and rear), overpressure, rise time, and impulse. Based on common assumptions, such as an isothermal atmosphere (a standard formula on how air temperature and density change with altitude), and premised on the supersonic area rule, they examined "how we can shape the equivalent body of revolution for the vertical plane to minimize a given signature parameter below the aircraft."[14] In addition to the obvious ways of reducing overpressure and impulse by lowering aircraft weight and improving efficiency (e.g., ratios of lift to drag and thrust to weight), they looked at specific aerodynamic design principles, citing key findings from the growing literature on the topic while presenting their own remedies.

By way of creating general rules to be considered, George and Seebass showed mathematically and graphically the relationships and tradeoffs between various aircraft-design features and sonic boom characteristics. For example, higher Mach numbers can somewhat lower impulse but not overpressure. They calculated and described how the proper combinations of shape, weight,

Albert George in 1978 (Cornell) and Richard Seebass in the early 1980s. (University of Colorado)

length, and altitude (lower than previously thought) can practically eliminate the explosive sound of the bow shock wave (but not the signature's total overpressure). "Thus, for example, a Mach 2.7, 600,000 lb., 300-foot aircraft can have a shock-free [but not silent] signature at altitudes below 30,000 ft."[15] They also mathematically examined various operational techniques, such as the transonic speeds that could prevent sonic booms from reaching the ground. At the other extreme, they predicted that hypersonic speeds might help with lowered shock waves but would not solve other problems—especially impulse. They concluded "that aircraft could be designed that would achieve overpressure levels just below $1/lb/ft^2$ (for both positive and negative phases of the pressure signature) and impulses of about $1/10$ lb/sec/ft². These numbers are not too different from the sonic boom generated by the SR-71, and experience with SR-71 overflights should give some indication of whether or not overpressures and impulses of this magnitude will prove acceptable."[16]

Unfortunately for the future of supersonic transports, the public's apparent tolerance of occasional sonic booms from the Air Force's small fleet of SR-71s did not transfer to the more frequent booms that scheduled supersonic airline traffic would generate along their routes. Although most of Seebass's and George's work at the time applied to large transports, their analysis of various types of airframe shaping to alter the formation and evolution of shock waves—such as slightly blunting the nose without significantly penalizing performance—would be relevant to smaller supersonic aircraft as well. Based on their insightful analyses, George and Seebass are considered the fathers of sonic boom minimization.

Meanwhile, other researchers under contract to NASA also continued to advance the state of the art. For example, Antonio Ferri of New York University in partnership with Hans Sorensen of the Aeronautical Research Institute of Sweden used new three-dimensional measuring techniques in Sweden's trisonic wind tunnel to more accurately correlate near-field effects with linear theory. Testing NYU's model of a 300-foot-long SST cruising at Mach 2.7 at 60,000 feet projected sonic booms of less than 1.0 psf.[17] Ferri's death in 1975 at the age of 63 left a big void in the field of supersonic aerodynamics, including sonic boom research.[18]

In addition to theoretical refinements and wind tunnel techniques, important new computer-modeling capabilities continued to appear in the early 1970s. In June 1972, Charles Thomas of the Ames Research Center published details on a computer program he called the waveform parameter method, which used new algorithms to extrapolate the evolution of far-field N-waves. This offered an alternative to using the F-function (the pattern of near-field shock waves emanating from an airframe) as required by the previously discussed program developed by Wallace Hayes and colleagues at Princeton's ARAP. Although both methods accounted for acoustical-ray tracing and would arrive at almost identical results, Thomas's code allowed for easier inputs of wind tunnel pressure measurements as well as such variables as Mach number, altitude, flightpath angle, acceleration, and atmospheric conditions for automated data processing.[19]

Both Thomas's waveform parameter program and Hayes's ARAP program remain relevant well into the 21st century. As explained 30 years after Thomas released his program by an expert in sonic boom modeling, "Both are full implementations of fundamental theory, accounting for arbitrarily maneuvering aircraft in horizontally stratified atmospheres with wind.... Moreover, virtually every full ray trace sonic boom program in use today is evolved in one way or another from one of these two programs."[20]

In June 1973, at the end of the AST program's first year, Harry Carlson, Raymond Barger, and Robert Mack of the Langley Research Center published a study on the applicability of sonic boom minimization concepts for an overland supersonic transport based on "the airplane design philosophy, most effectively presented by Ferri, in which sonic-boom considerations play a dominant role."[21] They examined two baseline AST designs and two reduced-boom concepts. The objective for all four was a commercially viable Mach 2.7 supersonic transport with a range of 2,500 nautical miles (nm) (the coast-to-coast distance across the United States). Applying the experimentally verified minimization concepts of George, Seebass, Hayes, Ferri, Barger, and L.B. Jones, and using the ARAP computer program, Carlson's team explored various ways to manipulate the F-function to project a quieter sonic boom signature. As with similar previous efforts, their options were limited by the lack of established signature characteristics (combinations of initial rise time, shock strength, peak overpressure, and duration) that people would best tolerate, both outdoors and especially indoors. Also, the complexity of aircraft geometry made measuring effects on tail shocks difficult. They therefore settled on lowering peak overpressure, with the goal being a plateau or flattopped signature.[22]

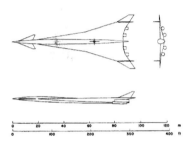

Figure 3-1. Aft arrow-wing configuration for low peak overpressure. (NASA)

Considering this objective along with numerous other parameters deemed necessary for a practical airliner, their study confirmed the advantages of highly swept wings located toward the rear of the fuselage with carefully designed twist and camber for sonic boom shaping. It also confirmed the use of canards (small airfoils used as horizontal stabilizers near the nose of rear-winged aircraft) and positive dihedral (angled up) wings to optimize lift distribution for sonic boom benefits. Although two designs (one with a delta wing and another with an arrow wing) showed bow shocks of less than 1.0 psf at an optimum cruising altitude of 53,000 feet to 59,000 feet, their report noted "that there can be no assurance at this time that [their] shock-strength values ... if attainable, would permit unrestricted overland operations of supersonic transports."[23]

In October 1973, Edward J. Kane of Boeing, who had been a key sonic boom specialist during the SST program, released the results of a similar NASA-sponsored study on the feasibility of a commercially viable low-boom transport using technologies projected to be available in 1985. Applying the latest theories (including the just-discussed Langley study), Boeing explored two longer range concepts: a high-speed (Mach 2.7) arrow-wing design that would produce a sonic boom of 1.0 psf or less at 55,000 feet and a medium-speed (Mach 1.5) highly swept wing design with a signature of 0.5 psf or less at 45,000 feet.[24] Ironically, these results were published just as the new FAA rule rendered them largely irrelevant. In retrospect, this study represented a final industry perspective on the prospects for boom minimization before the SCAR program dropped plans for supersonic cruise over land.

Obviously, the FAA's virtual ban on civilian supersonic flight in the United States dampened any enthusiasm by the major aircraft manufacturers to continue investing much capital in sonic boom research. Within NASA, funding for academic studies slowed to a trickle and many of its own employees with experience in sonic boom research redirected their efforts into other areas of expertise. Of the approximately 1,000 technical reports, conference papers, and articles by NASA and its contractors listed in bibliographies of the SCR program from 1972 to 1980, only eight dealt directly with the sonic boom.[25]

The Early Promise of Computational Fluid Dynamics

Even so, some foundations were being laid that would benefit future sonic boom research. Of great benefit to supersonic aerodynamics and other aeronautical endeavors, NASA was fostering the development and use of powerful new digital computer capabilities. In early March 1975, the Langley Research Center hosted a major conference, "Aerodynamic Analyses Requiring Advanced Computers," on the progress being made. In introductory remarks, J. Lloyd Jones, NASA's Deputy Associate Administrator for Aeronautics Technology, extolled the advances made since NASA's last aerodynamics conference in 1969. Rather than having to rely on simple shapes and various shortcuts to formulate solvable equations, usually involving only two dimensions, researchers were now able to calculate such phenomena as transonic mixed flows with embedded shock waves and flows around complex wing-body configurations.[26] The future promised much greater improvements. Dean R. Chapman of the Ames Research Center discussed the accelerating advances in the digital-modeling and analysis capabilities of computational fluid dynamics. CFD, he asserted, offered "tremendous potential for revolutionizing the way our profession has been doing business."[27] With the expected arrival of super computers, he envisaged CFD overcoming the inherent limitations of wind tunnels and, eventually, allowing timely processing of heretofore virtually unsolvable differential equations governing fluid flows over solid surfaces with the effects of friction.

Using advanced calculus, the brilliant 18th century mathematician Leonhard Euler had devised two nonlinear partial differential equations governing the momentum (velocity and pressure) and the continuity of flowing fluids. Euler's momentum equation did not account for the viscous effects of friction on fluid flow along the surface of an object (known later in aerodynamics as "boundary conditions"). This required highly complex, nonlinear partial equations developed independently in the first half of the 19th century by Claude-Louis Navier and George Gabriel Stokes that account for interdependent variables such as pressure, density, and velocity. These became known as the Navier-Stokes equations. During the late 19th century, advances in the field of thermodynamics would lead to an energy equation for the high-speed fluid flows that also would be needed later in aerodynamics. Unfortunately, the Euler and Navier-Stokes equations had no general analytical solutions when applied to practical problems (such as airflow over a wing) without simplifying selected factors to permit linear solutions. Future progress in computational fluid dynamics, however, would allow the partial derivatives or integrals in these equations to be replaced by discrete algebraic forms. Eventually, the data-processing capability of high-speed computers could repeatedly generate flow-field values for each variable at specific points in space and time, known as grid points, with results improving with each iteration. The end products, although not classic stand-alone mathematical solutions, would be of great practical use for aerodynamic design purposes, including (as will be seen) shock wave calculations.[28]

Although the revolutionary potential of CFD was still some years in the future, 2 of the 52 papers presented at this 1975 conference presented two evolutionary computer programs of value for sonic boom minimization. A NASA-sponsored paper by Richard Seebass and three others at Cornell reported on several recent advances in sonic boom theory and introduced an easy-to-use computer program for aerodynamic minimization calculations written by Joseph Liu Lung as his master's thesis.[29] The other paper by H. Harris Hamilton of Langley and Frank Marconi and Larry Yeager of the Grumman Aerospace Corporation reported a new technique for accurately and efficiently computing high-speed inviscid (frictionless) flows in three dimensions around real airframe configurations. Although this research was prompted by NASA's need to learn more about the aerodynamics of the Space Shuttle orbiter during return flights, their innovation could be applied to all supersonic and hypersonic vehicles.[30] NASA published full details on the procedure and its related computer code the following year.[31]

Progress in Prediction and Minimization Techniques

By 1975 Christine M. Darden of NASA Langley had developed an innovative computer code specifically for use in sonic boom minimization. She wrote it to convert Seebass and George's minimization theory, which was based on an isothermal (uniform) atmosphere, into a program that applied to the real (stratified) atmosphere. She did this with new equations for calculating pressure-signature changes, ray-tube areas, and acoustic impedance. Although reliance on an isothermal atmosphere allowed reasonable estimates of sonic boom signatures for many conditions, Darden's modification offered more accurate equivalent area distribution calculations, such as for flying at low Mach numbers and for designing better aerodynamics in the nose area (a goal she would continue to pursue).[32] Darden, who had earned a mathematics degree with highest honors from the Hampton Institute in 1962, taught high school before beginning her career at Langley in 1967. Even as she became NASA's top sonic boom expert, Darden continued her education—earning a master's in mathematics from Virginia State College in Petersburg in 1978 and a doctorate in mechanical engineering (specializing in fluid mechanics) from George Washington University in 1983.[33]

Christine Darden and Robert Mack presented a paper on current sonic boom research at the first SCAR conference, held at Langley from November 9 to 12, 1976. The conference took place after both the Concorde and the Soviet Tu-144 began scheduled supersonic flights that, because of their sonic booms, were restricted to routes over oceans and sparsely populated land areas.[34] Theirs was the only paper on the sonic boom issue among the 47 presentations at the conference.[35] In other areas, NASA and its industry partners were making significant advances over the Concorde and Tu-144 in the areas of engine noise, fuel consumption, lift-to-drag (L/D) ratio, airframe structure (using new titanium fabrication processes), and direct operating costs (estimated at 50-percent lower than the Concorde's). The major problem left unsolved for any second-generation supersonic transport was the sonic boom.[36]

One of the main obstacles to progress in sonic boom minimization was what Darden and Mack called the low-boom, high-drag paradox (figure 3-2). "Contrary to earlier beliefs," they explained, "it has now been found that improved efficiency and lower sonic boom characteristics do not always go hand in hand."[37] Both theory and experiments had shown that (as would be expected) an aerodynamically efficient sharp-nosed supersonic airframe generates a weaker bow shock than one with a less streamlined nose. Yet with a blunt-nose section, there is less propensity for the strong shock waves generated along the rest of

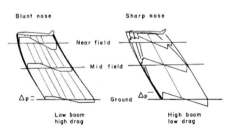

Figure 3-2. Low-boom, high-drag paradox. (NASA)

an airframe to merge with the bow shock and create the typical N-wave sonic boom at the surface. Unfortunately, the excess drag of a truly blunt-nosed supersonic aircraft would make it aerodynamically unacceptable.[38]

The two Langley researchers were exploring ways to deal with this dilemma, a full solution of which they said would require extensive tradeoff studies by engineering design teams. Meanwhile, they reported on preliminary results of their ongoing experiments using Darden's revision of the George-Seebass methodology to design lower-boom wind tunnel models that did not pay too great a penalty in aerodynamic efficiency. As for any progress on the still-critical question of what would be an acceptable sonic boom, the only research being done in North America was by the University of Toronto. Its Institute for Aerospace Studies was testing humans, animals, and materials with various sonic boom simulators. Other research there included focused booms, effects of turbulence, and signatures in the shadow zone.[39]

Another NASA contribution to understanding sonic booms came in early 1978 with the publication of Harry Carlson's "Simplified Sonic-Boom Prediction," a how-to guide on a relatively quick and easy method to determine sonic boom characteristics in a standard atmosphere. It could be applied to a wide variety of supersonic aircraft configurations as well as spacecraft at altitudes up to 76 kilometers (km) and cover the entire width of a boom carpet. Although his clever series of graphs and equations did not provide the accuracy needed for predicting booms from maneuvering aircraft or in designing airframe configurations, Carlson explained that "for many purposes (including the conduct of preliminary engineering studies or environmental impact statements), sonic-boom predictions of sufficient accuracy can be obtained by using a simplified method that does not require a wind tunnel or elaborate computing equipment. Computational requirements can in fact be met by hand-held scientific calculators, or even slide rules."[40] This procedure would be especially helpful to the armed services in preparing recently required environmental studies for areas where military aircraft flew supersonically.

Although it was drawing funds away from aeronautics, one aspect of NASA's Space Transportation System (STS) led to additional sonic boom research involving the full range of shock waves—from hypersonic speeds at the top of the atmosphere down to transonic speeds near the surface. In April 1978, NASA headquarters released its final environmental impact statement (EIS) for the Space Shuttle. It benefited greatly from the Agency's previous research on sonic booms, including the X-15 and Apollo programs as well as the adaptations of Charles Thomas's waveform-based computer program.[41] While the entire STS was ascending, the EIS estimated maximum overpressures of 6 psf (possibly up to 30 psf with focusing effects) about 40 miles downrange over open water. This would be caused by both its long exhaust plumes (which acted somewhat as a "phantom body") and its curving flight profile while accelerating toward orbit. During reentry of the orbiter, the sonic boom was estimated at a more modest 2.1 psf (comparable to the Concorde), which would affect about 500,000 people as it crossed the Florida peninsula or 50,000 people when landing at Edwards.[42] In the following decades, as populations in those areas boomed, millions more would be hearing the sonic booms of returning Shuttles, more than 120 of which would be monitored for their signatures.[43]

Some other limited experimental and theoretical work on sonic booms continued in the late 1970s—even if it was no longer based on an American supersonic transport. For example, Richard Seebass delved deeper into the tricky phenomena of caustics and focused booms, an area on which French researcher John Pierre Guiraud had written the governing equations and derived a related scaling law.[44] After a numeric solution of Seebass's ideas by one of his graduate students,[45] Kenneth Plotkin (who began working for Wyle Laboratories in 1972 after receiving his Ph.D. from Cornell) applied these techniques to analyzing predicted focused booms from the Shuttle for the Marshall Space Flight Center as part of the studies described in the previous paragraph.[46] At the end of the decade, Langley's Raymond Barger published a study on the relationship of caustics to the shape and curvature of acoustical wave fronts caused by aircraft maneuvers. To display these effects graphically, he programmed a computer to draw simulated three-dimensional line plots of acoustical rays in the wave fronts. Figure 3-3 shows how even a simple decelerating turn—in this case, from Mach 2.4 to Mach 1.5 in a radius of 23 km (14.3 miles)—can merge the rays into the kind of caustic that might cause a super boom.[47]

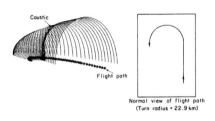

Figure 3-3. Acoustic wave front above a maneuvering aircraft. (NASA)

Unlike in the 1960s, there was little if any NASA sonic boom flight testing during the 1970s. As a case in point, NASA's YF-12 Blackbirds at Edwards AFB (where the Flight Research Center was renamed the Dryden Flight Research Center in 1976) flew numerous productive supersonic missions in support of the AST/SCAR/SCR program, but none of them were dedicated to sonic boom issues.[48] On the other hand, flight testing of the Concorde provided some new sonic boom data from a real supersonic transport. For example, an Anglo-American aeronautical conference in London in early June 1973 (just a few weeks after the FAA's new rule prohibited civilian supersonic flight in the United States) included an informative paper on sonic boom measurements and their effects on people, buildings, and wildlife during Concorde test flights along Great Britain's west coast.[49] Once these swift new airliners became operational, however, most of their supersonic flying was done over the open ocean, where there was presumably only limited opportunity for gathering sonic boom data.

One strange and unexpected discovery about secondary booms came after British Airways and Air France began regular Concorde service to the United States, first to Dulles International Airport near Washington, DC, in May 1976 and then (after much local opposition due mainly to jet-engine noise) to New York's Kennedy International Airport in November 1977.[50] Although the Concordes slowed to subsonic speeds while well off shore, residents along the Atlantic seaboard began hearing what were called the East Coast mystery booms. These were detected all the way from Nova Scotia to South Carolina, some of them measurable on seismographs.[51] Although a significant number of the sounds defied explanation, studies by the Naval Research Laboratory (NRL), the Federation of American Scientists, a committee of the DOD's JASON scientific advisory group, and the FAA eventually determined that most of the low rumbles heard in Nova Scotia and New England were secondary booms from the Concorde. These sounds were caused by distorted shock waves that were being bent or reflected by temperature variations high up in the thermosphere while Concordes were still about 75 miles to 150 miles offshore.[52] In July 1978, the FAA issued new rules prohibiting the Concord from creating sonic booms that could be heard in the United States. Although the FAA did not consider that this applied to secondary booms because of their low intensity, the affair apparently made the FAA even more sensitive to the sonic boom potential inherent in AST designs.[53] (As part of the NRL investigation, Harvey Hubbard and Domenic Maglieri determined that Aerospace Defense Command F-106s, scrambled from Langley AFB to intercept Soviet Tu-20 Bear long-range bombers flying along the Atlantic coast on their regular flights to or from Cuba, created many of the sonic booms heard farther south to the Carolinas.)[54]

At Langley, Christine Darden and Robert Mack continued to pursue their research on sonic boom minimization during the late 1970s. Using the Seebass-George procedure as modified for a stratified atmosphere, they followed up on the previously described studies by Kane's team at Boeing and, in particular, the studies by Carlson's team at Langley. They used wing analysis and wave-drag area rule computer programs, including viscous effects, to help design three specially shaped wing-body models with low-boom characteristics for comparison with two of Carlson's models that had been designed mainly for aerodynamic efficiency (figure 3-4). One of their models was configured for cruise at Mach 1.5 (the lowest speed for a truly supersonic transport) and two for cruise at Mach 2.7 (which approached the upper limit for applying near-field sonic boom theory). At 6 inches in length, these were the largest yet tested for sonic boom propagation within the confines of the Langley 4-foot-by-4-foot Unitary Plan Wind Tunnel—an improvement made possible by continued progress in applying the ARAP code to extrapolate near-field pressure signatures to the far field. These low-boom models showed much more reduced overpressure than the standard (unconstrained) delta-wing model and significantly lower overpressures than the unconstrained arrow-wing design, especially at Mach 1.5, where the tail shocks were softened as well. The low-boom models' pressure signatures also showed definite flattop characteristics. Darden and Mack first presented their findings at an AIAA conference in March 1979 and published them in a NASA technical paper that October as well as in an article in the *Journal of Aircraft* in March 1980.[55]

To Christine Darden, it had become obvious that "because extreme nose bluntness produces large drag, a method of relaxing the bluntness is needed to offer the opportunity for compromise between blunt-nose low-boom and sharp-nose low-drag configurations."[56] She wrote about this attempt in another NASA technical paper published in 1979 titled

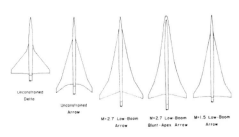

Figure 3-4. Models used in sonic boom minimization study. (NASA)

"Sonic Boom Minimization with Nose-Bluntness Relaxation." It focused on findings that "because the shape of the aircraft does influence the shape of a mid-field pressure signature, aircraft shaping has now become a powerful tool in reducing the sonic boom."[57] Using numerous equations and the previously mentioned NASA-sponsored computer code developed by Grumman for calculating supersonic and hypersonic inviscid flow around various configurations, she explored theoretical options for eliminating the bow shock, allowing unrestricted tail shock, and eliminating both shocks. Her calculations showed how to relax a blunt nose into a more conical shape to reduce drag. "Thus," she concluded, "the boom levels could be reduced significantly without prohibitive drag penalties by defining the proper ratio [of] y_f / l."[58] (In area distribution terms, this was the width of the shock wave spike along the front of the fuselage relative to the entire equivalent length of the airplane.)

Although premised on airliner-sized supersonic aircraft, Darden and Mack's rather lonesome work in the late 1970s on how carefully designed airframe shaping could in turn shape the signature of a sonic boom would help set the stage for future research on various-sized supersonic aircraft. Because of funding limitations, however, this promising approach could not be sustained beyond 1979.[59] It was apparently the last significant NASA experimentation on sonic boom minimization for almost another decade. Yet by validating the Seebass-George minimization theory and verifying design approaches for sonic boom reductions, their findings would serve as a point of departure

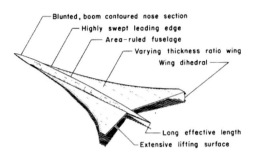

Figure 3-5. Features of low-boom models. (NASA)

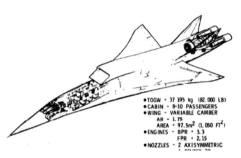

Figure 3-6. North American Rockwell SSBJ concept. (NASA)

when such research finally did resume. (Figure 3-5 illustrates the aerodynamic design characteristics for boom minimization confirmed by Darden and Mack's studies.[60])

The second conference on Supersonic Cruise Research, held at Langley in November 1979, was the first and last under its new name. More than 140 people from NASA, other Government agencies, and the aerospace industry attended. This time, there were no presentations on the sonic boom, but Robert Kelly from North American Rockwell did put forth the concept of a Mach 2.7 business jet for 8 to 10 passengers or possibly a military aircraft that could generate a sonic boom of only 0.5 psf. Because of the difficulty of developing a big supersonic airliner in one step, he proposed an alternate course: "to validate the critical supersonic technologies in a small research vehicle prior to the building of a full-size supersonic vehicle.... But," he asked, "would the research vehicle have only one use? Why not have the additional capability for military use or as a supersonic business jet?"[61] The concept he presented was a blended-variable camber arrow-wing/body design using fiber-reinforced titanium structures with superplastic forming and diffusion bonding (SPF/DB) and either of two different propulsion systems. The basic configuration is shown in figure 3-6.[62] Although it was not proposed as a business jet per se, Boeing had also submitted a study in 1977 on building a subscale (93-foot-long) Mach 2.4 SCAR demonstrator to test numerous characteristics and capabilities, including sonic boom acceptability and possible boom reducing modifications.[63] It would take another 20 years for ideas about either a low-boom demonstrator or a supersonic business jet (SSBJ) to go anywhere beyond paper studies.

Despite SCR's relatively modest cost versus its significant technological accomplishments, the program suffered a premature death in 1981. Reasons for this included the discouragingly high cost of Concorde operations, opposition to civilian R&D spending by some key officials in the new administration of President Ronald Reagan, and the growing Federal deficit. These factors, combined with cost overruns for the Space Shuttle, forced NASA to abruptly cancel Supersonic Cruise Research without even funding completion of many final reports.[64] As regards sonic boom studies, an exception to this was a compilation of useful charts for estimating minimum sonic boom levels for various combinations of aircraft length, weight, altitude, and Mach numbers by Christine Darden published in 1981.[65]

Endnotes

1. Stephen D. Ambrose, *Nixon: Triumph of a Politician* 2 (New York: Simon and Schuster, 1989), 433, cited in Conway, *High-Speed Dreams*, 153. For the background of the AST program, see Conway, 153–158.
2. F. Edward McLean, "SCAR Program Overview," in *Proceedings of the SCAR Conference Held at Langley Research Center, Hampton, Virginia, November 9–12, 1976*, pt. 1, NASA CP-001 (1976), 1–3; McLean, *Supersonic Cruise Technology*, 101–102.
3. Marvin Miles, "Hopes for SST are Dim but R&D Continues—Just in Case," *Los Angeles Times*, November 23, 1973, G-1, 11.
4. McLean, *Supersonic Cruise Technology*, 104–108; Sherwood Hoffman, "Bibliography of Supersonic Cruise Aircraft Research (SCAR)," NASA RP-1003 (November 1977), 1–5.
5. Chambers, *Innovation in Flight*, 39–40.
6. McLean, *Supersonic Cruise Technology*, 103. NASA photo of McLean with Driver and Mascitti provided courtesy of Joseph Chambers. Based on a drawing on page 38 in Spearman, "Evolution of the HSCT," NASA TM-109089, the model in the photo appears to be Lockheed's Mach 2.2 AST, no later than 1975.
7. FAA, "April 27, 1973," *FAA Historical Chronology, 1926–1996*, section 1973. The rule was included as Federal Aviation Regulation (FAR) section 91.817, Civil Aircraft Sonic Boom, effective September 30, 1973.
8. Miles, "Hopes for SST Are Dim," G1, 11; McLean, *Supersonic Cruise Technology*, 117–118; Conway, *High-Speed Dreams*, 176–180.
9. F. Edward McLean, "Some Non-Asymptotic Effects on the Sonic Boom of Large Airplanes," NASA TN D-2877 (June 1965).
10. Rudolph J. Swigart, "An Experimental Study in the Validity of the Heat-Field Concept for Sonic Boom Alleviation," NASA CR 2381 (March 1974).
11. For example, A.R. George and R. Seebass, "Sonic-Boom Minimization," *JASA* 51, no. 2, pt. 3 (February 1972): 686–694; A.R. Seebass and A.R. George, *Sonic Boom Minimization through Aircraft Design and Operation*, AIAA paper no. 73-241 (January 1973).
12. A.R. Seebass and A.R. George, "Design and Operation of Aircraft to Minimize their Sonic Boom," *Journal of Aircraft* 11, no. 9 (September 1974): 509–517.
13. Ibid., 509.

14. Ibid., 510.
15. Ibid., 513.
16. Ibid., 516.
17. Antonio Ferri, Huai-Chu Wang, and Hans Sorensen, "Experimental Verification of Low Sonic Boom Configuration," NASA CR 2070 (June 1973).
18. For a retrospective, see Percy J. Bobbitt and Domenic J. Maglieri, "Dr. Antonio Ferri's Contribution to Supersonic Transport Sonic-Boom Technology," *Journal of Spacecraft and Rockets* 40, no. 4 (July–August 2003): 459–466.
19. Charles L. Thomas, "Extrapolation of Sonic Boom Pressure Signatures by the Waveform Parameter Method," NASA TN D-6823 (June 1972).
20. Kenneth J. Plotkin, "State of the Art of Sonic Boom Modeling," *JASA* 111, no. 1, pt. 2 (January 2002): 532.
21. Harry W. Carlson, Raymond L. Barger, and Robert J. Mack, "Application of Sonic-Boom Minimization Concepts in Supersonic Transport Design," NASA TN D-7218 (June 1973), 2.
22. Ibid., 6-16, with figure 3-1 extracted from 15.
23. Ibid., 28.
24. Edward J. Kane, "A Study To Determine the Feasibility of a Low Sonic Boom Supersonic Transport," AIAA paper no. 73-1035 (October 1973); see also NASA CR 2332 (December 1973).
25. Sherwood Hoffman, "Bibliography of Supersonic Cruise Aircraft Research (SCAR) Program from 1972 to Mid-1977," NASA RP-1003 (November 1977); "Bibliography of Supersonic Cruise Research (SCR) Program from 1977 to Mid-1980," NASA RP-1063 (December 1980).
26. Langley Research Center, *Aerodynamic Analysis Requiring Advanced Computers, pt. 1*, NASA SP-347 (Washington, DC: National Technical Information Service, 1975), 3.
27. Ibid., 5.
28. For a layperson's introduction to this topic, see John D. Anderson, Jr., "NASA and the Evolution of Computational Fluid Dynamics," in Hallion, ed., *NASA's Contributions to Aeronautics* 1, 431–434.
29. J.L. Lung, B. Tiegerman, N.J. Yu, and A.R. Seebass, "Advances in Sonic Boom Theory," in *Aerodynamic Analysis Requiring Advanced Computers, pt. 2*, 1033–1047.
30. Frank Marconi, Larry Yeager, and H. Harris Hamilton, "Computation of High-Speed Inviscid Flows About Real

Configurations," in *Aerodynamic Analysis Requiring Advanced Computers, pt. 2*, 1411–1453.

31. Frank Marconi, Manuel Salas, and Larry Yeager, "Development of Computer Code for Calculating the Steady Super/Hypersonic Inviscid Flow Around Real Configurations," 1: Computational Technique, NASA CR 2675 (April 1976); 2: Computer Code, NASA CR 2676 (May 1976).

32. Christine M. Darden, "Minimization of Sonic-Boom Parameters in Real and Isothermal Atmospheres," NASA TN D-7842 (March 1975).

33. "Dr. Christine Mann Darden," accessed January 8, 2009, *http://www.rbc.edu/library/specialcollections/women_history_resources/vfwposter2002_darden.pdf*; "Dr. Christine M. Darden," accessed January 8, 2009, *http://blackhistorypages.net/pages/cdarden.php*. Darden, who was valedictorian of her graduating class at Allen High School in Asheville, NC, taught high school mathematics in Hampton for several years before being hired by NASA.

34. In January 1976, British Overseas Airways Corporation initiated Concorde flights between London and Bahrain, and Air France initiated Concorde service between Paris and Rio de Janeiro.

35. Christine M. Darden and Robert J. Mack, "Current Research in Sonic-Boom Minimization," in *Proceedings of the SCAR Conference* (1976), pt. 1, 525–541. Darden had discussed some of these topics in "Sonic Boom Theory–Its Status in Prediction and Minimization," AIAA paper no. 76-1, presented at the AIAA Aerospace Sciences Meeting, Washington, DC, January 26–28, 1976.

36. Craig Covault, "NASA Advances Supersonic Technology," *Aviation Week* (January 10, 1977): 16–18.

37. Darden and Mack, "Current Research in Sonic Boom Minimization," 526.

38. Ibid., 528–529. Figure 3-3 extracted from page 44.

39. Ibid., 530–532; J.J. Gottlieb, "Sonic Boom Research at UTIAS," *Canadian Aeronautics and Space Journal* 20, no. 3 (May 1974): 199–222, cited in Harvey H. Hubbard, Domenic J. Maglieri, and David G. Stephens, "Sonic Boom Research—Selected Bibliography with Annotation," NASA TM 87685 (September 1, 1986), 1.

40. Harry W. Carlson, "Simplified Sonic-Boom Prediction," NASA TP 1122 (March 1978), page 1 quoted.

41. Paul Holloway of Langley and colleagues from the Ames, Marshall, and Johnson Centers presented an early analysis in "Shuttle Sonic

Boom—Technology and Predictions," AIAA paper no. 73-1039 (October 1973).

42. Myron S. Malkin, "Environmental Impact Statement: Space Shuttle Program (Final)" NASA Headquarters (HQ) (April 1978), 106–116.

43. Including measurements in Hawaii with the Shuttle entering the atmosphere at 253,000 feet and decelerating from Mach 23; telephone interview, Domenic Maglieri by Lawrence Benson, March 18, 2009.

44. Plotkin and Maglieri, "Sonic Boom Research," AIAA paper no. 2003-3575, 5–6. For a description of Guiraud's findings in English, see J.P. Guiraud, "Focalization in Short Non-Linear Waves," NASA Technical Translation F-12,442 (September 1969); See also his previously cited "Acoustique géométrique bruit ballistique des avions supersoniques et focalisation," 215–267.

45. Peter M. Gill, "Nonlinear Acoustic Behavior at a Caustic," Ph.D. thesis, Cornell University, June 1974.

46. K.J. Plotkin and J.M. Cantril, "Prediction of Sonic Boom at a Focus," AIAA paper no. 76-2 (January 1976).

47. Raymond L. Barger, "Sonic-Boom Wave-Front Shapes and Curvatures Associated with Maneuvering Flight," NASA TP 1611 (December 1979). Figure 3-3 is from page 23.

48. James and Associates, ed., *YF-12 Experiments Symposium: A Conference Held at Dryden Flight Research Center, Edwards, California, September 13–15, 1978*, NASA CP-2054 (1978); Hallion and Gorn, *On the Frontier*, appendix P (YF-12 Flight Chronology, 1969–1978), 423–429. Dryden tested an oblique wing aircraft, the AD-1, from 1979 to 1982. Although this configuration might have sonic boom benefits at mid-Mach speeds, it was not a consideration in this experimental program. The AD-1 program is the subject of another study in this NASA book series by Bruce Larrimer.

49. C.H.E. Warren, "Sonic Bang Investigations Associated with the Concorde's Test Flying," paper presented at the 13th Royal Aeronautical Society (RAeS), AIAA, and Center for AeroSpace Information (CASI) Anglo-American Aeronautical Conference, London, June 4–8, 1973, AIAA paper no. 73-41174, cited in Hubbard et al., "Sonic Boom Research—Selected Bibliography," 1.

50. John L. McLucas et al., *Reflections of a Technocrat: Managing Defense, Air, and Space Programs During the Cold War* (Montgomery, AL: Air University Press, 2006), 265. McLucas was FAA administrator at the time.

51. "Second Concorde Noise Report for Dulles Shows Consistency," *Aviation Week* (July 19, 1976): 235; William Claiborne, "Mystery Booms Defy Expert Explanation," *Washington Post*, December 24, 1977, A1.

52. Recall that sound travels faster in warmer gases, as prevalent in the ozone layer.

53. Deborah Shapely, "East Coast Mystery Booms: A Scientific Suspense Tale," *Science* 199, no. 4336 (March 31, 1978): 1416–1417; "Concordes Exempted from Noise Rules," *Aviation Week* (July 3, 1978): 33; G.J. MacDonald et al., "Jason 1978 Sonic Boom Report," JSR-78-09 (Arlington, VA: SRI International, November 1978); Richard Kerr, "East Coast Mystery Booms: Mystery Gone but Booms Linger On," *Science* 203, no. 4337 (January 19, 1979): 256; John H. Gardner and Peter H. Rogers, "Thermospheric Propagation of Sonic Booms from the Concorde Supersonic Transport," Naval Research Laboratory Memo Report 3904 (February 14, 1979) (DTIC AD A067201).

54. Maglieri to Benson, e-mail message, August 18, 2011.

55. Robert J. Mack and Christine M. Darden, "Wind-Tunnel Investigation of the Validity of a Sonic-Boom-Minimization Concept," NASA TP-1421 (October 1979), with figure 3-4 extracted from page 22. The earlier paper was presented at an AIAA conference in Seattle on March 12–14, 1979, as "Some Effects of Applying Sonic Boom Minimization to Supersonic Cruise Aircraft Design," AIAA paper no. 79-0652, and later published in *Journal of Aircraft* 17, no. 3 (March 1980): 182–186.

56. Christine M. Darden, "Sonic Boom Minimization with Nose-Bluntness Relaxation," NASA TP 1348 (January 1979), 1.

57. Ibid., 4.

58. Ibid., 12.

59. Christine M. Darden, "Affordable/Acceptable Supersonic Flight: Is It Near?" 40th Aircraft Symposium, Japan Society for Aeronautical and Space Sciences (JSASS), Yokohama, October 9–11, 2002, 2.

60. Figure 3-5 extracted from Mack and Darden, "Wind-Tunnel Investigation of...Sonic-Boom-Minimization Concept," NASA TP-1421, 26.

61. Robert Kelly, "Supersonic Cruise Vehicle Research/Business Jet," in *Supersonic Cruise Research '79: Proceedings of a Conference Held at the Langley Research Center, Hampton, Virginia, November 13–16, 1979,* NASA CP 2108, pt. 2, 935–944, quotation from page 935.

62. Ibid. Figure 3-6 extracted from page 945.

63. The Boeing Company, "Supersonic Cruise Research Airplane Study," NASA CR 145212 (September 1977). This report does not appear in the NTRS database, but it is summarized by Domenic J. Maglieri in "Compilation and Review of Supersonic Business Jet Studies from 1963 through 1995," NASA CR 2011-217144 (May 2011). The proposal is also mentioned and illustrated in Spearman, "Evolution of the HSCT," NASA TM 109089, 8, 41–42.

64. Conway, *High-Speed Dreams*, 180–188; Chambers, *Innovations in Flight*, 48.

65. Christine M. Darden, "Charts for Determining Potential Minimum Sonic-Boom Overpressures for Supersonic Cruise Aircraft," NASA TP 1820 (May 1981).

F-16XL and SR-71 during in-flight shock wave measurements in 1995. (NASA)

CHAPTER 4

Resuming the Quest

High-Speed Research

For much of the next decade, the most active sonic boom research took place as part of the U.S. Air Force's Noise and Sonic Boom Impact Technology (NSBIT) program. This was a comprehensive effort begun in 1981 to study the noise resulting from military training and operations, especially those involving environmental impact statements and similar assessments. Although NASA was not intimately involved with NSBIT, Domenic Maglieri (just before his retirement from Langley) and the recently retired Harvey Hubbard compiled a comprehensive annotated bibliography of sonic boom research, organized into 10 major subject areas, to help inform NSBIT participants of the most relevant sources of information.[1]

One of the noteworthy achievements of the NSBIT program was building a detailed sonic boom database (known as Boomfile) on U.S. supersonic aircraft, first by flying them over a large array of newly developed sensors at Edwards AFB in the summer of 1987.[2] Called Boom Event Analyzer Recorders (BEARs), these unattended devices captured the full sonic boom waveform in digital format.[3] Other contributions of NSBIT were the long-term sonic boom monitoring of air combat training areas, continued assessment of structures exposed to sonic booms, studies on the effects of sonic booms on livestock and wildlife, and intensified research on focused booms (long an issue with maneuvering fighter aircraft).[4] Although Harry Carlson's simplified boom prediction program worked well for straight and level flights, the Air Force Human Systems Division at Wright-Patterson AFB attempted to supplement it in 1987 with a companion program called PCBoom to predict these focused booms. Its ray-tracing routines were adapted from a proposed mainframe computer program called BOOMAP2 to run on the basic desktop computers of the late 1980s.[5] Both programs were based on Albion D. Taylor's Tracing Rays and Aging Pressure Signatures (TRAPS) program, which adapted the ARAP code to account for caustics and focused booms.[6] Kenneth Plotkin later achieved this goal using the waveform parameter code of the Ames Research Center's Charles Thomas (described in the previous chapter) to create the widely used

PCBoom3. In one interesting application of sonic boom focusing that was first envisioned in the 1950s, fighter pilots were successfully trained to lay down super booms at specified locations.[7]

SST Reincarnated:
Birth of the High-Speed Civil Transport

By the mid-1980s, the growing economic importance of nations in Asia was drawing attention to the long flight times required to cross the Pacific Ocean from North America or to reach most of Asia from Europe. Meanwhile, in the face of growing competition from Europe and Japan for high-tech exports, the White House reversed its initial opposition to funding civilian aeronautical research. As part of this new policy, in March 1985, the Office of Science and Technology (OST) released a report, "National Aeronautical R&D Goals: Technology for America's Future," that included renewed support for a long-range supersonic transport.[8] Then, in his State of the Union address of January 1986, President Reagan ignited interest in the possibility of even a hypersonic transport—the National Aero-Space Plane (NASP)—dubbed the "Orient Express." In April the Battelle Memorial Institute established the Center for High-Speed Commercial Flight, which became a focal point and influential advocate for these proposals.[9]

NASA had been working behind the scenes with the Defense Advanced Research Projects Agency on the hypersonic technology that led to the NASP since the early 1980s. In February 1987, the OST issued an updated report on National Aeronautical R&D Goals subtitled "Agenda for Achievement," with an eight-point strategy for sustaining American leadership in aviation. It called for aggressively pursuing the NASP and developing the "fundamental technology, design, and business foundation for a long-range supersonic transport."[10] In response, NASA accelerated its hypersonic research and began a rejuvenated effort to help develop commercially viable supersonic technology. This started with contracts to Boeing and Douglas aircraft companies in October 1986 for market and feasibility studies, on what was now named the High-Speed Civil Transport (HSCT), accompanied by several internal NASA assessments. These studies soon ruled out hypersonic speeds (Mach 5 and above) as being impractical for passenger service because of technological, operational, and cost considerations. Eventually, NASA and its industry partners, after considering speeds from Mach 1.8 to Mach 3.2, settled on a fairly modest cruise speed of Mach 2.4 (which would allow the use of conventional jet fuel).[11] Although this would be only marginally faster than the Concorde, the HSCT was expected to double the Concorde's range and carry three times as many passengers. For its part, the high-risk, single-stage-to-orbit NASP survived until 1994 as a NASA-DOD experimental program (designated the X-30), with its sonic boom potential studied by current and former NASA and Air Force specialists.[12]

The contractual studies on the HSCT emphasized the need to resolve environmental issues, especially the restrictions on cruising over land because of sonic booms, before it could meet the goal of economically viable long-distance supersonic passenger service. As a first step toward this objective, Langley hosted a workshop on the status of sonic boom physics, methodology, and understanding on January 19 and January 20, 1988. Coordinated by Christine Darden, 60 representatives from Government, academia, and industry attended—including many who had been involved in the SST and SCR efforts and several from the Air Force's NSBIT program. Princeton's Wallace Hayes led a working group on theory, Cornell's Albert George led one on minimization, Pennsylvania State University's Allan D. Pierce led one on atmospheric effects, and Langley's Clemans A. Powell led a group on human response. Panels of experts from each of the working groups determined that the following areas most needed more research: boom carpets, focused booms, high-Mach predictions, atmospheric effects, acceptability metrics, signature prediction, and low-boom airframe designs. The report from this workshop served as a baseline on the latest knowledge about sonic booms and some of the challenges that lay ahead. As regards aerodynamics, it was recognized that the high-drag paradox (figure 3-2) would have to be resolved before a supersonic transport could be both quiet and efficient.[13]

Phase I of the High-Speed Research Program

While Boeing and Douglas were reporting on early phases of their HSCT studies, Congress approved an ambitious new High-Speed Research (HSR) program in NASA's budget for FY 1990. This effort envisioned Government and industry sharing the cost with NASA taking the lead for the first several years and industry expanding its role as research progressed. Because of the intermingling of sensitive and proprietary information, much of the work done during the HSR program was protected by a limited distribution system and some has yet to enter the public domain (or this book). Although the aircraft companies were making some progress on lower-boom concepts for the HSCT, they identified the need for more sonic boom research by NASA, especially on public acceptability and minimization techniques, before they could design a practical HSCT quiet enough to cruise over land without unacceptable performance penalties.[14]

Because solving environmental issues would be a prerequisite to developing the HSCT, NASA structured the HSR program into two phases. Phase I—focusing on engine emissions, noise around airports, and sonic booms as well as preliminary design work—was scheduled for 1990 to 1995. Among the objectives of Phase I were predicting HSCT sonic boom signatures, determining feasible reduction levels, and finding a scientific basis on which to set acceptability criteria. After, ideally, making sufficient progress on the environmental problems, Phase II would begin ramping up in 1994. With more industry participation and greater funding, it would focus on economically realistic airframe and propulsion technologies and, it was hoped, extend until 2001.[15]

NASA convened its first workshop for the entire High-Speed Research program in Williamsburg, VA, from May 14 to May 16, 1991. Because of the sensitivity and proprietary nature of much of the information, attendance was by invitation only. Thirteen separate sessions covered every aspect of high-speed flight with 86 of the papers presented published subsequently with limited distribution. Robert Anderson of NASA's Aeronautics Directorate opened the meeting by noting that the market for an environmentally acceptable, technically feasible, and economically viable HSCT might be as high as 300,000 passengers per day by 2000. But as a cautionary reminder on the challenges that lay ahead, he quoted from Ed McLean's portrayal of previous programs:[16]

> Past experience indicates that there will be little room for design compromises in the development of a successful SST. To meet the stringent environmental constraints of noise, sonic boom, and pollution in a safe, economically competitive SST will require the best possible combination of aerodynamic, structural, and propulsion technologies...integrated into a congruent airplane that meets all mission requirements.

A NASA Headquarters status report specifically warned that "the importance of reducing sonic boom cannot be overstated."[17] The stakes for the HSCT were high. One of the Douglas studies had projected that even by 2010, overwater-only routes would account for just 28 percent of long-range air traffic; but with supersonic overland cruise, the proposed HSCT could capture up to 70 percent of all such travel. Yet despite widespread agreement on the inherent advantages of a low-boom airliner, NASA's detailed program-management flowcharts included periodic decision points on whether or not to continue including sonic boom minimization as an essential criterion for the HSR designs. Based on previous efforts, the study admitted that research on low-boom designs "is viewed with some skepticism as to its practical application. Therefore an early assessment is warranted."[18]

As it is made evident by 15 of the presentations, NASA, its contractors, academic grantees, and the manufactures were already busy conducting a wide range of sonic boom research and minimization projects, including the long-postponed issue of human response. The main goals were to demonstrate a waveform shape that would be acceptable to the general public, to prove that a viable airplane could be built to generate such a waveform, to determine that such a shape would not be too badly disrupted during its propagation through the atmosphere, and to estimate that the economic benefit of overland supersonic flight would make up for any performance penalties imposed by a low-boom design.[19]

During the next 3 years, NASA and its partners went into a full-court press against the sonic boom.[20] They began several dozen major experiments and studies, results of which were published in reports and presented at several workshops dealing solely with the sonic boom. These were held at the Langley Research Center in February 1992,[21] the Ames Research Center in May 1993,[22] Langley in June 1994,[23] and again at Langley in September 1995.[24] These meetings, like the HSR's sonic boom effort itself, were organized into three major areas of research: (1) configuration design and operations (managed by Langley's Advanced Vehicles Division), (2) atmospheric propagation, and (3) human acceptability (both managed by Langley's Acoustics Division). The reports from these workshops were well over 500 pages long and included dozens of papers on the progress or completion of various projects, experiments, and research topics.[25]

The HSR program precipitated major advances in the design of supersonic configurations even for reduced sonic boom signatures. Many of these advances were made possible by the rapidly expanding field of computational fluid dynamics. With CFD, engineers and researchers were now able to use complex

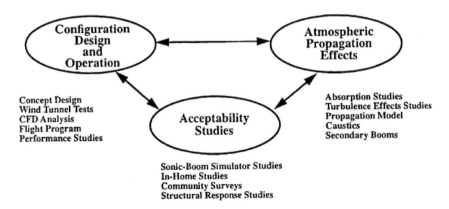

Figure 4-1. Structure and scope of HSR sonic boom studies. (NASA)

Christine Darden (second from left) and Robert Mack (right) examining a 12-inch low-boom HSCT model with Matthew Overholt and Kathy Needleman. (NASA)

computational algorithms processed by supercomputers and parallel computers to calculate the nonlinear aspects of near-field shock waves, including the Navier-Stokes equations, even at high Mach numbers and angles of attack. Results could be graphically displayed in mesh and grid formats that emulated three dimensions. (In simple terms: before CFD, the nonlinear characteristics of shock waves generated by a realistic airframe had involved far too many variables and permutations to calculate by conventional means.)

The great progress that had been made in recent years was already evident at the 1991 HSR Workshop's session on Sonic Boom and Aerodynamic Performance, which included minimization strategies. At Langley, Christine Darden, Robert Mack, and Raymond Barger—among the few to remain actively involved in sonic boom minimization ever since the demise of the SCR program—had recently been joined by talented new researchers, such as Peter G. Coen. With the help of a new computer program for predicting sonic booms devised by Coen,[26] Langley's 11-person design team applied two theoretical approaches and an iterative process of modifications to build two 12-inch wind tunnel models of HSCT-size airframes, including engine nacelles, intended to combine reduced sonic boom signatures with aerodynamic efficiency. One was designed for Mach 3 cruising at 65,000 feet and the other for cruising at Mach 2 and 55,000 feet.[27]

The finished models were first tested during October 1990 in the 9-foot-by-7-foot and 8-foot-by-7-foot supersonic sections of the Ames Research Center's very busy Unitary Plan Wind Tunnel, which allowed pressure measurements out to between five and six body lengths. With extrapolations confirmed for closer distances, they were then able to use the much tighter confines of Langley's 4-foot-by-4-foot Unitary Plan Wind Tunnel for additional experiments during the next 3 months with a specially made sting (the connecting device upon which a wind tunnel model is affixed) and angle-of-attack mounting mechanism. The tests found excellent agreement between the forward part of the extrapolated wind tunnel measurements and the predicted sonic boom signatures, and they validated the theory-derived design process, especially for the Mach 2 model. There were some disappointments. For example, even using 12-inch models, the openings to the nacelles proved too small to allow a sufficient airflow to pass through them at Ames, so they were removed for the testing at Langley.[28]

Exploiting the CFD Revolution

Although using the latest in wind tunnel technology, the limitations of these experiments also showed signs that a new era in sonic boom minimization research was dawning with the use of computational fluid dynamics. The wind tunnel tests revealed anomalies in the rear portions of the near-field pressure signatures caused by three-dimensional flows that could not be accounted for by the axisymmetric propagation methods. They also failed to reveal downstream shocks from the wings or capture the effects of exhaust plumes. CFD, on the other hand, indicated that these plumes might completely obscure the benefits of sonic boom shaping for Langley's Mach 3 design cruising at 60,000 feet.[29]

In addition to being unable to model complex airframe geometries and the effects of propulsion systems, traditional linear supersonic aerodynamics based on wind tunnel experimentation as well as quasi-linear acoustic propagation theory—sometimes referred to as modified linear theory (MLT)—had difficulty with analysis of nonlinear three-dimensional effects near the aircraft and higher Mach numbers or angles of attack. These limitations would seriously hamper designing the new HSCTs for optimum performance and low sonic booms. The Ames Research Center, with its location in the rapidly growing Silicon Valley area, had been a pioneer in applying CFD capabilities to aerodynamics, especially after establishing the Numerical Aerodynamic Simulation Facility with a powerful new Cray supercomputer in 1987.[30] At the 1991 HSR workshop, a report by the Ames sonic boom team led by Thomas Edwards and including modeling expert Samson Cheung predicted that "in many ways,

CFD paves the way to much more rapid progress in boom minimization…. Furthermore, CFD offers fast turnaround and low cost, so high-risk concepts and perturbations to existing geometries can be investigated quickly."[31]

The Ames researchers started off by using three previously tested supersonic models (a cone cylinder, a rectangular wing section, and a delta-winged airframe) to validate their new CFD codes for sonic boom predictions as well as those of several existing CFD programs. After obtaining good correlation of CFD and wind tunnel data, the Ames researchers concluded that "at this point it can be said without reservation that CFD can be used in conjunction with quasi-linear extrapolation methods to predict sonic booms in the near and far field accurately."[32] The experimentation showed the importance of using precise airframe geometry and adequate grid resolution. Results also indicated that inviscid Euler flow analysis (i.e., without the need to account for laminar flow) was sufficient for accurate sonic boom predictions. After validating their CFD codes, they next began applying them to the two Langley models tested in the Ames 9-foot-by-7-foot wind tunnel (see above), work that was still ongoing at the time of the HSR workshop. Both Ames and Langley would use the HSCT as a demonstration project for analyses on massively parallel computers under NASA's High Performance Computing and Communications Program (HPCCP).[33]

In a project sponsored by Christine Darden, Michael Siclari of the Grumman Corporate Research Center at Bethpage, NY, presented the results of applying a three-dimensional Euler code for multigrid-implicit marching, as modified to predict sonic boom signatures, from the two Langley HSCT models. (This new code accordingly was referred to as MIM3D-SB, or Multigrid Implicit Marching in Three Dimensions for Sonic Booms.) Stated as simply as possible, the code used a simple wave-drag geometry to input data, from which the computer calculated the propagation of shock waves from digitized replications of the models (with different length stings attached) in a series of more than 100 steps, with each new step calculated from the results of prior steps. The program recorded the propagation of shock waves on three-dimensional adaptive mesh grids, featuring denser grid points near the aircraft and (for faster processing) a coarser adaptive grid pattern farther out on which additional data points were progressively marched a specified distance from the airframe. From there, a waveform parameter code (derived from that of Charles Thomas) extrapolated the near-field signatures through atmospheric conditions to the ground. Figure 4-2 depicts a side view of one of the grid patterns used for Langley's Mach 3 model (just visible in the front apex of the grid system), swept back at the approximate angle of the shock waves to save computer time compared to a more complete grid system. Figure 4-3 shows the shock waves calculated on this grid network as pressure contours (something like they would appear in a wind tunnel shadowgraph). For visualizing the complex three-dimensional, cone-shaped expansion of the shock waves on paper, this early CFD application could also display slices of the pressure waves. Figure 4-4 shows the computed isobars in two vertical planes aft of the Mach 2 model, clearly indicating the complexity of its flow pattern—something not really possible with a wind tunnel. To achieve signatures at three body lengths from the aircraft axis (which at supersonic speeds meant 12 to 15 body lengths downstream of the aircraft) required approximately 2 million data points.[34]

Figure 4-2. Side view of MIM3DSB grid topology for Mach 3 HSCT model. (NASA)

Figure 4-3. Pressure contours from Mach 3 HSCT model. (NASA)

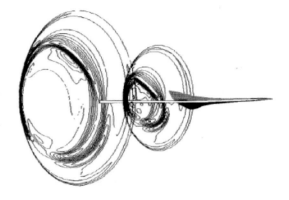

Figure 4-4. Isobars showing propagation of midfield pressure patterns downstream from Mach 2 HSCT model. (NASA)

At the 1992 sonic boom workshop, Darden and Mack admitted how recent experiments at Langley had revealed limitations in using near-field wind tunnel data for extrapolating sonic boom signatures.[35] During this and the two subsequent sonic boom workshops and at other venues, experts from Ames, Langley, and their contractors reported optimistically on the potential of new CFD computer codes to help design configurations optimized for constrained sonic booms and aerodynamic efficiency. In another potential application of CFD, former Langley researcher Percy "Bud" Bobbitt, who had joined Domenic Maglieri at Eagle Engineering, pointed out the potential of hybrid laminar flow control (HLFC) for both aerodynamic and low-boom purposes.[36]

Even the numbers-crunching capabilities of the supercomputers of that era were not yet powerful enough for CFD codes and the grids they produced to accurately depict effects much beyond the near field, but the use of massively parallel computing held the promise of eventually being able to do so. It was becoming apparent that, for most aerodynamic purposes, CFD was the design tool of the future, with wind tunnel models becoming more a means of verification. As predicted by Ames researchers in 1991, "the role of the wind tunnel in low-boom model design is to benchmark progress at significant intermediate stages and at the final design point of numerical model development."[37]

By the second sonic boom workshop in February 1992, there were already signs of progress in applying CFD methods for predicting sonic boom signatures. Both Susan Cliff of Ames and Michael Siclari of Grumman included the effects of engine nacelles in analyses of the Langley Mach 2 and Mach 3 configurations and a Boeing Mach 1.7 design. Cliff described lessons learned doing analyses with the Three-dimensional Euler/Navier-Stokes Aerodynamic Method (TEAM) and a faster Euler code-based program called AIRPLANE that relied on an unstructured tetrahedral mesh to calculate pressure signatures, including those from the nacelles that had defied wind tunnel measurements.[38] Siclari followed up on his earlier work using the efficient multiblock Euler marching code (MIM3D-SB) with Grumman's innovative mesh technology. It was able to calculate accurate three-dimensional pressure footprints at one body length (using 1.9 million grid points) and extrapolate them to the ground by using a linear waveform parameter method (derived from that of Charles Thomas). Besides the nacelles themselves, his modeling included an engine exhaust simulation to predict the effects of the plumes on the sonic boom signatures. As can be seen from the graphics printed out from one of these exercises in figure 4-5, the state of the art in CFD during the early 1990s was advancing rapidly.[39] Unfortunately, it would not yet progress enough to design a low-boom but also aerodynamically efficient supersonic transport.

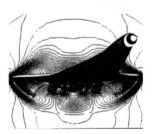

Among a dozen other aerodynamic papers at the 1992 workshop, the work by Samson Cheung and Thomas Edwards reported on progress in their CFD modeling using the UPS3D parabolized (simplified) Navier-Stokes code and a hyperbolic (curved in three dimensions) grid-generation scheme for minimization purposes. They were able to improve the lift-to-drag ratio for a simplified model of Boeing's baseline low-boom HSCT configuration (without nacelles or a complete tail assembly) by almost 4 percent while at the same time extrapolating a quieter flattop signature. To save expensive computer time, they relied on a course grid for their design work and only ran the end result on a fine grid to check for discrepancies.[40]

Despite the signs of rapid progress with CFD, designing low-boom characteristics into a practical airliner would not be easy. John Morgenstern

Figure 4-5. Computational grid and resulting pressure contours from the wings and nacelles on Langley's Mach 3 HSCT configuration. (NASA)

described McDonnell Douglas's strategy, after exploring numerous configurations, to optimize its HSCT for efficient Mach 2.4 cruising over water while slowing down to Mach 1.8 for reduced sonic boom over land.[41] In something of a reality check, veteran sonic boom specialist George Haglund and a Boeing colleague described analyses of their company's two low-boom designs: "Since L/D alone is not a good measure of airplane performance, each airplane was evaluated in sufficient depth to determine an operating empty weight...and maximum takeoff weight...for a 5000 n mi. mission [to allow] a meaningful performance comparison to a conventional baseline configuration." Although meeting some objectives, they found that "achieving a practical HSCT low-boom configuration with low drag, high payload, and good performance is a formidable design problem."[42]

Documentation of the work on configuration design and analysis presented at the Ames sonic boom workshop in 1993 is not publicly available; therefore, the papers presented at the Langley workshop in 1994 represent 2 years' worth of progress, especially in applying CFD techniques. By then, results were in from an Ames experiment comparing computational fluid dynamics with traditional, modified linear theory for predicting sonic boom signatures, something that would be essential for designing HSCTs that could shape such signatures in the near, mid, and far fields. Although modified linear theory was well established, fast, and efficient, with an inverse design capability, it had trouble modeling the effects of complex geometries on pressure signatures. The limitations of CFD were not yet fully understood, but it did have the capability to do complex geometrical modeling—at the cost of expensive computer time.

To compare CFD with MLT, the Ames Computational Aerospace Branch selected a modified Boeing arrow-wing, low-boom configuration as a test case. They then evaluated several CFD techniques—UPS3D, AIRPLANE, and HFL03 (a Euler time-relaxation code)—along with results contributed by Grumman with its very efficient MIM3D-SB code and Boeing's MLT techniques and ARAP-based extrapolations. Although calibration problems limited the use of wind tunnel data, the analysis found that all the CFD methods, although not consistent in their far-field pressure signatures, could more accurately predict the effects of lift and pitching as well as sonic booms as measured by perceived loudness (in decibels) at ground level. Measuring the effects of drag, however, was highly dependent on dense grid resolution. The results indicated that CFD predictions would continue to improve with experience.[43]

Eight more of the papers at the 1994 workshop described projects related to sonic boom minimization, most using CFD as well as wind tunnel analyses. As an example of the latter, Robert Mack reported some success in preventing the inlet shocks that had stymied previous experiments so as to obtain pressure signatures from four nacelles on a low-boom wind model in Langley's

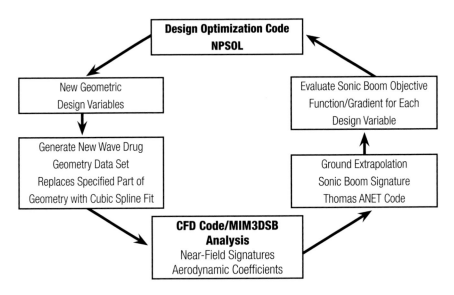

Figure. 4-6. CFD design process for sonic boom minimization. (NASA)

4-foot-by-4-foot Unitary Plan Wind Tunnel.[44] Meanwhile, Ames and Princeton researchers reported on using CFD to design an airframe that generated a type of multishock signature that might reach the ground with a quieter sonic boom than either the ramp or flattop wave forms that were a goal of traditional minimization theories.[45] (Although not part of the HSCT effort, Ames and its contractors also used CFD to continue exploring the possible advantages of oblique-wing aircraft, including sonic boom minimization.)[46] As an excellent case in point of how CFD was becoming more practical, Grumman's Michael Siclari described how his NASA-sponsored MIM3D-SB code and numerical optimization techniques, coupled with an aerodynamic code (in this case, one called NPSOL), could now analyze wing-body configurations in a matter of minutes rather than hours of supercomputer time, making it efficient and economical enough to be practical as a design tool. As examples, he showed results of this automated process (depicted in figure 4-6) with four HSCT configurations.[47]

Even with the advances being made in designing airframes for lower sonic booms, the issue of overall performance was still a critical concern for the High-Speed Research program. To get a better perspective on the relationship between sonic boom acceptability and other performance requirements, an eight-person team that included Donald Baize and Peter Coen from NASA Langley and former NASA intern Kathy Needleman from Lockheed Engineering & Sciences Company assessed eight of the current low-boom HSCT configurations against an unconstrained reference configuration. Predicated on some technologies projected to be available in 2005 (e.g., advanced composite materials, aeroelastic tailoring, mixed-flow turbofan engines, and multipurpose displays), the team evaluated such factors as L/D ratios, fuel capacity and consumption, passenger payload, takeoff distance, gross weight, and mission block time. Under these criteria, all of the designs achieved a total gross weight per passenger

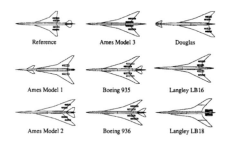

Figure 4-7. Eight low-boom HSCT configurations in the early 1990s. (NASA)

only slightly higher than the reference configuration, but all were heavier than originally assumed and would require at least another design cycle to ensure successful low-boom shaping. As with studies dating back to the SST, the most highly swept wing planforms did not have enough lift at low speeds. Reinforcing previous aircraft company projections, being able to fly supersonic over land areas—even on relatively short segments of routes—offered better block time and therefore economic advantages. Figure 4-7, showing the low-boom HSCT configurations studied in this project, offers an excellent idea of the various design options being explored during the first phase of the High-Speed Research program.[48]

Flight Tests and Acceptability Studies

Neither wind tunnels nor CFD could as yet empirically prove the physical persistence of a shaped waveform for more than a tiny fraction of the 200 to 300 body lengths needed to represent the distance from an HSCT to the surface. To fill this credibility gap, Domenic Maglieri and a team at Eagle Engineering looked at various options for performing relatively economical flight tests by using remotely piloted vehicles (RPVs) modified for low-boom characteristics, a proposal presented at the 1991 HSR workshop.[49] In 1992, they provided results of a feasibility study on the most cost-effective ways to verify design concepts with realistic testing. After exploring a wide range of alternatives, the team selected the Teledyne-Ryan BQM-34E Firebee II, which the Air Force and Navy had long used as a supersonic target drone. Four of these 28-foot-long RPVs, which could sustain a speed of Mach 1.3 at 9,000 feet (300 body lengths from the surface), were available as surplus. Modifying them with low-boom design features such as specially configured 40-inch nose extensions (shown in figure 4-8) could provide far-field measurements needed to verify the waveform shaping projected by CFD and wind tunnel models.[50]

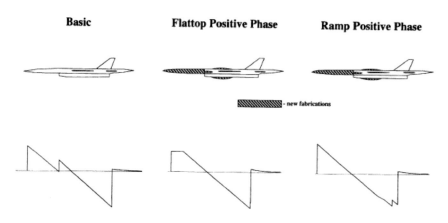

Figure 4-8. Proposed modifications and signatures of BQM-34E Firebee II. (NASA)

Meanwhile, a similar but more ambitious plan at the Dryden Flight Research Center led to NASA's first significant sonic boom testing there since 1970. SR-71 program manager David Lux, atmospheric specialist L.J. Ehernberger, aerodynamicist Timothy R. Moes, and principal investigator Edward A. Haering came up with a proposal to demonstrate CFD design concepts by having one of Dryden's SR-71s modified with a low-boom configuration. As well as being much larger, faster, and higher flying than the little Firebee (thereby more closely emulating the HSCT), an SR-71 would also allow easier acquisition of near-field measurements for direct comparison with CFD predictions.[51]

To lay the groundwork for this modification, Dryden personnel gathered baseline data from a standard SR-71 using one of its distinctive "cranked arrow" (double angle delta-winged) F-16XL aircraft (shown in a photograph preceding this chapter). Built by General Dynamics in the early 1980s for evaluation by the Air Force as a long-range strike version of the short-range F-16 fighter, the elegant F-16XL had lost out to the rival McDonnell Douglas F-15E Strike Eagle, which had even greater range and payload capability.

In tests at Edwards during July 1993, the F-16XL, flown by Dryden test pilot Dana Purifoy, probed as close as 40 feet below and behind an SR-71 cruising at Mach 1.8 to collect near-field pressure measurements.[52] Langley and McDonnell Douglas analyzed this data, which had been gathered using a standard flight-test nose boom. Both reached generally favorable conclusions about the ability of high-order CFD and McDonnell Douglas's proprietary MDBOOM program to serve as design tools.[53] Kenneth Plotkin and Juliet Page of Wyle Labs had developed MDBOOM from a focus version of the Thomas code that Plotkin and a colleague developed in 1976.[54] (This focus code was also adapted for PCBoom3, which replaced the original TRAPS-based PCBoom.)[55]

Based on these results, a team led by low-boom aerodynamicist John Morgenstern at McDonnell Douglas Aerospace West designed modifications to alter the bow and middle shock waves of the SR-71 by reshaping the front of the airframe with a nose glove and adding to the midfuselage cross section as partially illustrated in figure 4-9. (In this figure, M_∞ denotes Mach number and α denotes angle of attack.) An assessment of these modifications by Lockheed Engineering & Sciences Company found them feasible.[56] The next step—a big one—would be to obtain the considerable funding that would be needed for the modifications and testing.

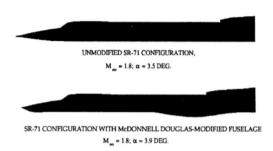

UNMODIFIED SR-71 CONFIGURATION,
$M_\infty = 1.8$; α = 3.5 DEG.

SR-71 CONFIGURATION WITH McDONNELL DOUGLAS-MODIFIED FUSELAGE
$M_\infty = 1.8$; α = 3.9 DEG.

Figure 4-9. Proposed SR-71 low-boom modification. (NASA)

In May 1994, Dryden used two of its fleet of F/A-18 Hornets to measure how near-field shock waves merged to assess the feasibility of a similar low-cost experiment in waveform shaping using two SR-71s. Flying at Mach 1.2 with one aircraft below and slightly behind the other, the first experiment positioned the canopy of the lower F/A-18 in the tail shock extending down from the upper F/A-18 (called a tail-canopy formation). The second experiment had the lower F/A-18 fly with its canopy in the inlet shock of the upper F/A-18 (an inlet-canopy formation). Ground sensor recordings revealed that the tail-canopy formation caused two separate N-wave signatures, but the inlet-canopy formation yielded a single modified signature, which two of the recorders measured as a flattop waveform. This low-cost technique, however, presented safety issues. Even with the excellent visibility from the F/A-18's bubble canopy (one pilot used the inlet shock wave as a visual cue for positioning the aircraft) and its responsive flight controls, maintaining such precise positions was still not easy. The pilots recommended against doing the same with SR-71s considering their larger size, slower response, and limited cockpit visibility.[57]

Atmospheric effects had long posed many uncertainties in understanding sonic booms, but advances in acoustics and atmospheric science since the SCR program promised better results. Not only did the way air molecules absorb sound waves need to be better understood but so did the old issue of turbulence. In addition to using the Air Force's Boomfile and other available material, Langley's Acoustics Division had Eagle Engineering, in a project led by Domenic Maglieri, restore and digitize data from the irreplaceable XB-70 records.[58]

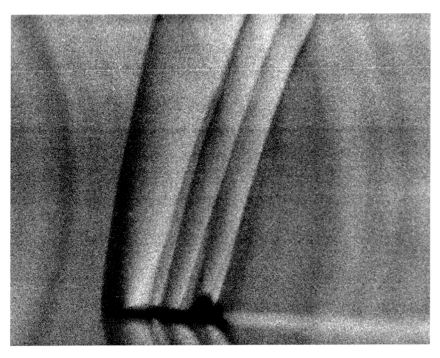

Historic schlieren photograph of shock waves from a T-38 flying Mach 1.1 at 13,000 feet. (NASA)

The Acoustics Division, assisted by Lockheed Engineering & Sciences Company, also took advantage of the NATO Joint Acoustic Propagation Experiment (JAPE) at the White Sands Missile Range in August 1991 to do some new flight testing. The researchers arranged for Air Force F-15, F-111, and T-38 aircraft and one of Dryden's SR-71s to make 59 supersonic passes over an extensive array of BEAR and other recording systems as well as meteorological sensors—both early in the morning (when the air was still) and during the afternoon (when there was usually more turbulence).[59] Although meteorological data was incomplete, results later showed the effects of molecular relaxation and turbulence on both the rise time and overpressure of bow shocks.[60] Henry Bass of the University of Mississippi, a key participant in the JAPE, was an important researcher on the acoustics of turbulence. Another academic researcher, David Blackstock of the University of Texas, and his graduate students also discovered more new effects of turbulence as well as other atmospheric instabilities on sonic booms, some of these by using innovative laboratory experiments.[61] Starting with the first HSR workshop, NASA and NASA-sponsored researchers, such as Allan D. Pierce of Penn State University, began producing a variety of papers on waveform freezing (persistence), measuring diffraction and distortion of sound waves, and trying to ascertain the complex relationship among molecular relaxation, turbulence, humidity, and other weather conditions.[62]

For better visualizing sonic booms, Leonard Weinstein of Langley even developed a way to capture stunning images of actual shock waves in the real atmosphere. He did this using a ground-based schlieren imaging system (a specially masked and filtered tracking camera with the sun providing back-lighting). As shown in the accompanying photo, this was first demonstrated in December 1993 with a T-38 flying just over Mach 1 at Wallops Island.[63]

All of the research into the theoretical, aerodynamic, and atmospheric aspects of sonic booms—no matter how successful—would not protect the High-Speed Research program from the Achilles' heel of previous efforts: the subjective responses of human beings. As a result, Langley, led by Kevin Shepherd of the Acoustics Division with researchers such as Brenda Sullivan, Jack Leatherwood, and David McCurdy, began a systematic effort to measure human responses to different strengths and shapes of sonic booms to help determine acceptable levels. As an early step, the division built an airtight, foam-lined sonic boom simulator booth (known as the boom box) derived from a similar apparatus at the University of Toronto. Using the latest in computer-generated digital-amplification and loudspeaker technology, it was capable of generating shaped waveforms up to 4 psf and 140 decibels (dB). Based on responses from subjects, researchers selected the perceived-level decibel (PLdB) as the preferred metric. For responses outside a laboratory setting, the NASA Langley team planned several additional acceptance studies.[64]

By 1994, early results had become available from two of these human-response projects. Langley and Wyle Laboratories had developed mobile boom-simulator equipment for what was called the In-Home Noise Generation/Response System (IHONORS). Depicted in figure 4-10, it consisted of computerized sound systems installed in 33 houses for 8 weeks at a time in a network connected by modems to a monitor at Langley. From February 1993 to December 1993, these households were subjected to almost 58,500 randomly timed sonic booms of various signatures for 14 hours a day. Although definitive analyses were not available until the following year, the initial results confirmed how the level of annoyance increased whenever subjects were startled or trying to rest.[65]

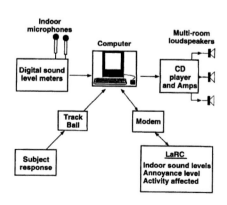

Figure 4-10. Schematic of the In-Home Noise Generation/Response System. (NASA)

Preliminary results were also in from the first phase of the Western USA Sonic Boom Survey of civilians who had been exposed to such sounds for many years. This part of the survey took place in remote desert towns and settlements located around the Air Force's vast Nellis combat training range complex in Nevada. Unlike previous community surveys, it correlated citizen responses to accurately measured sonic boom signatures (using BEAR devices) in places where booms were a regular occurrence yet where the subjects did not live on or near a military installation (where the economic benefits of the base to the local economy might influence their opinions). Although findings were not yet definitive, these 1,042 interviews proved more decisive than any of the many other research projects in determining the future direction of the HSCT effort. Based on a metric called day-night average noise level, the respondents found the booms much more annoying than previous studies on other types of aircraft noise even at the levels projected for low-boom designs. Their negative responses, in effect, dashed hopes that the HSR program might lead to an acceptable overland supersonic transport.[66]

HSR Phase II: Surrendering Again to the Sonic Boom

Well before the paper on this survey was presented at the 1994 Sonic Boom Workshop, its early findings had prompted NASA Headquarters to reorient the High-Speed Research program toward an HSCT design that would fly supersonic only over water. Just as with the AST program 20 years earlier, this became the goal of Phase II of the HSR program (which began with the help of FY 1994 funding left over from the canceled NASP).[67] Once again, public annoyance with the sonic boom had proved too big an obstacle for even a new generation of aeronautical technology to overcome. The revamped HSR program would continue intensive supersonic research for the rest of the decade, but after Boeing's absorption of McDonnell Douglas in 1996, this single company's continued willingness to invest in the program became crucial.

At the end of the 1994 workshop, Christine Darden discussed the progress and lessons learned to date as well as the next steps for sonic boom research. Regarding progress, she said, "tremendous advances in supercomputers, gridding schemes, and computational algorithms have allowed computational fluid dynamics...to become a new tool in the prediction of near field sonic-boom signatures."[68] Although major improvements were still needed in correlating this nonlinear data with the linear methodology used for effects in the mid- and far-fields, several achievements were already in evidence. In addition to the benefits of CFD for design concepts and analysis of near-field shock wave signatures, these achievements included improved F-function analysis and methods to predict inlet shocks, increased use of nonlinear corrections for modified linear theory techniques, minimization theories for cambered wing bodies, measurements of flow-through nacelles on wind tunnel models, and improving some performance criteria of low-boom concepts to within 3 percent of unconstrained baseline configurations.[69]

While the lower-boom design efforts had shown outstanding progress, management of this effort had not been ideal. Dispersal of the work among two NASA centers and two major aircraft manufacturers had resulted in communication problems as well as a certain amount of unhelpful competition (presumably among the contractors as well as between Langley and Ames). The milestone-driven HSR effort required concurrent progress in various technical and scientific areas, which is inherently difficult to coordinate and manage. And even if low-boom airplane designs had been improved enough to meet acoustic criteria, they would have been heavier and performed more poorly at slow speeds than unconstrained designs.[70]

Under the new HSR strategy, any continued minimization research was now aimed at lowering the sonic boom of the baseline overwater design while propagation studies would concentrate on predicting boom carpets, focused booms, secondary booms, and ground disturbances. In view of the HSCT's overwater mission, new environmental studies would devote more attention to the potential penetration of shock waves into water and the effects of sonic booms on the marine mammals and birds that might be affected.[71] Concorde operations had revealed no such problems, but since the HSCT would be about twice the weight but only 50 percent longer, the sonic boom overpressures generated by the baseline designs would tend to be about 50 percent higher. As a result, aerodynamicists such as Robert Mack of Langley, John Morgenstern of McDonnell Douglas, George Haglund of Boeing, and Michael Siclari of Grumman (which merged with Northrop Corporation in April 1994) turned their attention to minor modifications that could reduce this level with only minimal performance penalties.[72]

Although the preliminary results of the first phase of the Western USA Survey had already had a decisive impact, Wyle Laboratories completed the second phase with a similar polling of civilians in Mojave Desert communities exposed regularly to sonic booms, mostly from Edwards AFB and China Lake Naval Air Station. Surprisingly, this phase of the survey found the Californians there much more amenable to sonic booms than the less tolerant desert dwellers in Nevada, but they were still more annoyed by booms than by other aircraft noise of comparable perceived loudness.[73]

With the decision to end work on a low-boom HSCT, the proposed modifications of the Firebee RPVs and SR-71 had of course been canceled (postponing for another decade the first live demonstrations of boom shaping).

Nevertheless, some flight testing that would prove of future value continued to be conducted. From February 1995 through April 1995, the Dryden Flight Research Center conducted more SR-71 and F-16XL sonic boom flight tests. Led by Ed Haering, this experiment included an instrumented YO-3A light aircraft from Ames, an extensive array of various ground sensors, a network of new differential Global Positioning System receivers accurate to within 12 inches, and installation of a sophisticated new nose boom with four pressure sensors on the F-16XL.[74]

On eight long missions, one of Dryden's SR-71s flew at speeds between Mach 1.25 and Mach 1.6 at 31,000 feet to 48,000 feet while the F-16XL (kept aloft by in-flight refuelings) made numerous near- and midfield measurements of bow, canopy, inlet, wing, and tail shock waves at distances from 80 feet to 8,000 feet. Some of these showed that the canopy shock waves were still distinct from the bow shock after 4,000 feet to 6,000 feet. Comparisons of far-field measurements obtained by the YO-3A flying at 10,000 feet above ground level and the recording devices on the surface revealed effects of atmospheric turbulence. Analysis of the data validated two existing sonic boom propagation codes used for predicting far-field signatures (ZEPHYRUS and SHOCKN) and clearly showed how variations in the SR-71's gross weight, speed, and altitude and atmospheric phenomena such as molecular absorption caused differences in shock wave patterns and their coalescence into N-shaped waveforms.[75] Figure 4-11 depicts the participants and basic structure of these flight tests, which would serve as a precedent for others in the future.[76]

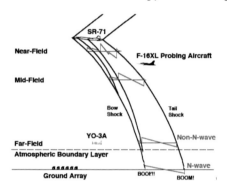

Figure 4-11. Measuring the evolution of shock waves from an SR-71. (NASA)

This innovative and successful experiment marked the end of dedicated sonic boom flight testing during the HSR program. Phase II testing focused on the many other issues involved in designing a practical, 320-foot-long, 300-passenger, Mach 2.4 HSCT with a range of 5,000 nm that would fly only subsonically over land. For example, NASA's creative partnership with Russia in using a Tu-144 as a supersonic laboratory from 1996 to 1999 did not include sonic boom measurements as originally planned.[77]

The last of the sonic boom workshops, held at Langley in September 1995, no doubt seemed rather anticlimactic for the 46 attendees in view of the new, less ambitious HSR goals for a high-speed civil transport. As with the SST and

SCR programs, however, their research—the latest of which would be published in a two-volume compendium—added greatly to the scientific knowledge and engineering skills that would be needed if and when another effort to develop a civilian supersonic airplane might be initiated.[78] Several papers indicated that the behavior of shock waves and acoustic rays under a wide range of atmospheric conditions were now well understood.[79] Yet the challenges in designing a practical airplane that could exploit this knowledge to control sonic boom signatures, especially in view of the disturbing new evidence collected on the sensitivity of human responses to them, were still daunting.[80] Even with the rapid progress with computational fluid dynamics, results so far indicated the need for much more computing power and new techniques. As Kenneth Plotkin explained, "due to a combination of computational costs and numerical algorithms losing resolution after many steps, CFD cannot be brought all the way to the ground or even very many body lengths away from the aircraft."[81]

Developing a high-speed civil transport ran into other barriers besides the sonic boom. By late 1998, the HSR program confronted a combination of economic, technological, political, and budgetary problems (including cost overruns for the International Space Station). The Boeing Company, now estimating that development of the HSCT would take $13 billion, cut its support, and the administration of President William J. Clinton, with the backing of NASA administrator Daniel S. Goldin, decided to terminate the HSR program at the end of FY 1999. Although other research programs picked up elements of the HSR, having to end it and a similar program for an Advanced Subsonic Transport deprived NASA of the focus these programs helped give to its aeronautical research.[82] Ironically, NASA's success in helping the aircraft industry develop quieter subsonic aircraft, which had the effect of moving the goal post for acceptable airport noise, was one of the factors convincing Boeing to drop plans for a supersonic airliner. Nevertheless, the High-Speed Research program was responsible for truly significant advances in technologies, techniques, and scientific knowledge, including a better understand of the sonic boom and ways to diminish it.[83]

To help identify areas for future research that might overcome the barriers to supersonic flight encountered by the HSR program, NASA in 2000 asked the National Research Council to conduct a comprehensive investigation of the relevant technologies that would be needed. The Council's Aeronautics and Space Engineering Board formed an expert 14-person committee on breakthrough technologies to perform this task. Released in 2001, its in-depth study focused on "high-risk, high-payoff technologies where NASA research could make a difference over the next 25 years."[84] While advising that NASA "should have its eye on the grand prize—supersonic commercial transports," the committee deemed it "quiet appropriate" for NASA to conduct sonic boom research related to supersonic business jets, which were increasingly seen as having a more realistic chance of meeting sonic boom requirements.[85] Their study concluded with the following admonition:

> If the United States intends to maintain its supremacy in the commercial aerospace sector, it has to take a long-term perspective and channel adequate resources into research and technology development. The technological challenges to commercial supersonic flight can be overcome, *as long as the development of key technologies is continued.* Without continued effort, an economically viable, environmentally acceptable, commercial supersonic aircraft is likely to languish.[86]

Endnotes

1. Harvey H. Hubbard, Domenic J. Maglieri, and David G. Stephens, "Sonic-Boom Research—Selected Bibliography with Annotation," NASA TM-87685 (September 1986).
2. Including F-4, F-14, F-15, F-16, F/A-18, F-111, T-38, and SR-71 aircraft.
3. J. Micah Downing, "Lateral Spread of Sonic Boom Measurements from U.S. Air Force Boomfile Flight Tests," in *High-Speed Research: Sonic Boom—Proceedings of a Conference Held at Langley Research Center, Hampton, Virginia, February 25–27, 1992* 1, NASA CR 3172, 117–129. For a description, see Robert E. Lee and J. Micah Downing, "Boom Event Analyzer Recorder: The USAF Unmanned Sonic Boom Monitor," AIAA paper no. 93-4431 (October 1993).
4. For example, see Kenneth J. Plotkin et al., "Sonic Boom Environment Under a Supersonic Military Operating Area," *Journal of Aircraft* 29, no. 6 (November–December 1992): 1069–1072. (Study conducted at White Sands Missile Range, NM.)
5. Dwight E. Bishop, "Noise and Sonic Boom Impact Technology: PCBOOM Computer Program for Sonic Boom Research Technical Report," 1, USAF, HSD-TR-88-014 (October 1988) (Defense Technical Information Center [DTIC] AD-A206290), 2–3; Plotkin and Maglieri, "Sonic Boom Research," 6.
6. Albion D. Taylor, "The TRAPS Sonic Boom Program," NOAA Technical Memorandum ERL ARL-87 (July 1980). In reviewing this chapter, Kenneth Plotkin pointed out Taylor's contribution.
7. Micah Downing, Kenneth Plotkin, Domenic Maglieri et al., "Measurement of Controlled Focused Sonic Booms from Maneuvering Aircraft," *JASA* 104, no. 1 (July 1998): 112–121.
8. Judy A. Rumerman, *NASA Historical Data Book Volume VI: NASA Space Applications, Aeronautics, and Space Research and Technology, Tracking and Data Acquisition/Space Operations, Commercial Programs, and Resources, 1979–1988*, NASA SP-2000-4012 (Washington, DC: NASA, 2000), 177–178.
9. Conway, *High-Speed Dreams*, 201–215; Paul Proctor, "Conference [sponsored by Battelle] Cites Potential Demand for Mach 5 Transports by Year 2000," *Aviation Week* (November 10), 1986, 42–46. Another potential use for a hypersonic transport was to economically deliver components of Reagan's Space Defense Initiative into low orbit.

10. Rumerman, *NASA Historical Data Book, Volume VI*, 178.

11. Conway, *High-Speed Dreams*, 218–228; Chambers, *Innovations in Flight*, 50.

12. Domenic J. Maglieri, Victor E. Sothcroft, and John Hicks, "Influence of Vehicle Configurations and Flight Profile on X-30 Sonic Booms," AIAA paper no. 90-5224 (October 29, 1990); Domenic J. Maglieri, "A Brief Review of the National Aero-Space Plane Sonic Booms Final Report," USAF Aeronautical Systems Center TR 94-9344 (December 1992).

13. Christine Darden et al., *Status of Sonic Boom Methodology and Understanding; Proceedings of a Workshop Sponsored by the National Aeronautics and Space Administration and Held at NASA Langley Research Center, Hampton, Virginia, January 19–20, 1988*, NASA CP 3027 (June 1989), 2–7.

14. Boeing Commercial Airplanes, "High-Speed Civil Transport Study; Final Report," NASA CR 4234 (September 1989); Douglas Aircraft Company, "1989 High-Speed Civil Transport Studies," NASA CR 4375 (May 1991) (published late with an extension). For a summary of Boeing's design process, see George T. Haglund, "HSCT Designs for Reduced Sonic Boom," AIAA paper no. 91-3103 (September 1991).

15. Allen H. Whitehead, ed., *First Annual High-Speed Research Workshop; Proceedings of a Workshop Sponsored by the National Aeronautics and Space Administration, Washington, DC, and Held in Williamsburg, Virginia, May 14–16, 1991*, NASA CP 10087 (April 1992), pt. 1, 5–22, 202 (hereafter cited as *1991 HSR Workshop*). Later sonic boom workshop titles will be similarly abbreviated with the year conducted.

16. Robert E. Anderson, "First Annual HSR Program Workshop: Headquarters Perspective," in Whitehead, *1991 HSR Workshop*, pt. 1, 7, 20, quotation taken from McLean, *Supersonic Cruise Technology*, 6.

17. George Unger, "HSR Community Noise Reduction Technology Status Report," in Whitehead, *1991 HSR Workshop*, pt. 1, 272.

18. Ibid., 275.

19. Session V, "Sonic Boom (Aerodynamic Performance)," 665–810; Session IX, "Sonic Boom (Human Response and Atmospheric Effects," in Whitehead, *1991 HSR Workshop*, pt. 3, 1199–1366.

20. As examples of recent work at Langley, see Robert J. Mack and Kathy E. Needleman, "A Methodology for Designing Aircraft to

Low Sonic Boom Constraints," NASA TM 4246 (February 1, 1991); Kathy E. Needleman, Christine M. Darden, and Robert J. Mack, "A Study of Loudness as a Metric for Sonic Boom Acceptability," AIAA paper no. 91-0496 (January 1991).

21. Christine M. Darden, ed., *High-Speed Research: Sonic Boom; Proceedings of a Conference Held at Langley Research Center, Hampton, Virginia, February 25–27, 1992*, parts 1 and 2, NASA CR 3172 (October 1992).

22. Thomas A. Edwards, ed., "High-Speed Research: Sonic Boom," in *Proceedings of a Conference Sponsored by the National Aeronautics and Space Administration, Washington, DC, and Held at the Ames Research Center, Moffett Field, California, May 12–14, 1993*, NASA CP 10132, 1. (The second volume, on configurations and design, including sonic boom minimization, has never been publicly released on the NTRS.)

23. David A. McCurdy, ed., "High-Speed Research: 1994 Sonic Boom Workshop, Atmospheric Propagation and Acceptability Studies, in *Proceedings of the Third High-Speed Research Sonic Boom Workshop, Hampton, Virginia, June 1–3, 1994*, NASA CP 3209; "High-Speed Research: 1994 Sonic Boom Workshop: Configuration, Design, Analysis, and Testing," in *Proceedings of the Third High-Speed Research Sonic Boom Workshop, Hampton, Virginia, June 1–3, 1994*, NASA CP 209669 (December 1999).

24. Daniel G. Baize, *1995 NASA High-Speed Research Program Sonic Boom Workshop: Proceedings of a Workshop Held at Langley Research Center, Hampton, Virginia, September 12–13, 1995*, NASA CP 3335 (July 1996); "Configuration, Design, Analysis, and Testing" 2, NASA CP 1999-209520 (December 1999).

25. For guidance in helping to choose which of the many research projects to cover, the author referred to Christine M. Darden, "Progress in Sonic-Boom Understanding: Lessons Learned and Next Steps," *1994 Sonic Boom Workshop*, 269–292. Figure 4-1 is copied from page 270.

26. Peter G. Coen, "Development of a Computer Technique for Prediction of Transport Aircraft Flight Profile Sonic Boom Signatures," NASA CR 188117 (March 1991) (based on M.S. thesis, George Washington University, Washington, DC.).

27. Christine M. Darden et al., "Design and Analysis of Low Boom Concepts at Langley Research Center," *1991 HSR Workshop*, pt. 2, 676–679. Photo with HSCT model provided to author by Dr. Darden.

28. Ibid., 680–698.
29. Ibid., 685, 687, 693.
30. Kinney, "NASA and the Evolution of the Wind Tunnel," in Hallion, *NASA's Contributions to Aeronautics* 2, 346.
31. Thomas A. Edwards et al., "Sonic Boom Prediction and Minimization Using Computational Fluid Dynamics," *1991 HSR Workshop*, pt. 2, 728–732, quotation from 732.
32. Ibid., 732. The other CFD codes used were TRANAIR (a full potential code with local mesh refinement capability), TEAM (an Euler/Navier-Stokes code with versatile zonal grid capability), AIRPLANE (an unstructured-grid Euler solver), and UPS (a parabolized Euler/Navier-Stokes code).
33. Ibid., 736.
34. M.J. Siclari, "Sonic Boom Predictions Using a Modified Euler Code," *1991 HSR Workshop*, pt. 2, 760–784. Figures 4-2, 4-3, and 4-4 extracted from 762, 766, and 772.
35. Robert J. Mack and Christine M. Darden, "Limitations on Wind-Tunnel Pressure Signature Extrapolation," *1992 Sonic Boom Workshop* 2, 201–220.
36. Percy J. Bobbitt, "Application of Computational Fluid Dynamics and Laminar Flow Technology for Improved Performance and Sonic Boom Reduction," *1992 Sonic Boom Workshop* 2, 137–144.
37. Edwards et al., "Sonic Boom Prediction Using Computational Fluid Dynamics," *1991 HSR Workshop*, 732.
38. Susan E. Cliff, "Computational/Experimental Analysis of Three Low Sonic Boom Configurations with Design Modifications," *1992 Sonic Boom Workshop* 2, 89–118.
39. M.J. Siclari, "Ground Extrapolation of Three-Dimensional Near-Field CFD Predictions for Several HSCT Configurations," *1992 Sonic Boom Workshop* 2, 175–200, with figure 4-5 copied from 192.
40. Samson H. Cheung and Thomas A. Edwards, "Supersonic Airplane Design Optimization Method for Aerodynamic Performance and Low Sonic Boom," *1992 Sonic Boom Workshop* 2, 31–44.
41. John M. Morgenstern, "Low Sonic Boom Design and Performance of a Mach 2.4/1.8 Overland High Speed Civil Transport," *1992 Sonic Boom Workshop* 2, 55–63.
42. George T. Haglund and Steven S. Ogg, "Two HSCT Mach 1.7 Low Sonic Boom Designs," *1992 Sonic Boom Workshop* 2, 65–88, quotations from 66 and 72.

43. Eugene Tu, Samson Cheung, and Thomas Edwards, "Sonic Boom Prediction Exercise: Experimental Comparisons," *1994 Sonic Boom Workshop: Configuration Design, Analysis, and Testing* 2, 13–32.

44. Robert Mack, "Wind-Tunnel Overpressure Signatures from a Low-Boom HSCT Concept with Aft-Mounted Engines," *1994 Sonic Boom Workshop: Configuration Design, Analysis, and Testing* 2, 59–70.

45. Susan E. Cliff et al., "Design and Computational/Experimental Analysis of Low Sonic Boom Configurations," *1994 Sonic Boom Workshop: Configuration Design, Analysis, and Testing* 2, 33–57. For a review of CFD work at Ames from 1989–1994, see Samson Cheung, "Supersonic Civil Airplane Study and Design: Performance and Sonic Boom," NASA CR-197745 (January 1995).

46. Christopher A. Lee, "Design and Testing of Low Sonic Boom Configurations and an Oblique All-Wing Supersonic Transport," NASA CR-197744 (February 1995).

47. Michael J. Siclari, "The Analysis and Design of Sonic Boom Configurations Using CFD and Numerical Optimization Techniques," *1994 Sonic Boom Workshop* 2, 107–128, figure 4-6 extracted from 110.

48. Donald G. Baize et al., "A Performance Assessment of Eight Low-Boom High-Speed Civil Transport Concepts," *1994 Sonic Boom Workshop* 2, 149–170, figure 4-7 copied from 155.

49. Domenic J. Maglieri, Victor E. Sothcott, Thomas N. Deffer, and Percy J. Bobbitt, "Overview of a Feasibility Study on Conducting Overflight Measurements of Shaped Sonic Boom Signatures Using RPV's," *1991 HSR Workshop*, pt. 2, 787–807.

50. Domenic J. Maglieri et al., "Feasibility Study on Conducting Overflight Measurements of Shaped Sonic Boom Signatures Using the Firebee BQM-34E RPV," NASA CR-189715 (February 1993). Figure 4-8 is copied from page 52, with waveforms based on a speed of Mach 1.3 at 20,000 feet rather than the 9,000 feet of planned flight tests.

51. David Lux et al., "Low-Boom SR-71 Modified Signature Demonstration Program," *1994 Sonic Boom Workshop: Configuration, Design, Analysis and Testing* 2, 237–248.

52. Interview, Dana Purifoy by Lawrence Benson, Dryden Flight Research Center, April 8, 2011.

53. Wyle Laboratories developed MDBOOM to meld CFD with sonic boom prediction theory. J.A. Page and K.J. Plotkin, "An Efficient

Method for Incorporating Computational Fluid Dynamics into Sonic Boom Theory," AIAA paper no. 91-3275 (September 1991).

54. Plotkin and Cantril, "Prediction of Sonic Boom at a Focus," AIAA paper no. 76-2.

55. Kenneth Plotkin, Juliet Page, and J. Micah Downing, "USAF Single-Event Sonic Boom Prediction Model: PCBoom3," *1994 Sonic Boom Workshop* 1, 171–184.

56. Edward H. Haering et al., "Measurement of the Basic SR-71 Airplane Near-Field Signature, *1994 Sonic Boom Workshop: Configuration, Design, Analysis, and Testing*, 171–197; John M. Morgenstern et al., "SR-71A Reduced Sonic Boom Modification Design," *1994 Sonic Boom Workshop* 2, 199–217; Kamran Fouladi, "CFD Predictions of Sonic-Boom Characteristics for Unmodified and Modified SR-71 Configurations," *1994 Sonic Boom Workshop* 2, 219–235. Figure 4-9 is copied from 222.

57. Catherine M. Bahm and Edward A. Haering, "Ground-Recorded Sonic Boom Signatures of F/A-18 Aircraft in Formation Flight," *1995 Sonic Boom Workshop* 1, 220–243.

58. J. Micah Downing, "Lateral Spread of Sonic Boom Measurements from US Air Force Boomfile Flight Tests," *1992 Sonic Boom Workshop* 1, 117–136; Domenic J. Maglieri et al., "A Summary of XB-70 Sonic Boom Signature Data, Final Report," NASA CR 189630 (April 1992).

59. William L. Willshire and David Chestnut, eds., "Joint Acoustic Propagation Experiment (JAPE-91) Workshop," NASA CR 3231 (1993).

60. William L. Willshire and David W. DeVilbiss, "Preliminary Results from the White Sands Missile Range Sonic Boom Propagation Experiment," *1992 Sonic Boom Workshop* 1, 137–144.

61. Bart Lipkens and David T. Blackstock, "Model Experiments to Study the Effects of Turbulence on Risetime and Waveform of N Waves," *1992 Sonic Boom Workshop* 1, 97–108; Robin O. Cleveland, Mark F. Hamilton, and David T. Blackstock, "Effect of Stratification and Geometrical Spreading on Sonic Boom Rise Time," *1994 Sonic Boom Workshop* 1, 19–38; Richard Raspet, Henry Bass, Lixin Yao, and Wenliang Wu, "Steady State Risetimes of Shock Waves in the Atmosphere," *1992 Sonic Boom Workshop* 1, 109–116. The author was alerted to the contributions of Bass and Blackwood by Kenneth Plotkin.

62. Gerry L. McAnich, "Atmospheric Effects on Sonic Boom—A Program Review," *1991 HSR Workshop*, pt. 1, 1201–1207; Allan D. Pierce and Victor W. Sparrow, "Relaxation and Turbulence Effects on Sonic Boom Signatures," *1991 HSR Workshop*, pt. 1, 1211–1234; Kenneth J. Plotkin, "The Effect of Turbulence and Molecular Relaxation on Sonic Boom Signatures," *1991 HSR Workshop*, pt. 1, 1241–1261; Lixin Yao et al., "Statistical and Numerical Study of the Relation Between Weather and Sonic Boom," *1991 HSR Workshop*, pt. 3, 1263–1284.

63. Leonard M. Weinstein, "An Optical Technique for Examining Aircraft Shock Wave Structures in Flight," *1994 Sonic Boom Workshop, Atmospheric Propagation and Acceptability* 1, 1–18. The following year, Weinstein demonstrated improved results using a digital camera: "An Electronic Schlieren Camera for Aircraft Shock Wave Visualization," *1995 Sonic Boom Workshop* 1, 244–258.

64. Kevin P. Shepherd, "Overview of NASA Human Response to Sonic Boom Program," *1991 HSR Workshop*, pt. 3, 1287–1291; Shepherd et al., "Sonic Boom Acceptability Studies," *1991 HSR Workshop*, pt. 3, 1295–1311.

65. David A. McCurdy et al., "An In-Home Study of Subjective Response to Simulated Sonic Booms," *1994 Sonic Boom Workshop: Atmospheric Propagation and Acceptability* 1, 193–207; McCurdy and Sherilyn A. Brown, "Subjective Response to Simulated Sonic Boom in Homes," *1995 Sonic Boom Workshop* 1, 278–297, with figure 4-10 copied from 279.

66. James M. Fields et al., "Residents' Reactions to Long-Term Sonic Boom Exposure: Preliminary Results," *1994 Sonic Boom Workshop: Atmospheric Propagation and Acceptability* 1, 209–217.

67. Conway, *High-Speed Dreams*, 253.

68. Christine M. Darden, "Progress in Sonic-Boom Understanding: Lessons Learned and Next Steps," *1994 Sonic Boom Workshop, Configuration, Design, and Testing* 2, 272.

69. Ibid., 272–274.

70. Ibid., 275.

71. Ibid. 289–290.

72. *1995 Sonic Boom Workshop: Configuration, Design, Analysis, and Testing* 2, NASA CP 1999-209520, 47–175.

73. James M. Fields, "Reactions of Residents to Long-Term Sonic Boom Noise Environments," NASA CR-201704 (June 1997).

74. Edward A. Haering, L.J. Ehernberger, and Stephen A. Whitmore, "Preliminary Airborne Experiments for the SR-71 Sonic Boom Propagation Experiment," *1995 Sonic Boom Workshop* 1, 176–198.

75. Ibid., Stephen R. Norris, Edward A. Haering, and James E. Murray, "Ground-Based Sensors for the SR-71 Sonic Boom Propagation Experiment," *1995 Sonic Boom Workshop* 1, 199–218; Hugh W. Poling, "Sonic Boom Propagation Codes Validated by Flight Test," NASA CR 201634 (October 1996).

76. Figure 4-11 copied from Edward A. Haering and James E. Murray, "Shaped Sonic Boom Demonstration/Experiment Airborne Data: SSBD Final Review," PowerPoint presentation, August 17, 2004, slide no. 3.

77. Robert J. Mack, "A Whitham-Theory Sonic-Boom Analysis of the TU-144 Aircraft at Mach Number of 2.2," *1995 Sonic Boom Workshop* 2, 1–16. For a complete account of this cooperative project by one of the test pilots, see Robert A. Rivers, "NASA's Flight Test of the Russian Tu-144 SST," in Hallion, *NASA's Contributions to Aeronautics* 2, 914–956.

78. Volume 1 of the *1995 NASA High-Speed Research Program Sonic Boom Workshop*, published in July 1996 (without a subtitle) as CP-3335, covered theoretical and experimental sonic boom propagation, while the second volume, *Configuration Design, Analysis, and Testing*, was published in December 1999 as CP 1999-209520.

79. Ibid., 1, eight papers in Session 1 on atmospheric propagation effects, 1–175.

80. Ibid., 1, five papers in Session 3 on human response, 278–332.

81. Kenneth J. Plotkin, "Theoretical Basis for Finite Difference Extrapolation of Sonic Boom Signatures," *1995 Sonic Boom Workshop: Configuration, Design, Analysis, and Testing* 1, 55.

82. Graham Warwick, "Cutting to the Bone," *Flight International* (July 17, 2001), accessed ca. June 15, 2011, *http://www.flightglobal.com/articles/2007/07/17/134122/cutting-to-the-bone.html*.

83. Chambers, *Innovations in Flight*, 61–62; Conway, *High-Speed Dreams*, 286–300; James Schultz, "HSR Leaves Legacy of Spinoffs," *Aerospace America* 37, no. 9 (September 1999): 28–32. The Acoustical Society held its third sonic-boom symposium in Norfolk from October 15–16, 1998. Because of HSR distribution limitations, many of the presentations could be oral only, but a few years later the ASA was able to publish some of them in a special edition of its journal. For a status report on one major spinoff at the end of

the HSR, see Kenneth J. Plotkin, "State of the Art of Sonic Boom Modeling," *JASA* 111, no. 1, pt. 3 (January 2002): 530–536.

84. National Research Council, *Commercial Supersonic Technology: The Way Ahead* (Washington, DC: National Academies Press, 2001), 1, accessed ca. June 15, 2011, *http://www.nap.edu/openbook. php?record_id=10283*.

85. Ibid.

86. Ibid., 43.

Image of Northrop Grumman's QSP design concept. (NGC)

CHAPTER 5

The Quiet Supersonic Platform
Innovative Concepts and Advanced Technologies

With NASA's High-Speed Research program having once again revealed how difficult it would be to design and produce a full-size airliner with a sonic boom quiet enough to fly over land, the alternative of small- or medium-size supersonic aircraft for civilian passengers began attracting more attention. One of the world's top sonic boom experts was among those looking into this option. In November 1998, as the HSR program was winding down, Richard Seebass presented two papers on supersonic flight and the sonic boom at NATO's von Kármán Institute in Belgium.[1] One of his papers examined the general problems and prospects for commercial supersonic transports,[2] while the other traced the history and current status of sonic boom minimization theory.[3] In each of these and subsequent publications, he concluded by endorsing a less ambitious but more pragmatic way than the HSCT to surmount the sonic boom barrier: a supersonic business jet (SSBJ).

Making the Case for a Supersonic Business Jet

Although no company had yet to begin actual development of an SSBJ, the idea itself was not new. Fairchild Swearingen, McDonnell Douglass, Lockheed-California, and British Aerospace had all seriously looked at the possibility in the mid-1980s, and the fractional-ownership company NetJets had come to believe that an SSBJ would fit well with its business model.[4] Because smaller supersonic aircraft would inherently have a weaker sonic boom, Seebass was among those who became most interested in pursuing this concept.

At a 1995 NASA Langley workshop on transportation technologies beyond 2000, Seebass helped Randall Greene, president of Aeronautical Systems Corporation, make a presentation on the market for and configuration of a proposed supersonic corporate jet. At 91 feet long and 66,000 pounds with a cranked-arrow wing similar to the F-16XL, this aircraft was designed to carry 8 to 10 passengers 3,350 nautical miles at Mach 1.8. Greene predicted sales of $1 billion a year for the plane, which he optimistically hoped (with enough outside financial support) to bring to market in 2000. Using Seebass's minimization techniques, the design was projected to have a sonic boom overpressure as low as 0.4 psf, although locally focused booms during acceleration would still be a problem. FAA certification, especially for the sonic boom and jet noise near airports under the FAA's Stage 3 standard of 1978, was identified as a potential "show stopper."[5] Greene's presentation emphasized the importance of Government help in developing such an aircraft. Although the United States had failed to be first to develop an SST, he argued that "it is NASA's role to make the US first in business jets."[6]

Based on the ongoing experience of the Concorde and related market analyses, Richard Seebass's subsequent presentations in 1998 were decidedly pessimistic about the viability of a large supersonic passenger plane in the foreseeable future. With 350,000 mostly supersonic flying hours during the Concorde's first 14 years of reliable operation, Seebass did consider the Concorde "a great technical success."[7] Economically, however, the case for another SST had yet to be made. The British and French governments paid for most of the Concorde's development and production (essentially donating the last five of them to their national airlines). This allowed their small fleet of 12 aircraft to attract enough passengers willing and able to pay a high fare for the two airlines to break even on operations, even at a fuel-cost-per-passenger mile several times that of a Boeing 747. But because such a U.S. Government subsidy was highly unlikely in the future, Seebass observed how "the development of a supersonic transport that can be operated at a profit by the airlines and sold in sufficient numbers for the airframe and engine manufacturers to realize a profit as well remains a challenge."[8] Specifically, "the challenge is to build, certify, and operate an SST at marginally increased fares while providing the airlines a return on investment comparable to a similar investment in subsonic aircraft."[9]

As shown by repeated studies, generating sufficient passenger loads to justify the expense of a supersonic airliner would most likely require overland supersonic routes from a large number of airports. This meant solving the acoustic issues of jet noise, especially when taking off, and the sonic boom when accelerating and cruising. Recent NASA HSR data indicated that adequate sound suppression of 15 to 20 perceived noise decibels would add about 6,500 pounds per engine, or the equivalent weight of 90 passengers.[10] As researched by NASA as far back as the SCAR program of the 1970s (see chapter 3), a variable-cycle engine that could switch from a quieter high-bypass ratio during takeoff and landing to low-bypass ratio to limit drag during cruise could be needed. Although engine noise was an intimidating challenge, it might be potentially solvable with some future technical breakthroughs.

When it came to sonic boom minimization, some immutable laws of physics posed even more intractable problems. One of these, warned Seebass, was that "the sonic boom due to lift cannot be avoided. The aircraft's weight must be transmitted to the ground."[11] In general, as he verbally explained a key equation, the minimum achievable sonic boom is related to the aircraft's weight divided by three-halves the power of its length. In addition to the easier-said-than-done goal of reducing weight, the main way to alleviate the effects of lift was to find acceptable tradeoffs in designing an airframe (i.e., the aircraft's volume) to shape the sonic boom signature in a manner tolerable to listeners but not too detrimental to aerodynamic performance.

As for the old problem of determining what would be acceptable to the public, the HSR's human-response surveys and NASA Langley's simulator experiments along with related research in Canada and Japan had improved ways to measure the apparent loudness of variously shaped sonic boom signatures. Although about 5 percent of people might find any sonic boom they can discern as unacceptable, some of the results indicated that a perceived level of 68 decibels outdoors would be acceptable to 95 percent of those exposed to it. This, Seebass predicted, could be achieved by a signature with an initial shock-pressure rise of 0.25 psf—if a maximum pressure of 1 psf is delayed 20 milliseconds after the front shock arrives and then begins to recede 20 milliseconds before the onset of the rear shock. Still to be determined, however, were the effects of such waveform shaping on the longstanding issue of acceptable sonic boom vibrations indoors. This would need to be determined by flight testing with an aircraft designed for this purpose.[12]

Seebass was already convinced, however, that it would not be possible for an SST-size airplane's elongated N-wave signature to avoid causing the structural vibrations that annoy people indoors, thereby continuing to restrict it to intercontinental routes over water and some unpopulated regions.[13] The one possible exception, at least in theory, might be a supersonic oblique-wing transport. As Seebass explained in his sonic boom minimization paper, "The aerodynamic optimum supersonic aircraft [is] an elliptic wing flying obliquely, which we note is unusual in that its maximum sonic boom does not occur directly below the aircraft."[14] As regards market potential, "it appears that an oblique flying wing could provide a Mach 1.4–1.6 transport that operates with no surcharge over future subsonic transports and compete with them over land as well."[15] Such an unconventional configuration, with its long wingspan, would of course require some airport modifications, but even more daunting, it would require a very expensive R&D effort. One can also assume that passenger acceptance of such a strange-looking airplane and its interior accommodations might also pose a challenge.[16]

Since building vibrations were not an inherent problem for a properly designed smaller airplane, Seebass asked a hypothetical question: "Could a 100 ft. long, Mach 1.6 supersonic business jet, cruising at an altitude of 40,000 feet and weighing 60,000 lbs.[,] have an acceptable sonic boom?"[17] The beneficial effects of vibrational relaxation from small aircraft "were well understood many years ago, but we did not consider them in sonic boom minimization because they are not important in the sonic boom of transport-sized aircraft." The shock waves from a much smaller, slender-bodied supersonic airplane, however, could be so weak as to be "nearly inaudible" while also containing "less energy in the frequencies important in structural response and indoor annoyance."[18] The fact that business jets do not follow scheduled routes might also help in the certification of supersonic versions, since they would not create the repetitive sonic booms of supersonic airliners. This, Seebass concluded, "leads us to conclude that a small, appropriately designed supersonic business jet's sonic boom may be nearly inaudible outdoors and hardly discernible indoors."[19] Such an airplane, he further stated, "appears to have a significant market ... if ... certifiable over most land areas."[20]

Previous SSBJ Studies and Proposals

Even though developing a supersonic business jet never became a goal of either the Supersonic Cruise Research or High-Speed Research programs, it had long been considered by some as a realistic possibility. The idea of building a small supersonic jet for general aviation, technology demonstrations, or potential military purposes had inspired a limited number but wide range of concepts in the past. After the HSR program dropped plans for a full-size overland supersonic airliner, Domenic Maglieri of Eagle Aeronautics—who by then had been involved with sonic boom research for 40 years—drafted a study for NASA in which he summarized all known proposals involving small supersonic aircraft intended mainly for business passengers.[21]

Between 1963 and 1995, there had been a total of at least 22 such studies or projects on developing small supersonic civilian airplanes. Academic institutions performed six of them, sometimes as student projects or theses, and all during either the 1960s or the early 1990s. The aircraft industry initiated eight more, starting with one by Spain's Construcciones Aeronauticas Sociedad Anonima (CASA), which at the time was building supersonic Northrop F-5s under license.[22] Although Boeing internally examined a supersonic 10-passenger plane concept in 1971, which was delta winged like its canceled 2707-300 SST, Fairchild Swearingen conducted the first serious design project published by an American company. It started with a feasibility study involving several

major airframe and engine manufacturers. The company then drew up preliminary designs for four two-engine configurations with a range of 4,000 miles
and an ability to cruise subsonically as well as supersonically using a modified
version of the Concorde's proven Rolls-Royce Snecma Olympus 493 engine.
By 1985, concerns about weight, the FAA's Stage 3 noise restrictions, and the
sonic boom brought the project to an end.[23]

In early 1988, while Douglas and Boeing were engaged with NASA in studies for the HSCT, Gulfstream Aerospace began studying market and technical
criteria for an SSBJ.[24] (Grumman had started Gulfstream in 1958 as part of
a diversification strategy into civilian aircraft but divested itself of the brand
in 1972.[25]) The company, which catered to the high end of the executive jet
market, drew up plans for a 125-foot, 100,000-pound, Mach 1.5 airplane
with ogive-delta wings (i.e., with their trailing edges angled forward, much as
the leading edges swept back). In a preliminary attempt at sonic boom
minimization, the designers were able to lower its predicted overpressure from
1.0 psf to 0.6 psf but only at the expense of some increased wave drag.[26]

Meanwhile, the sudden ending of the Cold War (and the unraveling of state
funding for the Russian aircraft industry) led the Sukhoi Design Bureau, which
had been studying a 114-foot, cranked-arrow wing SSBJ (the Su-51), to seek an
international partner. At the 1989 Paris Air Show, Sukhoi's chief designer and
Gulfstream's chairman agreed to explore joint development, taking advantage
of the former's expertise with supersonic fighters and the latter's expertise with
successful business jets.[27] The companies aimed at a speed of Mach 2 and range
of 4,000 miles as they considered design options, but the problems of weight
versus performance requirements proved to be beyond current technologies.
Although variable-cycle and ejector-mixer engine designs might partly mitigate
the level of jet noise, Gulfstream concluded that a concerted effort by the FAA,
NASA, industry, and academia would be needed to solve the problem of sonic
boom acceptability.[28] Even after the two companies parted ways in 1992, Sukhoi
continued pre-prototype design work in the hopes of forming another partnership in the future.[29] As will be shown in later sections, Gulfstream too remained
interested in a supersonic SSBJ, including sonic boom minimization technology.

In addition to the university and company projects, NASA conducted or
sponsored eight SSBJ-related studies between 1977 (4 years after it dropped
sonic boom minimization from the SCAR program) and 1986 (just as it initiated studies on the HSCT). The first, by Vincent Mascitti of Langley, explored
five possible configurations for an eight-passenger, Mach 2.2 supersonic executive aircraft based on the latest SCAR research findings and technological
advances. Although reduced engine noise was an objective, none of the options
were designed with the expressed goal of sonic boom minimization, so a trans-
atlantic range of 3,200 nautical miles was one of the criteria.[30] Also in 1977,

Boeing completed a feasibility study for NASA on a subscale SCAR demonstrator followed in 1979 by North American Rockwell's proposed supersonic business jet presented at the last Supersonic Cruise Research Conference (both described in chapter 3).

The next NASA study, left unpublished in 1981 as a possible casualty of the Reagan administration's abrupt cancellation of the SCR program, was the first phase of what had been planned as a three-phase market survey for supersonic business jets.[31] The same year, however, also marked the completion of the first of four SSBJ studies performed for Langley by the local technology division of Kentron International (later PRC Kentron). Each of the studies applied the latest technical advances to various SSBJ concepts during the period between the SCR and HSCT programs. Kentron's 1981 report presented concepts for an advanced droop-nose, two-engine Mach 2.7 business jet carrying eight passengers a distance of 3,200 nautical miles. Reflecting advances since Mascitti's study in 1977, the researchers assumed the use of the latest titanium- and superelastic-formed diffusion bonded materials to reduce its weight from 74,000 pounds to 64,000 pounds and a scaled down version of the GE 21/J11 variable-cycle turbofan engine for propulsion. As regards its sonic boom, the predicted overpressure of 1.0 psf at the start of cruise and 0.7 psf at the finish (due to reduced fuel weight) would still prohibit overland operations.[32]

The next study, completed in 1983, examined the use of a more fuel-efficient turbofan engine, the smallest possible eight-passenger compartment, and only one pilot to reduce takeoff weight to only 51,000 pounds. The result was a 103-foot-long, arrow-winged Mach 2.3 executive jet with a range of 3,350 nautical miles at Mach 2.3. Using Carlson's simplified overpressure prediction method with additional area-rule calculations, former NASA supersonic aerodynamicist A. Warner Robins hoped the combination of low wing loading, high cruise altitude, and modified flight profiles for climb and acceleration would alleviate the sonic boom problem on cross-country flights. The plane was also designed to fly 2,700 nautical miles at Mach 0.9 if necessary when cruising over land.[33]

In 1984, the same Kentron researchers completed the concept for a 114-foot-long executive jet with variable-sweep wings for better low-speed performance, which would eliminate the need for a droop nose as on the previous configuration. Although such adjustable wings had been found infeasible for the SST in the 1960s, the researchers hoped lower weight materials and advances in stability and control technology would make them more practical (which subsequent analysis proved overly optimistic). This latest design (figure 5-1) would have a ramp weight of 64,500 pounds with eight passengers and a two-person crew. Its performance included a range of almost 3,500 nautical miles at Mach 2.0 and over 5,000 miles at Mach 0.9 with takeoff and landing

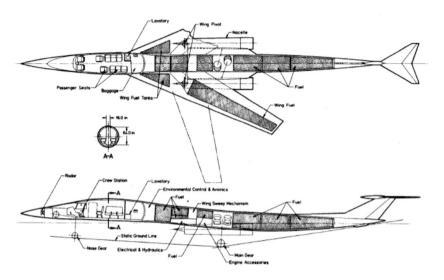

Figure 5-1. Kentron's concept for a swing-wing SSBJ. (NASA)

distances of less than 5,000 feet. Using the same prediction method as before, the overpressures at Mach 1.2 and Mach 2.0 varied from 0.9 psf to 2.0 psf depending on weight and altitude, making speeds no higher than Mach 0.9 mandatory for overland cruise.[34]

The last of the NASA studies was completed in 1986. For possible expansion of the customer base, the Kentron design team assessed the feasibility of an eight-passenger, long-range SSBJ with a planform similar to the 1981 and 1983 studies that could be converted into a missile-carrying interceptor (presumably for foreign sales). With a takeoff weight of 61,600 pounds for the civilian version and 63,246 pounds for the military version, its low-bypass-ratio turbofan engines would give it a range of more than 3,600 nautical miles or a combat radius of more than 1,600 nautical miles, both at Mach 2.0. Takeoffs would require a 6,600-foot runway. By flying an optimum profile for climb and acceleration, sonic boom overpressure was calculated at 1.0 psf, but the plane could also cruise transonically for 3,780 nautical miles at Mach 0.96.[35]

Although NASA and the major aircraft manufacturers focused on the HSCT for the next decade, the idea of a small supersonic plane continued to intrigue many in the small airplane manufacturing and general aviation communities. The rapidly growing corporate jet market appeared to have room for higher speeds, perhaps using more fractional ownership arrangements. The main roadblocks were the complex technology and considerable resources that would be required to develop, test, and produce such an advanced aircraft. This made Government support and partnerships among competing companies appear necessary. Overseas, France's Dassault Aviation explored developing a supersonic version of its popular Falcon business jet in the 1990s before scaling back its effort because of the lack of a suitable engine.[36]

Sometime in the mid 1990s, Lockheed's legendary but secretive Skunk Works (officially titled its Advanced Development Company), which 20 years earlier began work on the first jet airplane to have a very low radar signature, became interested in learning how to design airframes with a low sonic boom signature. (In March 1995, Lockheed merged with Martin Marietta Corporation to become Lockheed Martin.) Obviously, there would also be military advantages for some air vehicles—such as aircraft designed for high-speed reconnaissance— not to betray their presence by laying down a loud sonic boom carpet. To help in this effort, the Skunk Works hired McDonnell Douglas aerodynamicist John Morgenstern, who had been that company's lead boom analyst for the HSCT (see chapter 4). He was among those involved in designing a patented control surface near the nose of an airplane that could be extended to reduce the pressure and slope of the shock waves as a way of shaping its sonic boom.[37] To further add to its expertise, the Skunk Works also brought in none other than Richard Seebass as a consultant.[38] By 1998, Lockheed Martin had made enough progress on sonic boom minimization that it teamed up with Gulfstream to work on ways to develop a low-boom SSBJ.[39] (General Dynamics acquired Gulfstream Aerospace Corporation in 1999 as a wholly owned subsidiary.[40])

Meanwhile, the market for business jets was booming. It grew about 400 percent from 1995 through 2000, much of this captured from the scheduled airlines' business and first-class passenger categories. Furthermore, a good portion of this growth was in new models of more sophisticated and expensive corporate jets, whether privately purchased or under fractional ownership arrangements. Progress in understanding how to deal with the sonic boom reinforced a conviction that customers would be willing to pay the premium required to develop and produce an SSBJ. "With the advent of new technologies, and a travel market that increasingly desires time above all else, the business case is clearing emerging for new, fast transports."[41] The National Research Council's study on "the way ahead" for commercial supersonic flight (described at the end of chapter 4) found that airframe manufacturers believed customers would be willing to pay about twice as much for a plane that could fly twice as fast as current business jets and estimated the potential market for such an SSBJ to be at least 200 aircraft over a 10-year period.[42]

Unlike the Skunk Works' highly classified stealth technology, which did not have civilian applications, reducing the sonic boom could obviously benefit the private sector as well as have potential military advantages. The Department of Defense, however, had no current operational requirement to develop a new supersonic bomber, let alone one with a quiet sonic boom. Indeed, the Air Force's "bomber roadmap," released in March 1999, focused on sustaining its current mix of B-1B Lancers, B-2 Spirits, and B-52H Stratofortresses for decades to come with fielding of a new bomber postponed until the 2030s.[43] Internally, however, many in the Air Force were still interested in advanced strike concepts.

Birth of the QSP Program

One DOD agency is not bound by the pull of formal requirements. Instead, the mission of the Defense Advanced Research Projects Agency is to push innovative new technologies that might be of value in the future—including many with dual military and civilian uses. In February 2000, DARPA listed Supersonic Aircraft Noise Mitigation as a new program in its portion of the Department of Defense's FY 2001 budget estimates.[44] The new DARPA program resulted, at least in part, from lobbying by Lockheed Martin and Gulfstream (which had previously approached NASA for support)[45] and an earmark from Senator Ted Stevens of Alaska.[46] The Defense appropriation became law in August 2000.[47] This allowed DARPA to begin funding projects for supersonic noise mitigation, which it had since renamed the Quiet Supersonic Platform, in October 2000. For fiscal years 2001 and 2002, the QSP was allocated approximately $35 million.[48] A DARPA budget submission described the program—funded under Program Element (PE) 0603285E, Advanced Aerospace Systems—as follows:

> The Quiet Supersonic Platform (QSP) program is directed toward development and validation of critical technology for long-range advanced supersonic aircraft with substantially reduced sonic boom, reduced takeoff and landing noise, and increased efficiency.... Improved capabilities include supersonic flight over land without adverse sonic boom consequences with boom overpressure rise less than 0.3 pounds per square foot, increased unrefueled range approaching 6,000 nmi [nautical miles], gross takeoff weight approaching 100,000 pounds, increased area coverage, and lower overall operational cost. Highly integrated vehicle concepts will be explored to simultaneously meet the cruise range and noise level goals. Advanced airframe technologies will be explored to minimize sonic boom and vehicle drag including natural laminar flow, aircraft shaping, plasma, heat and particle injection, and low weight structures.[49]

DARPA initially identified three potential military roles for quiet, efficient supersonic aircraft: a reconnaissance vehicle, a medium bomber, and a high-speed transport that could quickly deliver vital spares and other equipment to forward-operating locations (the function most related to an SSBJ). To manage the QSP program, DARPA chose Richard W. Wlezien, a researcher from NASA Langley recently assigned to DARPA to manage a program on microadaptive flow controls. His specialty, the manipulation and control of shear flows, was a good match for overseeing technologies relevant to the QSP program.[50] In seeking participants from both industry and academia, Wlezien made sure to cast a wide net.

As DARPA's initial step in disseminating information about the program, it hosted an Advanced Supersonic Platform Industry Day in Alexandria, VA, on March 28, 2000. The announcement for this event, released 1 month earlier, informed interested parties that "it is our desire to facilitate the formation of strong teams and business relationships in order to develop competitive responses to a forthcoming DARPA Request for Information (RFI) and any subsequent solicitation."[51] Although encouraging the participation of small technology companies and academic institutions with specialized expertise, DARPA needed major aerospace corporations to assess and assimilate the wide range of airframe and engine technologies that would be required for the type of quiet, long-range supersonic aircraft desired. With the consolidations in the defense industry after the end of the Cold War, the three corporations with the required expertise and resources to be these system integrators were Boeing, Lockheed Martin, and Northrop Grumman.

For help in formulating the program's sonic boom strategy, Richard Wlezien received briefings from experts in the field such as Peter Coen of NASA Langley and Domenic Maglieri and Percy Bobbitt of Eagle Aeronautics.[52] The latter two planted some seeds for a sonic boom demonstration to eventually become part of the QSP by reviewing their Firebee proposal from the early HSR program and pointing out the continued value of physically proving sonic boom minimization predictions with an actual airframe in the real atmosphere.[53] (Through a Lockheed Martin contract, DARPA later had them prepare a survey on the findings of previous sonic boom research as background information for QSP participants.[54])

In August 2000, the DARPA Tactical Technology Office issued its formal solicitation for QSP systems studies and technology integration to include seeking detailed proposals for fostering new technologies sufficient to mitigate the sonic boom for unrestricted supersonic flight over land. Phase I of the program was expected to last 12 months. Phase II contracts, to be awarded later through a down-select process, would extend through the second year. The solicitation informed interested participants that "the program is designed to motivate approaches to sonic boom reduction that bypass incremental 'business as usual' approach and is focused on the validation of multiple new and innovative 'breakthrough' technologies for noise reduction that can ultimately be integrated into an efficient quiet supersonic vehicle."[55]

The initial goal of the QSP program was "to develop and validate critical technology for long range advanced sonic boom, reduced take-off and landing noise, and increased efficiency relative to current-technology supersonic aircraft."[56] The only firm requirement at the start of the program, mentioned in the solicitation and succinctly put into context by Richard Wlezien, was a concept that would reduce the overpressure of the sonic boom to 0.3 psf—a level that "won't rattle your windows or shake the china in your cabinet."[57] It was hoped a signature this low would allow unrestricted operations over land, although a sonic boom with 0.5 psf might be permissible in designated corridors.

System goals (less firm than the sonic boom requirement) included a speed of Mach 2.4, a gross weight of 100,000 pounds (about one-quarter that of the Concorde), a range of 6,000 miles, a 20-percent payload capacity, and meeting the FAA's Stage 3 noise restrictions.[58] Derived goals included a lift-to-drag ratio of 11 to 1, an engine-thrust-to-weight ratio of 7.5 to 1, a specified fuel-consumption rate, a 40-percent fuel fraction, and a 40-percent empty-weight fraction (both relative to gross takeoff weight). The concept aircraft was also expected to have adequate subsonic performance. As explained by Wlezien, "We have worked with NASA and the US Air Force to come up with numbers which make sense and are self-consistent. In our view, the numbers are reasonable given the state of the technologies, but still well off the projected trend lines."[59] Even so, meeting these multiple goals would not be easy. This was made clear by David Whelan, director of the DARPA Tactical Technology Office. "We do not see any 'silver bullet' solution.... But it might be possible to make improvements in many different areas that add up to a real net improvement."[60]

Achieving these goals would require the R&D capabilities of major aircraft and engine manufacturers, scientific and technical ideas from university engineering departments and specialized contractors, and the support and facilities of Government agencies. The needed NASA contributions would include modeling skills, wind tunnel facilities, and eventual flight-test operations. NASA administrator Dan Goldin strongly approved the QSP's approach. "Rather than a big point-design program that characterized HSR, [it] is a precompetitive study addressing core issues—efficiency, engine jet noise, sonic boom overpressure, and emissions.... Once we have sufficiently explored a broad range of promising technologies, we will work to develop and fund a more substantial industrial partnership."[61] The QSP emphasized potential military uses, but the sonic boom was currently a bigger problem for civilian aviation. Military aircraft had always been able to fly supersonic in designated airspace in the United States, so DARPA's goal of a validated concept for boom minimization could be of greatest benefit to the development of an SSBJ.

QSP Phase I: Defining Concepts and Technologies

DARPA awarded initial 1-year QSP contracts in November 2000 to the three systems integrators that would perform the large-scale design studies. Northrop Grumman's Air Combat Systems Integrated Systems Sector (ISS) received the first, on November 7, for $2.5 million.[62] Shortly thereafter, DARPA awarded contracts to the other two systems integrators, Lockheed Martin's Skunk Works and Boeing's Phantom Works. (The predecessor of the Phantom Works had been a part of McDonnell Douglas before that company's merger with Boeing in December 1996.) Neither Northrop nor Grumman had been among the aircraft manufactures that submitted designs for the SST, AST, or HSCT in the past, but almost as soon as DARPA announced the new program, the Northrop Grumman Corporation (NGC) decided to participate. As explained by Charles W. Boccadoro, NGC's future strike systems manager at the ISS Western Region in El Segundo, CA, upon award of the contract: "We started liking the capability offered by a long-range, efficient supersonic flight platform."[63]

On March 1, 2000, the day after DARPA issued the invitation to its industry day, Boccadoro had flown to Washington, DC, to meet with Richard Wlezien. There, Boccadoro went over a study he had presented at Headquarters Air Force in January on concepts for next generation supersonic strike aircraft. This detailed report involved capabilities directly relevant to DARPA's newly announced program.[64] At the end of the month, four other Northrop Grumman officials attended DARPA's industry day, including Steve Komadina. Shortly thereafter, Boccadoro was given the additional duty of program manager of Northrop Grumman's participation in what became the QSP program, with Komadina becoming the chief engineer and later the deputy program manager.[65]

A graduate of MIT and the von Kármán Institute for Fluid Dynamics, Charles Boccadoro had been hired in 1980 by Northrop, where he worked on such state-of-the-art aircraft programs as the B-2 Spirit stealth bomber and the YF-23 advanced tactical fighter.[66] In May 2000, he appointed an experienced systems engineer, Joseph W. Pawlowski, to lead a small team to develop the company's strategy for responding to DARPA's solicitation. Pawlowski's duties included coordinating developmental activities among a number of subcontractors, which would be good experience for the QSP endeavor. To look for help from outside the company, NGC hosted its own industry day on July 13, 2000. The following month, in what would turn out to be a very shrewd move, Northrop Grumman hired Eagle Aeronautics and Wyle Laboratories—with their long experience in sonic boom analysis—as subcontractors. It also teamed up Raytheon Corporation as a cost-sharing QSP partner.[67] Northrop Grumman submitted its response to DARPA's QSP solicitation on September 29, 2000.[68]

After the announcement of Northrop Grumman's QSP contract, Boccadoro provided some insight into the company's team-oriented approach. Because of its lack of commercial airplane experience, NGC sought out Raytheon, specifically the Raytheon Aircraft Company subsidiary that made Beechcraft and Hawker corporate jets, as its primary subcontractor.[69] "They will be working principally the civil applications, and we'll be working principally the military applications," he explained.[70] For help on engine technology and concepts, Northrop Grumman would be working with Pratt & Whitney, General Electric, and MIT's Gas Turbine Laboratory (all awarded their own QSP contracts) as well as General Motors' Allison Transmission and the Air Force Propulsion Laboratory.[71] In addition to having the sonic boom expertise of Wyle Laboratories, Eagle Aeronautics, and Stanford University, Northrop Grumman's own scientists and engineers had also gained some relevant knowledge in previous decades. As Boccadoro put it, "We understand the physics of boom mitigation."[72]

By January 2001, all 16 of the QSP Phase I contracts had been announced. Many of them focused on engine technologies, where major innovations were considered essential. To study concepts for advanced propulsions systems using high-bypass turbofans to meet the QSP goals, DARPA selected General Electric and Pratt & Whitney. Other contracts called for analyses of specific propulsion subcategories: Aerodyne for a vaporization-cooled turbine blade; Honeywell for ceramic components and compressor flow control, Techsburg (of Blacksburg, VA) for controlling the boundary-layer thickness of engine passageways; and MIT's Gas Turbine Laboratory for a two-stage, counter-rotating aspirated compressor.[73]

Most of the other QSP contracts involved innovative or even radical technologies for sonic boom mitigation. Gulfstream would follow up on some of its previous work by looking at integrated, top-mounted supersonic inlets that (being above the wings) could counter the contribution of inlet nacelle shocks to the sonic boom signature. Weidlinger Associates of New York City was engaged to investigate the previously dismissed theory of increasing virtual body length to spread out shock waves using the heat from a thermal keel or ramjet. Directed Technologies of Arlington, VA, in partnership with Reno Aeronautics, would assess using foamed metallic surfaces to promote natural laminar flow over a thin unswept wing (similar in shape to that of the F-104 Starfighter). Laminar flow, which is easier to achieve at supersonic speeds than at subsonic speeds, would greatly decrease the boundary layer turbulence and friction that causes aerodynamic drag by keeping air adjacent to the surface in a thin, smoothly shearing layer. (Active laminar flow requires the use of airflow devices creating suction to draw air into tiny holes in a special material covering a wing's surface.) In January 2000, NASA Dryden had tested a scale model of Reno Aeronautics Corporation's natural laminar flow wing attached to the center pylon of an F-15B on four supersonic flights with "remarkable results."[74]

Universities were the recipients of the remaining sonic boom research contracts. DARPA chose Stanford (by then the university doing the most advanced sonic boom research) to develop an efficient boom propagation tool optimized for multidisciplinary design techniques, Princeton for integrating aircraft shaping with energy-generated ionization of plasmas to prevent shock wave strengthening, and Arizona State University to demonstrate and develop design tools for using distributed roughness to inhibit crossflow instabilities on natural laminar flow over moderately swept wings. Finally, the University of Colorado received a contract for a more conventional assessment of aircraft-shaping techniques with a three-dimensional propagation tool to prevent shock waves from coalescing into the sonic boom.[75] Sadly, Richard Seebass, chair of the University of Colorado's Department of Aerospace Engineering Sciences until May 1999, passed away in November 2000 at the age of 64—just as the Quiet Supersonic Platform was getting ready to put his and Albert George's longstanding sonic boom minimization theory into practice.[76]

These selections reflected DARPA's policy to encourage smaller businesses and academic organizations to participate. As Richard Wlezien put it, "We are trying to get the traditional players to think out of the box and to bring in people with new ideas on an equal footing."[77] Not only did the QSP program aim to promote innovative technologies, it also employed an innovative management philosophy to get its contractors—including those who were traditional competitors—to work together. Although some of their techniques, findings, and data remained proprietary, DARPA required the major aircraft and engine companies to assess and integrate the impact of all the technologies under consideration. In Phase I of the QSP (which lasted through 2001), the three systems integrators developed conceptual airplane designs intended to meet the aforementioned sonic boom requirement and performance goals with promising technologies and configurations. In addition to relevant findings by the QSP technology contractors, the designs relied heavily on tools and methods developed during the HSR and previous NASA programs while incorporating the latest computational and optimization techniques, especially increasingly powerful CFD capabilities. Even with improved modeling and prediction of sonic boom propagation, however, the value of actually demonstrating the persistence of a reduced sonic boom signature through the atmosphere became increasingly apparent as the program continued.[78]

The three systems integration contractors, their partners, and all the technical and propulsion contractors worked intensely but quietly for the next year with relatively little about their progress appearing in the aerospace trade press or other media. The first task of Northrop Grumman, Boeing, and Lockheed

Martin was to perform 3-month studies on system scoping for their conceptual aircraft designs and technology assessments on developing and validating sonic boom mitigation measures. At the same time, the specialized sonic boom mitigation contractors worked on technology scoping studies of their own. Meanwhile, the advanced propulsion contractors worked on 6-month studies.[79]

By the end of the QSP program's first 3 months, the technology and propulsion contractors provided their findings to date to the system integration teams, which also shared the results of their own sonic mitigation studies among themselves. For the remaining 9 months of Phase I, the three major contractors worked on their conceptual supersonic aircraft designs while completing technology evaluation reports on sonic boom mitigation. The sonic boom contractors also completed technology evaluation reports on their assigned areas while the propulsion contractors, upon completing their 6-month scoping studies, moved on to integrating technologies into conceptual designs.[80] The progress being made to address the sonic boom problem using computational fluid dynamics was somewhat encouraging. "It doesn't require new science," said Richard Wlezien, "it requires good engineering."[81]

Although most of the QSP program went pretty much according to plan, two major changes involving the sonic boom occurred toward the end of its first year. Despite the progress being made on minimization, the sonic boom requirement of 0.3 psf was downgraded to be just one of the goals, equivalent to such other goals as long-range and low takeoff weight. This reflected a course adjustment to move the program more in the direction of military missions (perhaps at least partly a response to the terrorist attacks of September 11). By then, however, QSP management had also decided that the most pressing issue involving the sonic boom was to actually demonstrate the persistence of a shaped signature through the atmosphere. This would be consistent with the

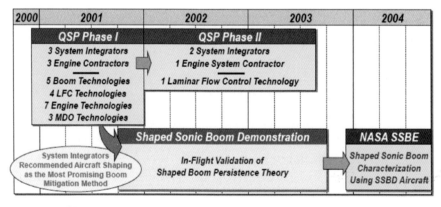

Figure 5-2. QSP timeline and major program activities. (DARPA)
Key: LFC = laminar flow control; MDO = multiple discipline optimization.

recently published report on supersonic technology by the National Research Council, which recommended proceeding to a "system/subsystem model or prototype demonstration in a relevant environment."[82]

The Northrop Grumman QSP team had already made preliminary plans on how to do this in the 3 months before April 2001, when DARPA formally solicited proposals for this demonstration.[83] After the NGC proposal was selected, this spinoff of the QSP program became known as the Shaped Sonic Boom Demonstration. The SSBD is the subject of the next three chapters, with the third of these also covering a follow-on project sponsored by NASA known as the Shaped Sonic Boom Experiment. Figure 5-2 depicts the final structure of the QSP program as it evolved after these changes.[84] Although there had been some hopes for a full-scale QSP Phase III that would have continued work on the design concepts or lead to development of a truly low-boom X-plane,[85] the SSBE was later considered by some sources to have been Phase III of the Quiet Supersonic Platform.

QSP Phase II: Refining Concepts and Technologies

By the end of 2001, each of the three system integrators had completed their Phase I studies defining their preferred design concepts and identifying the technologies needed to produce a real airplane. The Northrop Grumman team had submitted its study on December 12. As the QSP moved into Phase II in January 2002, Richard Wlezien gave some hints on how the research was going. "We have changed the face of supersonics.... We have simultaneously looked at long range and low boom and found that they are not mutually exclusive."[86] Although using various configurations, the preliminary design concepts all featured long, thin aircraft with lift distributed along their length, low wave drag, and highly integrated propulsion systems. Because of the light weight of the airframes relative to their volume, advanced composites would be essential for strength and stiffness. To achieve the goal of a high lift-to-drag ratio, ways of achieving supersonic laminar flow also emerged as key factors. "The question is how to integrate laminar flow into a real vehicle,"[87] cautioned Wlezien. Taken together, the QSP goals were a tremendous challenge. "Contractors have told us this is the toughest program they have ever worked, and we are surprised they have come up with ways to get there."[88]

At about this same time, Charles Boccadoro revealed some details about Northrop Grumman's overall design concept. It featured top-mounted, mixed compression inlets for the engines, which Gulfstream's computational analysis of three engine positions in four basic configurations showed could significantly reduce the sonic boom by shielding the flow field below the aircraft from the inlet shock.[89] Boccadoro's team also found that an above-airframe engine position resulted in less spillage as well as external compression and expansion fields. To achieve lower drag, the team was using Arizona State's distributed roughness concept to enable laminar flow on the plane's lifting surfaces while using natural laminar flow on some of its other surfaces. As for applying any of the more revolutionary methods, "a key finding of our studies was that the QSP goals could be achieved without active or exotic sonic boom reduction technologies."[90]

More details on the QSP concepts came out during the annual AIAA meeting in Reno, NV, during mid-January. After studying 12 design concepts, Northrop Grumman and Raytheon came up with a preferred dual-relevant concept appropriate for civilian as well as military purposes. They expected this configuration would meet the QSP's sonic boom mitigation goal with a slightly slower cruise speed of Mach 2.2 and a takeoff distance of 7,000 feet, which would be about halfway between the shorter business jet distance and the longer allowance for a military strike aircraft. The design featured a strut-braced (or joined-wing) configuration. A single vertical tail extended above the two engine nacelles nested on the rear of the aircraft.[91] Steve Komadina, chief engineer on Northrop's QSP team, later said this configuration is a design "we think can be evolved into a strike aircraft or business jet."[92] It could accommodate

FOR PUBLIC RELEASE

QSP Phase I & II System Studies

Phase I (CY 00/01)
Concept Study

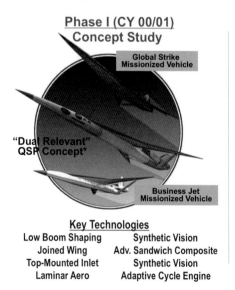

Global Strike
Missionized Vehicle

"Dual Relevant"
QSP Concept*

Business Jet
Missionized Vehicle

Phase II (CY 02/03)
System Validation

Focused On
Strike Concept

Key Activities
Definitive CONEMP Study
Detailed Vehicle & Subsystem Definition
Six Wind Tunnel Tests
High Fidelity CFD
Adv. Composite Manufacturing Demo
F-5 Shaped Sonic Boom Flight Test

Key Technologies

Low Boom Shaping	Synthetic Vision
Joined Wing	Adv. Sandwich Composite
Top-Mounted Inlet	Synthetic Vision
Laminar Aero	Adaptive Cycle Engine

Figure 5-3. Northrop Grumman's QSP program. (NGC)

either two 27-foot-long weapons bays or a 22-foot passenger cabin.[93] A briefing slide released later (figure 5-3) depicts Northrop Grumman's concept as it evolved during both phases of the QSP.[94]

Advance news of DARPA's selection of Northrop Grumman and Lockheed Martin to continue developing their concepts under Phase II of the QSP first leaked out in early March 2002.[95] Lockheed Martin's concept had a slender fuselage (described as sinuous), highly contoured swept wings with engines beneath, and a V tail while Boeing, which later published a paper with details on its design effort, had probably the most radical configuration. It featured two sets of thin, unswept wings (with natural laminar flow) fore and aft and a swiveling main wing that could be stowed along the top of the fuselage during cruise.[96]

DARPA officially awarded its Phase II contracts in May 2002. Northrop Grumman's Integrated Systems Sector received $2.7 million to validate the QSP concepts defined during the program's first phase. This would include wind tunnel testing of its preferred aircraft configuration and work with Raytheon on the fabrication and testing of a structural component made with an advanced composite core. At the same time, DARPA also awarded the NGC Integrated Systems Sector a $3.4 million contract for what became the Shaped Sonic Boom Demonstration.[97] Other contract awards included Lockheed Martin's Skunk Works for its design concept, General Electric for its advanced propulsion system, and Arizona State University for its distributed-roughness laminar flow research. By the time these contracts were awarded, DARPA had decided to make a priority of the long-range supersonic bomber for the QSP's military mission with the more liberal sonic boom goal of 0.5 psf overpressure, and it decided to place more emphasis on such factors as survivability.[98] This reduced boom might allow the aircraft to fly in new supersonic corridors beyond the limited confines of military training airspace without causing the public relations problems experienced by the Air Force's last midrange Mach 2 bomber, the B-58 Hustler (described in chapter 1).

On September 26, 2002, Northrop Grumman unveiled more about the preferred system concept of its QSP team, including an image of the sleek plane in flight (as pictured in front of this chapter). Its joined wing airframe was 156 feet long with a wingspan of 58 feet, a speed somewhat higher than Mach 2, and a range of 6,000 nautical miles. As had been a consideration with the Concorde, this speed would allow the use of lower cost materials, especially aluminum. The main wings were highly swept but thin and narrow for lower drag and better laminar flow, which would be easier to sustain with less turbulence across a shorter chord (wing width). These high-aspect-ratio cranked-arrow wings were braced by two much smaller wings swept forward from the rear of the aircraft. The concept also featured a dual top-mounted isentropic inlet (designed for smooth and steady airflow), extensive laminar flow aerodynamics, and an adaptive leading edge on its wings. Team members from Raytheon Aircraft Company designed an SSBJ variant.[99] Further refinements of the military concept gave it a cruise speed of Mach 2.2.[100] To supplement its extensive CFD modeling, Northrop Grumman tested a scale model of its final QSP configuration at Mach 2.2 in the 9-foot-by-7-foot section of the NASA Ames Supersonic Wind Tunnel for 33 hours in April 2003.[101]

Richard Wlezien moved to NASA Headquarters in the early fall of 2002. He was replaced as QSP manager by Steven H. Walker, who had been assigned to DARPA from Defense Research and Engineering in the Pentagon.[102] Walker later explained that even though the sonic boom goal had been relaxed, "What we ended up finding out was that if you improve lift and drag, if you improve specific fuel consumption, if you reduce your empty weight, all these things lend themselves to lower sonic boom as well."[103]

Except for the ongoing Shaped Sonic Boom Demonstration, DARPA phased out the QSP program during the first half of 2003. Its biennial budget estimate submitted in February included $4.8 million for FY 2003 but nothing for FY 2004.[104] Northrop Grumman's QSP team submitted extensive documentation of its work on May 22, 2003. Results of its and Lockheed Martin's QSP concepts went to the Air Force for use in its ongoing long-range-strike platform study.[105] No longer comfortable with the rather unambitious projections in its 1999 bomber roadmap, the Air Force leadership was seeking the latest ideas on long-range strike from the aerospace industry. It eventually examined more than 20 proposals from Northrop Grumman, Lockheed Martin, and Boeing.[106] None of these, however, led to a follow-on program like the QSP that could continue refining and demonstrating other supersonic technologies, such as reduced boom designs, as had once been contemplated.

Even so, the QSP participants had learned much and documented a great deal of data that could be of potential value in the future. The program had explored and evaluated a wide range of cutting-edge technologies, advancing the state of the art in aeronautics, propulsion, and related fields. For Northrop Grumman's engineers, who had not had any major supersonic projects after developing the YF-23 and the supersonic inlets for the Boeing F/A-18E/F Super Hornet, the QSP afforded valuable experience and new skills. This helped sustain the company's aerodynamic design capabilities for future projects, such as DARPA's Switchblade oblique-wing study, and advanced work with the Air Force Research Laboratory (AFRL).[107] The most publicized aspect of the QSP program, however, was the opportunity it provided for Northrop Grumman and its partners to make aviation history by being the first to demonstrate the creation of less intense sonic booms.

Endnotes

1. Seebass, who had been born and raised in Denver before becoming a professor at Cornell and later the University of Arizona, moved in 1981 to the University of Colorado at Boulder, where he served as dean of its College of Engineering until 1994.

2. Richard Seebass, "History and Economics of, and Prospects for, Commercial Supersonic Transport," Paper 1, NATO Research and Technology Organization, *Fluid Dynamics Research on Supersonic Aircraft* (proceedings of a course held in Rhode Saint-Genèse, Belgium, May 25–29, 1998), Research and Technology Organization (RTO)-EN-4 (November 1998).

3. Richard Seebass, "Sonic Boom Minimization," Paper 6, NATO Research and Technology Organization, *Fluid Dynamics Research on Supersonic Aircraft* (proceedings of a course held in Rhode Saint-Genèse, Belgium, May 25–29, 1998), RTO-EN-4 (November 1998). A slightly expanded version was also published with Brian Argrow as "Sonic Boom Minimization Revisited," AIAA paper no. 98-2956 (November 1998).

4. Domenic Maglieri to Lawrence Benson, "Comments on QSP Chapter," e-mail message, August 23, 2011.

5. Randall Greene and Richard Seebass, "A Corporate Supersonic Transport," in *Transportation Beyond 2000: Technologies Needed for Engineering Design, Proceedings of a Workshop Held in Hampton, Virginia, September 26–28, 1995*, NASA CP-10184 (February 1996), pt. 1, 491–508.

6. Ibid., 506.

7. Seebass, "Prospects for Commercial Supersonic Transport," I-1.

8. Ibid., I-3, I-5.

9. Ibid.

10. Ibid., I-4; National Research Council, *U.S. Supersonic Aircraft: Assessing NASA's High-Speed Research Program* (Washington, DC: National Academies Press, 1997), 46–49.

11. Seebass, "Sonic Boom Minimization Revisited," 6.

12. Ibid., 8–9. As described in chapter 9, NASA Dryden later developed a way to repeatedly create reduced sonic booms using existing fighter aircraft.

13. Ibid., 8.

14. Ibid., 3.

15. Seebass, "Prospects for Commercial Supersonic Transport," I-5.

16. Author's speculation.

17. Seebass, "Sonic Boom Minimization" (NATO version), VI-8. Although Seebass still attributed the structural vibrations specifically to the signature's impulse (which he defined as "the integral of the positive phase of the pressure with respect to time"), other structural engineering factors related to the ultralow frequencies of the sound waves caused by the extended N-waves of larger aircraft were also found to be involved.

18. Seebass, "Sonic Boom Minimization Revisited," 8–10.

19. Ibid., NTRS Abstract.

20. Seebass, "Prospects for Commercial Supersonic Transport," I-6.

21. The study, originally completed in 1997, was finally published just in time for use in this book. See Domenic J. Maglieri, "Compilation and Review of Supersonic Business Jet Studies from 1963 through 1995," NASA CR 2011-217144 (May 2011).

22. CASA became part of the European Aeronautic Defence and Space Company N.V. (EADS) in 1999.

23. Maglieri, "Compilation and Review of SSBJ Studies," NASA CR 2011-217144, 13.

24. "Gulfstream Studies Development of Supersonic Business Jet," *Aviation Week* (September 12, 1988): 47.

25. "The History of Gulfstream, 1958–2008," accessed November 13, 2011, *http://www.gulfstream.com/history/*.

26. Maglieri, "Compilation and Review of SSBJ Studies," NASA CR 2011-217144, 14.

27. Richard DeMeis, "Sukhoi and Gulfstream Go Supersonic," *Aerospace America* 28, no. 4 (April 1990): 40–42.

28. H.S. Bruner, "SSBJ—A Technological Challenge," *ICAO Journal* 46, no. 8 (August 1991): 9–13.

29. "Sukhoi Goes Supersonic," *Aviation Week* (September 20, 1993): 41; Graham Warwick, "Sonic Dreams," *Flight International* (May 6, 2003); 34.

30. Vincent R. Mascitti, "A Preliminary Study of the Performance and Characteristics of a Supersonic Executive Aircraft," NASA TM 74055 (September 1977).

31. Maglieri, "Compilation and Review of SSBJ Studies," 21, summarizes from a draft copy of this study.

32. Roy A. Da Costa et al., "Concept Development Studies for a Mach 2.7 Supersonic Cruise Business Jet," NASA CR 165705 (April 1981). This report is not available through the NTRS, but Maglieri's compilation, 21–22, summarizes its abstract.

33. F.L. Beissner, W.A. Lovell, A. Warner Robins, and E.E. Swanson, "Effects of Advanced Technology and a Fuel-Efficient Engine on a Supersonic-Cruise Executive Jet with a Small Cabin," NASA CR 172190 (August 1983).

34. F.L. Beissner, W.A. Lovell, A. Warner Robins, and E.E. Swanson, "Application of Near-Term Technology to a Mach 2.0 Variable Sweep Wing Supersonic Cruise Executive Jet," NASA CR 172321 (March 1984).

35. F.L. Beissner, W.A. Lovell, A. Warner Robins, and E.E. Swanson, "Effects of Emerging Technology on a Convertible, Business/Interceptor, Supersonic Cruise Jet," NASA CR 178097 (May 1986).

36. Graham Warwick, "Sonic Dreams," *Flight International* (May 6, 2003): 34–36.

37. John M. Morgenstern, low sonic boom shock control/alleviation surface. US Patent 5,740,984, filed September 22, 1994, and issued April 21, 1998. Accessed November 12, 2011, *http://www.patents.com/us-5740984.html*.

38. Univ. of Colorado, Biography of A. Richard Seebass, accessed April 27, 2011, *http://www.colorado.edu/aerospace/ARichardSeebass.html*. Seebass, who had stepped down from being dean of the University of Colorado's College of Engineering and Applied Sciences to become chair of the Department of Aerospace Engineering Sciences in 1995, had created a Lockheed Martin Engineering Management Program there.

39. Bill Sweetman, "Whooshhh!" *Popular Science*, posted July 30, 2004, accessed February 20, 2009, *http://www.popsci.com/military-aviation-space/article/2004-07/whooshhh*.

40. "The History of Gulfstream, 1958–2008," accessed November 13, 2011, *http://www.gulfstream.com/history/*.

41. Richard Aboulafia, "The Business Case for Higher Speed," *Aerospace America Online*, accessed July 8, 2010, *http://www.aiaa.org/aerospace/Article.cfm?issuetocid=ArchiveIssueID=16*.

42. National Research Council, *Commercial Supersonic Technology: The Way Ahead* (Washington, DC: National Academies Press, 2001), 10.

43. John Tirpak, "The Bomber Roadmap," *Air Force Magazine* 82, no. 6 (June 1999): 30–36.

44. "Department of Defense FY 2001 Budget Estimates: Research, Development, Test and Evaluation Defense-Wide," 1, Defense Advanced Research Projects Agency, February 2000, Program Element 0602702E, 87–88, 107–109. (Page 88 notes that it was a "congressional add.")

45. Guy Norris, "Back to the Future," *Flight International* (December 18, 2001), accessed November 13, 2011, *http://www.flightglobal.com/articles/2001/12/18/140226/back-to-the-future.html*.

46. Bill Sweetman, "Back to the Bomber," *Jane's International Defence Review* 37, no. 6 (June 2004): 55.

47. Department of Defense Appropriations Act 2001, Pub. L. No. 106-259, 114 Stat. 656 (August 9, 2001), accessed November 13, 2011, *http://www.gpo.gov/fdsys/pkg/PLAW-106publ259/content-detail.html*.

48. Sweetman, "Back to the Bomber," 54–59.

49. Department of Defense FY 2002 Amended Budget Submission: Research, Development, Test, and Evaluation Defense-Wide, 1, Defense Advanced Projects Agency, June 2001, 132.

50. Biography, Rich Wlezien, accessed ca. April 15, 2011, *http://www.richwelzein.com/Personal/Wlezien.html*.

51. DARPA Special Notice 00-17, "Advanced Supersonic Program—Industry Day," March 28, 2000, Alexandria, VA (February 29, 2000).

52. Chambers, *Innovations in Flight*, NASA SP-2005-4539, 65.

53. D.J. Maglieri, P.J. Bobbitt, and H.R. Henderson, "Proposed Flight Test Program to Demonstrate Persistence of Shaped Sonic Booms Signatures Using BQM 34E RPV," Eagle Aeronautics response to DARPA RFI-Advanced Supersonic Program, April 19, 2000, cited in Joseph W. Pawlowski, David H. Graham, Charles H. Boccadoro, Peter G. Coen, and Domenic J. Maglieri, "Origins and Overview of the Shaped Sonic Boom Demonstration Program," AIAA paper no. 2005-5, 6, 14.

54. Domenic J. Maglieri and Percy J. Bobbitt, "History of Sonic Boom Technology Including Minimization," Eagle Aeronautics, Hampton, VA, November 1, 2001. This comprehensive study totaled 373 pages.

55. "Quiet Supersonic Platform (QSP) Systems Studies and Technology Integration," *Commerce Business Daily*, SOL RA 00-48, posted on CBDNet, August 16, 2000, accessed August 18, 2011, *http://www.fedmine.us/freedownload/CBD/CBD-2000/CBD-2000-18au00.html*.

56. Richard Wlezien (DARPA) and Lisa Veitch (IDA), "Quiet Supersonic Platform Program," AIAA paper no. 2002-0143, 40th Aerospace Sciences Meeting, Reno, NV, January 14–17, 2002, 4. IDA is the acronym for Institute for Defense Analysis.

57. Robert Wall, "DARPA Envisions New Supersonic Designs," *Aviation Week* (August 28, 2000), 47.

58. Wlezien and Veitch, "Quiet Supersonic Platform Program," 5–6.

59. Graham Warwick, "Supersonic and Silent," *Flight International* (December 9, 2000), accessed November 13, 2011, *http://www.flightglobal.com/articles/2000/09/12/120156/supersonic-and-silent.html.*

60. Ibid.

61. Cited in Warrick, "Cutting to the Bone," *Flight International* (July 17, 2001), accessed November 13, 2011, *http://www.flightglobal.com/articles/2001/07/17/134122/cutting-to-the-bone.html.*

62. Pennington Way, "Northrop Grumman Awarded First Contract to Study Quiet Supersonic Platform," *Defense Daily* (November 14, 2000), *http://findarticles.com/p/articles/mi_6712/is_30_208/ai_n28800344/.*

63. Robert Wall and William Scott, "Northrop Grumman Gets Quiet Supersonic Work," *Aviation Week* (November 13, 2000): 32.

64. Telephone interview of Charles H. Boccadoro by Lawrence Benson, August 20, 2011.

65. Ibid.; Joseph Pawlowski to Lawrence Benson, "Re: More SSBD Questions," e-mail message, August 17, 2011. Also representing Northrop Grumman at the industry day were Tony Springs, Rob Chapman, and Jay Trott.

66. Biography of Charles Boccadoro, in *Decadal Survey of Civil Aeronautics: Foundation for the Future* (Washington, DC: The National Academies Press, 2006), 179. ISS headquarters was in Dallas.

67. Pawlowski to Benson, "Re: More SSBD Questions," August 17, 2011.

68. Benson/Boccadoro interview, August 20, 2011. The comprehensive NGC plan had eight sections on the various aspects it envisioned for the program.

69. In 2007, Raytheon sold its aircraft company to outside investors, who renamed their new entity Hawker Beechcraft Corporation.

70. Wall and Scott, "Northrop Grumman Gets Quiet Supersonic Work," 32.

71. Way, "Northrop Grumman Awarded First Contract."

72. Wall and Scott, "Northrop Grumman Gets Quiet Supersonic Work," 32.

73. Robert Wall, "New Technologies in Quest of Quiet Flight," *Aviation Week* (January 8, 2001): 61–62; Graham Warwick, "DARPA Cash to Fund Studies into Sonic Boom Reduction," *Flight International* (January 9, 2001): 17.

74. NASA Dryden, "Extensive Supersonic Laminar Flow Attained Passively in Flight," News Release 00-13, January 26, 2000.

75. Wall, "New Technologies in Quest of Quiet Flight," 61–62; Warwick, "DARPA Cash to Fund Studies"; Warwick, "Supersonic and Silent."

76. University of Colorado, News Center, "Former Dean of Engineering Richard Seebass Dies Tuesday," November 14, 2000, accessed ca. July 15, 2011, *http://www.colorado.edu/news/ r/4c5668f27e7454651c98568ed3ae0f5.html.*

77. Warwick, "Supersonic and Silent."

78. Pawlowski et al., "Origins and Overview of SSBD Program," 4.

79. Wlezien and Veitch, "Quiet Supersonic Platform Program," 7.

80. Ibid.

81. Bill Sweetman, "Quiet Supersonics in Sight: Technology Currently Under Development Could Form the Basis for Bizjet, Strike/ Reconnaissance Aircraft," at *Interavia Business & Technology* (November 1, 2001): accessed ca. July 20, 2011, *http://www. highbeam.com/doc/1G1-80743274.html.*

82. National Research Council, *Commercial Supersonic Technology: The Way Ahead* (Washington, DC: National Academy Press, 2001), recommendation number 2, 42, which references conducting a "technology demonstration" as defined on page 7.

83. Pawlowski et al, "Origins and Overview of SSBD Program," 5; Benson/Boccadoro interview, August 20, 2011.

84. Source for figure 5-2: Peter G. Coen, David H. Graham, Domenic J. Maglieri, and Joseph W. Pawlowski, "QSP Shaped Sonic Boom Demo," PowerPoint presentation, 43rd AIAA Aerospace Sciences Meeting, Reno, NV, January 10, 2005, slide 3.

85. Guy Norris, "Preferred Concepts Unveiled for Strike and Business QSP Versions," *Flight International* (October 1, 2002), 4, quoting Charles Boccadoro.

86. Graham Warwick, "Super Striker," *Flight International* (January 15, 2002), accessed November 13, 2011, *http://flightglobal.com/ articles/2002/01/15/141264/super-striker.html.*

87. Ibid.

88. Ibid.

89. Guy Norris, "Overwing Engine Placing Shows Low Boom Promise," *Flight International* (January 29, 2002): 29.

90. As quoted in Warwick, "Super Striker."

91. Guy Norris, "Manufacturers Unveil Dual Relevant QSP Configuration," *Flight International* (January 21, 2002): 6. No AIAA

papers were published on the QSP information revealed at this meeting.

92. Paul Lowe, "Answers Sought for SSBJ Questions," *Aviation International News* (June 2002), accessed ca. July 20, 2011, *http://www. ainonline.com/aviation-news/aviation-international-news/2007-10-03/ answers-sought-ssbj-questions.*

93. Ibid.

94. Northrop Grumman, "Quiet Supersonic Platform (QSP) Shaped Sonic Boom Demonstrator (SSBD) Program," presentation, Washington Press Club, September 3, 2003, slide 3.

95. "DARPA Selects Two To Develop Supersonic Aircraft," *Jane's Defence Weekly* (March 8, 2002), accessed ca. July 30, 2011, *http://www. articles.janes/Janes-Defence-Weekly-2002/DARPA.*

96. Warwick, "Super Striker." For a description of Boeing's design effort by leaders of its QSP team, see Peter M. Hartwich, Billy A. Burroughs, James S. Herzberg, and Curtiss D. Wiler, "Design Development Strategies and Technology Integration for Supersonic Aircraft of Low Perceived Sonic Boom," AIAA paper no. 2003-0556, 41st Aerospace Sciences Meeting, Reno, NV, January 6–9, 2003. Neither of the other two system integrators published similar reports on their design process.

97. Northrop Grumman, "Northrop Grumman Awarded Additional Contracts To Continue Work on Quiet Supersonic Platform," news release, May 14, 2002.

98. Robert Wall, "Bomber Becomes Focus of Quiet Aircraft Effort," *Aviation Week* (May 6, 2002): 28.

99. Northrop Grumman, "Northrop Grumman Unveils Concept for Quiet Supersonic Aircraft," news release, September 26, 2002, with the illustration used in front of this chapter, accessed ca. July 30, 2011, *http://www.irconnect.com/noc/press/pages/news_releases. html?d=32118;* Sweetman, "Back to the Bomber," 56; "Northrop Grumman Unveils Concept for Quiet Supersonic Flight," *Space Daily* (October 2, 2002, accessed ca. July 30, 2011, *http://www. spacedaily.com/new/plane-sonic-02b.html.*

100. Robert Wall, "Noise Control," *Aviation Week* (August 4, 2003): 23–24.

101. Benson-Boccadoro interview, August 20, 2011.

102. Biography, Rich Wlezien; U.S. Air Force Biography of Dr. Steven H. Walker, accessed November 13, 2011, *http://www.af.mil/information/ bios/bio.asp?bioID=13291.*

103. Laura M. Colarusso, "Fly Fast With No Boom," *Air Force Times*, September 22, 2003, accessed ca. August 5, *http://www.airforcetimes. com/legacy/new/0-AirPaper-2209084.php*.

104. Department of Defense, "Fiscal Year (FY) 2004/FY 2005 Biennial Budget Estimates, Research, Development, Test, and Evaluation Defense-Wide, 1—Defense Advanced Research Projects Agency," February 2003, 248.

105. Colarusso, "Fly Fast With No Boom"; Benson-Boccadoro interview, August 20, 2011. NGC's QSP II report and related data was submitted on a DVD.

106. Bill Sweetman, "Back to the Bomber," *Jane's International Defence Review* 37, no. 6 (June 2004): 54–59; Adam J. Herbert, "Long-Range Strike in a Hurry," *Air Force Magazine* 87, no. 11 (November 2004): 27–31.

107. Benson-Boccadoro interview, August 20, 2011. The Switchblade project is summarized in chapter 9.

Navy F-5E Tiger II fighters off the California coast. (NGC)

CHAPTER 6

Planning and Starting the SSBD Project

With most of the Quiet Supersonic Platform program consisting of engineering studies, computer models, and laboratory experiments, its most tangible legacy became the Shaped Sonic Boom Demonstration (SSBD). This innovative project used an actual airplane—the Shaped Sonic Boom Demonstrator (also SSBD)—to finally put theory into practice. Yet despite all the confidence that decades of peer-reviewed articles, wind tunnel experiments, and computational fluid dynamics had conferred on the basic principles of the Seebass-George-Darden sonic boom minimization theory, showing that it would actually work with a real airplane in the real atmosphere was anything but easy.

Selecting a Demonstrator

In June 2001, Charles Boccadoro picked Joseph W. Pawlowski, who was in charge of systems engineering for the QSP effort, to manage Northrop Grumman's sonic boom demonstration proposal.[1] Pawlowski was a versatile engineer who had worked on a wide variety of systems since being hired by Northrop in 1973. He and another veteran engineer, aerodynamicist David H. Graham from NGC's Advanced Air Vehicle Design office, had gone on fact-finding trips in late summer of 2000 to garner some of the latest information on sonic boom mitigation. Helping the pair bond for the challenging project that lay ahead, on their first flight Graham offered the much taller Pawlowski his first-class seat, which had been reserved using frequent flyer miles. In Hampton, VA, they, along with Charles Boccadoro and Steve Komadina, visited Eagle Aeronautics, where Domenic Maglieri described his ideas for a low-cost sonic boom demonstrator. Later, while at the Georgia Institute of Technology in Atlanta, Pawlowski and Graham met Wyle Laboratory's sonic boom specialist Ken Plotkin (there on a visit from his office in Arlington, VA). Plotkin went over some of his thoughts on sonic boom minimization with the two NGC engineers. The conversation continued when Plotkin gave Pawlowski a ride to the airport. Northrop Grumman engaged both Eagle and Wyle to become members of its QSP team in August 2000.[2]

Although DARPA had not planned for the sonic boom demonstration to be an initial part of the QSP program, Boccadoro's QSP team was interested almost from the beginning in Maglieri's long-standing proposal to use a supersonic Ryan BQM-34E remotely piloted vehicle as a relatively low-cost sonic boom demonstrator (described in chapter 4). The Firebee's modular construction, performance characteristics, and interchangeable components as well as previous wind tunnel data continued to make it an attractive option, at least in theory. Northrop Grumman's recent purchase of Teledyne Ryan perhaps added to the team's incentive to explore this opportunity. In anticipation of a future sonic boom demonstration contract, Northrop Grumman acquired all of the Navy's usable BQM-43E components except for engines that were still being used in subsonic models of the Firebee. The airframes and spare parts were trucked from the Naval Air Weapons Station at Point Mugu, CA, to one of NGC's facilities along Aviation Boulevard in El Segundo.[3]

By early 2001, the Northrop Grumman QSP team began to reconsider its concept for the demonstration. Analysis by NASA indicated that the Firebee's airframe might not have been long enough to demonstrate a definitive shaped boom signature. CFD modeling also raised concern about effects of the shock waves from the jet-engine inlet located under the airframe. Furthermore, NGC technicians had found that the Firebee fuselages and parts obtained at Point Mugu, where they had been stored outdoors in the salty air, had deteriorated significantly since the mid-1990s. So the team decided to put the Firebee option on the back burner.[4]

As a possible long-shot alternative, David Graham pointed out that Northrop Grumman's own F-5E fighter had two variations: the two-seat F-5F trainer and the RF-5E reconnaissance version with noses up to 42.5 inches longer and of different shapes than the basic F-5E (figure 6-1). Perhaps flying each of these aircraft supersonically at short intervals over an array of pressure sensors under the right conditions could show enough difference in their sonic booms to demonstrate the effect of airframe shaping—all at very little cost. However, some preliminary analysis in February 2001 by Graham and NGC colleague Hideo Ikawa and more detailed sonic boom modeling by Eagle Aeronautics revealed that all the signatures would still be typical N-waves. This had been predicted 2 months earlier by Domenic Maglieri, who determined that the longer noses did not have the smooth equivalent area distribution needed to produce a flattop or ramp-type signature. As a potential solution, Maglieri thought the F-5 would be an excellent candidate for using a new, properly designed nose extension to reshape its initial pressure rise into a flattop signature—something like what had been proposed for the Firebee.[5] Of all

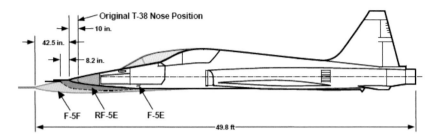

Figure 6-1. Profile comparison of F-5 Tiger II variants. (NGC)

the supersonic fighters in the U.S. inventory, the F-5E was uniquely suitable for such a modification.[6]

The Northrop F-5 Story

Back in 1954, when the United States Air Force and Navy were seeking supersonic combat aircraft of increasing size and sophistication, a U.S. Government study warned that many of America's Cold War allies needed smaller and simpler yet high-performance fighters. Thomas Jones and others at Northrop Corporation immediately saw the potential export market in being able to offer a fast but economical jet fighter. Jones, a chief engineer who succeeded founder Jack Northrop as company president in 1959, wanted a plane that was not only a relative bargain to buy but one that would also be cost effective throughout its life cycle.[7]

This approach featured lots of doors and removable panels for ready access by maintenance personnel and easily replaceable components. With that in mind, Northrop's design team came up with a concept designated the N-102. Helped by General Electric's development of the small but powerful J85 turbojet, the design evolved by 1955 into the N-156, a lightweight aircraft, which Northrop hoped might also be suitable for the U.S. Navy's small escort-type aircraft carriers. The Navy soon announced the retirement of these ships, but the Air Force released a requirement in 1955 for a supersonic trainer. Applying the area rule to its current design to improve transonic acceleration, Northrop created the TZ-156, which began flight tests at Edwards AFB in April 1959 as the YT-38. These tests went so well that the Air Force placed its first order for 50 T-38 Talons in October—the start of a production run of almost 1,200 aircraft lasting until 1972. The sleek T-38, easily capable of speeds up to about 820 mph, remains in use for undergraduate pilot training, introduction to fighter fundamentals, and a variety of special purposes, including some by NASA.[8]

Encouraged by progress with the TZ-156, Northrop's design team contin-ued working on its N-156F lightweight fighter version, using corporate funds to build the first prototype in early 1958. The Air Force soon agreed to buy two more prototypes, the first of which made its maiden flight at Edwards AFB on July 30, 1959—going supersonic without its engine yet having an afterburner. The early flight tests went so well that the Air Force stopped work on the third N-156F, which was eventually completed as an Air Force YF-5A that first flew in May 1963. While retaining as many T-38 structures as pos-sible, Northrop spent the next few years weaponizing its NF-156F design with internal guns, bomb racks, missile pylons, fuel tanks, and other features needed in a rugged combat aircraft. The result was the F-5A and the two-seat F-5B. After the Air Force awarded its first F-5 production contracts in October 1962, Northrop built them alongside T-38s on its highly efficient assembly line in Hawthorne, CA. The F-5B entered operational service as a trainer with the Air Force in April 1964, followed 4 months later by the F-5A.[9] In view of the F-5's intended international role during the Cold War, the Air Force named it the Freedom Fighter.

In October 1965, the Air Force deployed a unit of 12 F-5As, modified for aerial refueling and armored against small-caliber antiaircraft weapons, to South Vietnam for a 6-month combat evaluation code named Skoshi Tiger (Little Tiger). Although the F-5As did not fly enough missions over North Vietnam among their 2,664 sorties to test their air-to-air capabilities, they acquitted themselves well in air-to-ground operations considering their limited range and payload compared to the F-4 Phantom II and even the older F-100 Super Sabre. Maintenance personnel hours per flying hour were slightly better than with the F-100 and much better than with the big, complicated F-4. After completion of Skoshi Tiger, the F-5As were used to help form a commando fighter squadron and later transferred to South Vietnam's Air Force in 1967. By 1972, 15 nations had received F-5As, F-5Bs, and RF-5As under the U.S. Government's military assistance program or foreign military sales program while others were built under license in Canada and Spain.[10]

Based on the Vietnam deployment and feedback from other nations using the initial models of the F-5 Freedom Fighter, Northrop began testing an improved version, the F-5-21, which could better engage the latest models of the MiG-21.[11] Rather than accept Northrop's unsolicited bid for this to become the F-5A/B's replacement, the Air Force decided to sponsor what it called the International Fighter Aircraft competition. Lockheed, McDonnell Douglas, and Ling-Temco-Vought submitted modified versions of existing fighters as other candidates. In November 1970, the Air Force declared Northrop's entry the winner, with an initial contact for 340 aircraft. One month later, the Air Force gave it the designations F-5E and (for the two-seat version) F-5F. The

Air Force also tried to bestow the generic name International Fighter Aircraft on the F-5E/F but eventually renamed it the Tiger II, an informal nickname it had picked up largely in memory of Skoshi Tiger. The RF-5E reconnaissance version was later called the Tigereye. The maiden flights of the F-5E and F-5F were in August 1972 and September 1974 respectively. In June 1973, an F-5E put on a spectacular display of its agility for potential customers at the Paris Air Show.[12] Later in the 1970s, Northrop developed the RF-5E at its own expense, using three interchangeable nose pallets for various camera systems. (Joe Pawlowski spent the first 10 years of his career at NGC working on the F-5E and RF-5E, so he was already intimately familiar with the aircraft chosen for the SSBD project.) Northrop's Hawthorne facility would eventually build 792 F-5Es and 140 F-5Fs plus 12 RF-5Es while factories in Switzerland, Korea, and Taiwan would build more than 500 additional variants of these aircraft under license.[13]

Compared to the F-5A, the F-5E incorporated GE's more powerful J85-21 engines with 20 percent higher thrust, 9 percent more wing area, maneuvering flaps, bigger leading edge extensions that extended all the way to the wing roots, a larger fuselage with more internal fuel storage, and a two-stage nose gear strut that can be raised almost 12 inches for a better angle of attack during takeoff. As regards military utility, the biggest improvement over its predecessor (basically a day fighter) was the installation of an Emerson AN/APQ-153 radar as part of an integrated fire control system with a computing gunsight, a missile-launch computer, air-data inputs, and a gyroscopic platform. The F-5E has a maximum takeoff weight of 25,350 pounds, a takeoff run of 5,100 feet at that weight, a combat ceiling of 52,000 feet, and a ferry range (with three external tanks) of over 1,550 miles. Mach 1.64 is generally listed as the plane's maximum speed at 36,000 feet above sea level.[14] (This is where the coldest temperatures of the tropopause generally begin—allowing supersonic aircraft to achieve their highest Mach numbers.)

The relatively heavy losses suffered by large U.S. fighter-bombers from the hit-and-run tactics of smaller North Vietnamese MiG-17 and MiG-21 interceptors led to more realistic combat training of American aircrews. In 1969, the Navy's Fighter Weapons School at Naval Air Station (NAS) Miramar, near San Diego, CA, began a training regimen known as Top Gun. A key aspect of Top Gun was flying small and agile aircraft, such as the Douglas A-4 Skyhawk, against larger fighters, a practice known as Dissimilar Air Combat Tactics (DACT). The Navy's success eventually prompted the USAF Tactical Air Command to form an adversary training squadron to provide DACT to Air Force aircrews. Because Air Force fighter wings in the early 1970s were mostly equipped with F-4s, the proponents of DACT wanted the new squadron to fly Northrop F-5s since its size, performance, and smokeless jet engines were a close match to the MiG-21. Unfortunately, the Air Force owned no F-5s when its 64th Aggressor Squadron was being formed at Nellis AFB, NV, in 1972, but its Air Training Command had plenty of Northrop T-38s. Equipped with some of these Talons, the squadron soon proved its value in training "road shows" to other fighter bases.[15]

Although the T-38s proved to be a worthy adversary, they had not been designed or built for such strenuous maneuvers and began to suffer premature wear and tear. Fate intervened with the sudden North Vietnamese conquest of Saigon in May 1975. Seventy brand new F-5Es earmarked for the South Vietnamese Air Force suddenly became available to equip the 64th and two new aggressor squadrons in Nevada and England with a fourth soon activated in the Philippines.[16] These F-5Es (painted in a wide variety of camouflage schemes) helped hone the skills of USAF aircrews for the rest of the Cold War—skills that were demonstrated during Desert Storm air operations in early 1991. By then, the Air Force had inactivated its aggressor squadrons because of force reductions in Europe and the Pacific, the aging of their F-5Es, and the option of flying F-16s (which more closely emulated some current Russian-built fighters) against larger F-4s, F-15s, and F-111s for DACT training. During the 1970s and 1980s, the U.S. Navy had supplemented and eventually replaced its A-4s with F-5Es, surplus Air Force F-16As (refurbished as F-16Ns), and Israeli Kfir C.1 fighters (redesignated as F-21s) for Navy and Marine Corps aggressor squadrons.[17] The Navy and Marines have continued to use F-5Es well into the 21st century.

In 1996, the Navy moved its Top Gun program from NAS Miramar to the Naval Strike and Air Warfare Center at NAS Fallon in northwestern Nevada. As part of this realignment, Composite Fighter Squadron Thirteen (VFC-13), a Naval Reserve unit recently equipped with F/A-18 Hornets, converted to F-5E Tiger IIs to help provide adversary training at Fallon.[18] The Marine Corps had a similar unit, Marine Fighter Training Squadron (VMFT) 401, stationed since 1987 at its Marine Corps Air Station (MCAS) in Yuma, AZ.[19] As was the case with the Air Force, however, the Navy and Marine F-5Es began showing their age after years of hard use. To help keep them flying as long as possible, the Navy contracted with Northrop Grumman in 1999 to perform phased depot-level maintenance on its F-5s at the NGC East Coast manufacturing center in St. Augustine, FL. Northrop Grumman and selected subcontractors also continued to provide maintenance support, spares, and modifications for the F-5s still being flown by foreign nations, many of them increasingly being used as a lead-in jet trainer rather than a frontline combat aircraft.[20]

Besides its supersonic speed, the F-5E has many features that made it an attractive choice for serving as a sonic boom demonstrator. These included its light weight, a high fineness ratio, a blended canopy, and a relatively long forebody with engine inlets located farther back than most other fighter air-craft—all of which could help diminish the contributions of secondary shock waves to the planned demonstrator's bow shock. Northrop Grumman's experience in producing larger F-5F and RF-5E versions by simply adding forebody extensions, and the extensive analytical and flight-test data collected when doing so, also boded well for the planned sonic boom modifications. From a financial standpoint, the costs of operating an F-5 were relatively low.[21] Ironically, however, Northrop Grumman had no F-5s of its own to modify.

Forming the SSBD Working Group

On March 12, 2001, key members of NGC's QSP team, including Domenic Maglieri and Percy Bobbitt, gave Richard Wlezien the pros and cons of three options for demonstrating the persistence of a shaped sonic boom signature through the atmosphere. The options were (1) flying two F/A-18s in close over-and-under formations to generate a flattop signature, which would follow up on a flight test performed by NASA Dryden in 1994 (described in chapter 4); (2) testing Eagle Aeronautics's longstanding idea of comparing results from a modified and unmodified Firebee; and (3) comparing the signatures of a baseline and specially modified F-5E.[22] Except for the cost, the third option was obviously the preferred alternative.

Some members of Northrop Grumman's QSP team were already making plans for the demonstration. Joe Pawlowski, who had become Northrop Grumman's systems engineer for the QSP, had been appointed as project manager for the demonstration. David Graham was selected as its chief aerodynamicist. His expertise in dealing with wave drag on the company's supersonic and stealthy YF-23 advanced tactical fighter prototype afforded him good experience for dealing with the shock waves that cause sonic booms. From outside the company, Ken Plotkin and Juliet Page of Wyle Laboratories as well as Domenic Maglieri and Bud Bobbitt of Eagle Aeronautics offered many years of expertise in sonic boom research. These and several other NGC personnel became core members of what would be become known at first as the Shaped Boom Demonstration (SBD) Working Group.

One of the key participants in planning for the demonstration was M.L. "Roy" Martin, chief test pilot at the Integrated Systems Sector's Western Region. Martin was a veteran of more than 200 F-4 combat missions in Vietnam, a distinguished graduate of the USAF Test Pilot School, and a holder of a master of science degree in aeronautics and astronautics from Stanford. During his time with the Air Force, Air Force Reserve, and (starting in 1980) Northrop, he had flown several dozen types of aircraft, logging 6,000 flight hours in the

T-38/F-5 family and with what Northrop had hoped would be its high performance offspring, the single-engine F-20 Tigershark. Most recently, he had flown a series of high-stress flight tests of the F-5 for the Navy in 2000. He also had many contacts throughout the civilian and military test pilot community that would be helpful in arranging support for the project.[23]

In what would prove to be an advantageous move, Charles Boccadoro got Roy Martin involved early in the QSP program. In the summer of 2000, Boccadoro called him into his office along with David Graham to talk about the possible options for a sonic boom demonstration. Martin was less than enthusiastic about the proposal to use Firebee RPVs. When Graham asked about modifying an F-5, Martin thought that might be a good solution and said he was scheduled to go up to NAS Fallon for a safety day and would begin checking on the availability of Navy F-5Es stationed there. While at Fallon, Martin ate lunch with Mike Ingalls, a manager at the NGC facility in St. Augustine that performed depot maintenance for the Navy, including that for F-5s. He informed Martin that the Navy was reviving a previous proposal for the U.S. Government to buy back surplus Swiss F-5Es with low flying hours to replace its stable of heavily used F-5Es. Once the Navy got congressional approval and funding for the deal, obtaining and modifying one of its tired Tiger IIs presumably would not be too difficult.[24]

In April 2001, DARPA released its formal solicitation for proposals from all three systems integrators on how best to show the persistence of a shaped sonic boom. DARPA added an unusual twist to this QSP minicompetition. The winner would not only have to propose the best plan in terms of technology and cost but would also have to propose the best plan for incorporating design reviews by the other two systems integrators and for sharing data collected among all QSP participants.[25] Northrop Grumman quickly responded that same month with a proposal structured as a cooperative effort involving other companies and Government agencies. As for the vehicle to be used, the company submitted its preliminary design for modifying an F-5E with a specially shaped nose extension (shown in the next chapter). The cover of its proposal featured a CFD-generated image showing the modified aircraft's hoped-for shock wave pattern.[26] Because of the Dryden Flight Research Center's sonic boom experience, resources, and credibility, NGC proposed making NASA responsible for data collection. Based on a comparison of this and the other proposals, DARPA awarded a $3.4 million contract (MDA 972-01-2-0017) to the NGC Integrated Systems Sector in late July 2001 to begin preparing its flight demonstration of a shaped sonic boom signature using F-5Es.[27] Northrop Grumman officially came under contract in mid-August when it received the first payment.[28]

Several weeks before the contract award, DARPA invited attendees from Northrop Grumman, the other two system integrators, NASA, the Air Force, and some of the other contractors and universities to a shaped sonic boom workshop on July 10, 2001, in Valencia, CA (just north of Santa Clarita).[29] Consistent with DARPA's rule that anyone who participates in the demonstration would share in the data, the workshop examined how the various attendees and their organizations could help in the project. Some of the invitees who soon became key members of Northrop Grumman's Sonic Boom Demonstration Working Group (SBDWG) included Peter Coen of NASA Langley, Edward Haering of NASA Dryden, Mark Gustafson of the Air Force Research Laboratory (representing DARPA as its technical agent), and John Morgenstern, Lockheed Martin's veteran sonic boom specialist. The working group also included the sonic boom experts from Wyle and Eagle who were already part of the QSP team.

Right after the contract award was announced, Joe Pawlowski began formally establishing the SBDWG. Because the working group's members were located all across the United States, and getting them all together at one time and place would be difficult, he set up a special Web site for sharing information and ideas.[30] (Later, this information was also posted on the DARPA Web site for access by Government participants.) As the weeks went by, additional experts from NASA Langley, NASA Dryden, Boeing, Lockheed Martin, Gulfstream, and Raytheon joined the working group.[31] Eventually, more than 30 people served as members, with 9 of them composing the SSBD's program management team: Richard Wlezien and then Steven Walker of DARPA, Charles Boccadoro and Joe Pawlowski of Northrop Grumman, Mark Gustafson of the AFRL, Peter Coen of NASA Langley, and Ed Haering and David Richwine of NASA Dryden.[32]

At first, Northrop Grumman optimistically predicted being able to conduct the flight tests—projected to require about 18 sorties—at Edwards AFB in the summer of 2002.[33] But finding an F-5E to modify was only one of the many challenges confronting the SBD project. Before DARPA would let Northrop Grumman obtain an aircraft to modify, the company would have to complete an approved design. And to do this meant overcoming some unanticipated technical obstacles. One of the first to be identified was the need to better understand the effects of shock waves generated in front of jet-engine inlets.

NASA sonic boom specialist Ed Haering, who had gained valuable experience from NASA Dryden's supersonic probing and measurement of shock waves during flight experiments in the 1990s (described in chapter 4), became involved in the project when he was invited to the Valencia workshop. He introduced himself in advance to David Graham by alerting him to the GPS, telemetry, recording, and other equipment that would need to be installed in the participating F-5Es to ensure the collection of accurate data—the topic of a presentation he later made at the workshop. He wrote that he was also "concerned that inlet-wing shocks will be different than predicted, adversely affecting the signature."[34]

Following up on his presentation at the QSP workshop, Haering provided more details about his proposal to conduct a preliminary F-5E flight test to accurately measure the shock waves and expansions (regions of decreasing pressure) as they came off the various body parts of an F-5E, especially the shock waves that tended to spill out from around its engine inlets. He proposed using one of Dryden's F-15s or F/A-18s as a probe aircraft, perhaps as early as December.[35] These secondary shock waves coalesced through the atmosphere to merge with and reinforce the strength of the front and rear shock wave of the typical sonic boom. As the first prerequisite for conducting this test, Roy Martin went to NAS Fallon in mid-August to begin arranging for VFC-13 to deploy one of its F-5Es to Edwards, hopefully in January if not December.[36] He also continued investigating ways to obtain F-5Es later for the shaped sonic boom demonstration itself.

The first meeting of the SBDWG was held at the NGC Advanced Systems Development Center in El Segundo on August 22, 2001. Those who could not attend in person participated via the new Web site and special telephone connections. The initial intent was to meet for approximately 2 hours every other Thursday, but with scheduling conflicts and other events, the working group would normally hold these meetings somewhat less often. Joe Pawlowski described the SBDWG ground rules as follows: "Remember that this is a working group meeting and that team interaction and brainstorming are the desired products. Your ideas and support are appreciated. As stated at the Boom Workshop in Valencia, our goal is to reach technical consensus among team members in the formulation and execution of the SBD Program. The working group meetings are [also] designed to support critical decisions at the scheduled milestone reviews."[37] After implementation of the DARPA contract and establishment of the SBDWG, work to design the most effective possible F-5 modifications (described in chapter 7) greatly accelerated.

By February 2002, some of the members began brainstorming a new name and acronym for the working group as well as the SBD project itself, which occasionally was referred to in jest as "silent but deadly."[38] Among the more creative suggestions were Boom Shaping Technology (BooST) and the Boom Aerodynamic and Atmospheric Attenuation Demonstration (BAAAD).[39] The final choice was less colorful but very descriptive: the Shaped Sonic Boom Demonstration (SSBD).[40] (The abbreviation of the working group thus became SSBDWG.)

The First Flight Test:
Inlet Spillage Shock Measurements

One of the early priorities of the SBDWG was to begin planning the in-flight shock wave measurements of a standard F-5E recommended by Ed Haering. The main purpose of this probing, which Haering gave the name Inlet Spillage Shock Measurement (ISSM), was to collect data that could be incorporated into the CFD models needed to fully design the modifications for the shaped boom demonstrator.[41] To help prepare for the ISSM probes and check on some preliminary CFD modeling by the NGC design team, Haering had Dana Purifoy, a Dryden test pilot who did the in-flight measurements of the SR-71 with an F-16XL in the 1990s, use an F-15B to probe the shock waves from a Dryden F/A-18 flown by fellow NASA test pilot Jim Smolka in late September 2001.[42]

Some of the issues examined by various working group members—especially Ed Haering, Dave Graham, and Roy Martin—involved the attributes of the Navy's F-5E aggressor configuration, such as the effects that its permanently installed wingtip missile-launch rails and the normally carried Tactical Air Combat Training System (TACTS) telemetry pod might have on the ISSM test.[43] Such details could potentially affect the propagation of pressure waves. Another seemingly simple requirement illustrates just how complicated test planning could become. This involved the installation of some special telemetry equipment in the F-5E's cockpit—a GPS antenna, receiver, modem, and transmitter—that would be needed to obtain useful data.[44] As Ed Haering explained, "Measuring the sonic boom signature is comprised of two parts, the measured pressure, and the relative position of the two aircraft."[45] A discrepancy of several inches could make a difference when measuring the behavior of shock waves, so the more accurate the GPS units, the better. Even Edwards AFB's sophisticated radars could not be depended on to track aircraft locations to within 20 feet. During the F-16XL-SR-71 tests (which relied heavily on recently available differential GPS equipment), anomalies in the radar returns sometimes resulted in errors of 1,000 feet for the location of the F-16XL when the two planes were actually flying in close formation.[46]

Martin and Haering spent much of their time over the next month researching and testing GPS units and associated equipment that would be accurate enough for data collection yet small enough to be carried in the cockpit, especially without interfering with the ejection sequence—a major safety issue.[47] They eventually determined that the preferred GPS unit, an Ashtech Z-12 with an ultrahigh frequency (UHF) modem and power supply, would fit in the right-hand console of an F-5E with some adjustments, including the removal of a duplicate radio and use of a slightly larger map case to be built by Northrop

Grumman technicians.[48] When Martin took the proposed changes to VFC-13, officers in the echelon above the squadron recommended a "top-down" approval process to get clearance.[49] As a result, the proposal was elevated to Naval Air Systems Command (NAVAIR) at NAS Patuxent River in Maryland, where specialists gave the proposed cockpit installation a thorough review.[50] Northrop Grumman's Electrical and Data Systems Design office responded to NAVAIR's concerns with a detailed layout for the GPS and modem setup, including its wiring, that was delivered to NAVAIR as part of a data package on December 21, 2001.[51]

While awaiting Navy approval for the F-5 deployment, Ed Haering planned the routes and profiles that the F-5E and F-15B would need to follow during the in-flight measurements as well as location of the ground sensors. He worked closely with David Richwine, Dryden's F-15B manager, and Tim Moes, the F-15B chief engineer. Other members of the SBDWG helped in planning for optimal data collection, but Dave Graham alluded to Haering's key role as follows: "If in doubt use Ed's values. This is his test; we are all just very interested observers."[52] A flight safety specialist at Fallon thought the basic flight plan looked acceptable,[53] and VFC-13's operations officer assured Roy Martin on the qualifications of its pilots for the test. "They are all second tour, fleet fighter pilots who are highly experienced in formation flying and multi-plane operations."[54]

Early in January, scheduling conflicts involving Dryden's three F-15B pilots and other factors caused the test to be postponed until mid-February, which also allowed more time to get the Navy's approval.[55] On January 23, having done a fit test of the custom-built map case with an F-5E at Fallon, Martin delivered the GPS and associated equipment from Dryden to El Segundo for installation into the custom-built map case. After sending an apparent approval for the installation of the GPS equipment on January 30, a NAVAIR official apologetically informed Joe Pawlowski that the command still needed to check its electrical connections and possible effects on instrumentation.[56] With Pawlowski having key Northrop Grumman employees who worked at Patuxent River help to expedite the process,[57] NAVAIR's final clearance for the cockpit equipment came through on February 8.[58]

This approval came just in time to conduct the ISSM tests the following week. NASA Dryden activated a previously planned schedule of events. Dana Purifoy, its F-15B probe pilot, flew Northrop technicians and Ed Haering up to Fallon in one of Dryden's small passenger planes on Monday, February 11. Purifoy briefed VFC-13's chosen pilot, Lt. Commander Edgar "Sting" Higgins, on operating procedures at Edwards while the GPS equipment was being installed. Navy technicians then checked the equipment for any electromagnetic interference.[59] Haering used a receiving radio modem to ensure the GPS

NASA F-15B and Navy F-5E during Inlet Spillage Shock Measurement Test, February 2002. (NASA)

package was transmitting data properly. He and Purifoy also asked that the F-5E's centerline tank be removed since its shock waves would be incompatible with the design of the SSBD.[60]

Higgins flew the F-5E down to Edwards the following morning. As he entered the base's airspace, Purifoy met him in an F/A-18 to familiarize Higgins with the test area and for the Dryden control room to check data reception from the F-5E's GPS equipment. After landing for fuel, Higgins's F-5E and Purifoy's accompanying F-15B took off for the first of two flight tests on February 12, 2002, flying at about Mach 1.4 in both directions through the Edwards supersonic corridor. The pair completed two more similar test sorties on February 13, with Higgins returning the F-5E to Fallon later in the day.[61] The F-15B, using its special nose boom with sensitive pressure instrumentation, gathered 56 supersonic shock wave signatures from the F-5B at distances from 60 feet to 1,355 feet while various sensors on the ground collected plentiful data in their sonic boom carpets.[62]

After the first sorties on February 12, Ed Haering immediately sent members of the working group some encouraging preliminary data just in time for a previously scheduled Interim Design Review (IDR) the next day in El Segundo. He also passed along an intriguing atmospheric phenomenon. A

NASA employee at one of the sensor sites "could see the shock waves in the clouds for one pass. The shocks hit a sundog, increasing the brightness about ten times, and this lasted for 5–10 seconds, then started to fade. He is guessing the shocks may have crystallized the water vapor, or melted the ice crystals, or something. Of course he did not have a camera."[63]

In accordance with earlier arrangements, Northrop Grumman reimbursed VFC-13 for its expenses and Peter Coen arranged for Dryden to get some special NASA supersonic research funding to help cover its portion of the test.[64] In expressing NGC's and its partners' gratitude for the Navy's support, Joe Pawlowski acknowledged that "it took a series of small miracles to pull this off, and I want to thank everyone involved for their support."[65] The precedents set in working with the Navy boded well for having VFC-13 deploy an F-5E again in the future. In fact, the entire ISSM flight-test project served as an "excellent dry run" for the Shaped Sonic Boom Demonstration.[66]

Integrating Flight Data with Computational Fluid Dynamics

After refinement and validation of the flight-test data, including rectifying some rather inexact GPS readings, the working group's engineers and scientists went to work analyzing the data and applying it to CFD models. The major participants in this effort were Northrop Grumman's David Graham, Keith B. Meredith (CFD lead), John A. Dahlin (deputy program manager), and Michael Malone (CFD specialist); Juliet Page and Ken Plotkin of Wyle Labs; and Ed Haering of NASA Dryden. Ultimately, they selected four pressure signatures for CFD validation based on those probes that had the smallest and

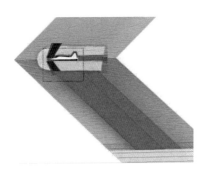

Figure 6-2. Near-field (inner box) and midfield CFD grids used for post-ISSM analysis of shock wave signatures. (NGC)

most stable vertical separation distances (ranging from 77 feet to 105 feet), the straightest flight paths, and the steadiest speeds (ranging from Mach 1.396 to Mach 1.448). They then used Northrop Grumman's Generalized Compressible Navier-Stokes Finite Volume (GCNSfv) CFD program, a structured implicit finite volume code (derived from the Ames Research Center's ARC3D code). It was first applied in an inviscid mode, since viscous effects from the F-5E's surfaces were assumed to be negligible at the distances measured. Later CFD analyses of midfield pressure signatures using both inviscid and viscous modes did verify negligible differences. The GCNSfv program included an extensive number of boundary conditions useful for measuring inlet flows. Its grid originally contained 8 million data points in the near field (within 0.5 body lengths of the aircraft) and another 4 million out to the midfield (3.0 body lengths).[67]

The analysts did repeated calculations and adjustments of GCNSfv's predictions based on the actual ISSM probing data (which varied in the number of usable data points collected). By repeatedly refining the results and increasing the number of grid points to 14.2 million, concentrated mostly along the angle followed by the shock waves under the aircraft (shown in figure 6-2), the analysts were eventually able to validate the accuracy of the CFD solutions out to three body lengths.[68]

The analysts also developed a process "to interpolate the CFD solutions onto the actual relative flight paths between the F-5E and the F-15B during each probing to accurately simulate the pressure measurement signature."[69] Among the lessons learned, they found that determining the two aircraft's relative speeds and flight paths, both vertically and horizontally, was critical for accurate correlations. Even knowing the exact location of the F-5E's GPS antenna down to almost the centimeter and the plane's precise angles of attack with decreasing fuel levels was important.[70] Taking all these data into account, the CFD code's postprocessed pressure signatures compared very well with those collected during the in-flight probes. "The comparison between the final CFD computed pressure signatures and the four selected flight test measurement[s] provided excellent correlation."[71] Figure 6-3 shows the close correlation of the pressure readings collected by the F-15B on February 13 during its 47th probe from about 94 feet beneath the F-5E compared with the postprocessed CFD prediction using the same flight conditions.[72]

For continuing design work on the modified F-5E, ISSM results found that the CFD estimates of F-5E inlet performance were acceptable. "Indeed, the entire process of the ISSM flight test and CFD correlation was successful, providing all the necessary procedures and confidence in predictive tools needed for the SSBD program."[73]

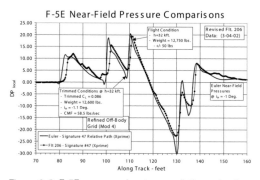

Figure 6-3. F-5E pressure signature prediction using the Euler-based CFD code (solid line) compared to in-flight measurements (line with dots) by F-15B. (NGC)

Endnotes

1. With the participants being organized as an integrated product team (IPT), his formal title became QSP Shaped Sonic Boom Demo IPT Lead.
2. Interview of Joseph W. Pawlowski and David H. Graham by Lawrence R. Benson, Northrop Grumman Space Park, Redondo Beach, CA, April 12, 2011; Biographical Sketch and Resume, Joseph W. Pawlowski, as of April 2011; David Graham to Lawrence Benson, "Biographical Information," e-mail, April 28, 2011; Joseph Pawlowski to Lawrence Benson, e-mail, August 17, 2011.
3. Joseph W. Pawlowski, David H. Graham, Charles H. Boccadoro, Peter G. Coen, and Domenic J. Maglieri, "Origins and Overview of the SSBD Program," Aerospace Sciences Meeting, Reno, NV, Jan 10-13, 2005, AIAA paper no. 2005-5, 4, 6, (also published with briefing slides by the Air Force Research Laboratory as AFRL-VA-WP-2005-300 [January 2005]).
4. Pawlowski et al., "Origins and Overview of the SSBD Program," AIAA paper no. 2005-5, 6.
5. Ibid., 7; Benson, Pawlowski/Graham interview, April 12, 2011; Domenic J. Maglieri by Lawrence R. Benson, telephone interview, March 18, 2009; David Graham to Lawrence Benson, "SSBD Design Questions," August 4, 2011.
6. Source for figure 6-1: David Graham and Roy Martin, "Aerodynamic Design and Validation of SSBD Aircraft," Northrop Grumman PowerPoint presentation, August 17, 2004, slide no. 3.
7. Marcelle S. Knaack, Office of Air Force History, *Post–World War II Fighters, 1945–1973* (Washington, DC: GPO, 1986), 287; Frederick A. Johnsen, *Northrop F-5/F-20/T-38*, 44, Warbird Tech Series (North Branch, MN: Specialty Press, 2006), 5–6; William G. Stuart, *Northrop F-5 Case Study in Aircraft Design* (AIAA, September 1978), 5-9. Stuart provides an engineer's perspectives on the design process.
8. Johnsen, *Northrop F-5/F-20/T-38*, 7-8, 53–56; "2011 USAF Almanac: T-38 Talon," *Air Force Magazine* 94, no. 5 (May 2011): 93–94.
9. Knaack, *Post–World War II Fighters*, 288–290; Johnsen, *F-5/F-20/T-38*, 8–22.
10. Johnsen, *F-5/F-20/T-38*, 39-51, 91; Knaack, *Post–World War II Fighters*, 288-289.
11. Johnsen, *F-5/F-20/T-38*, 62–64, 71–76.

12. John L. McLucas with Kenneth J. Alnwick and Lawrence R. Benson, *Reflections of a Technocrat: Managing Defense, Air, and Space Programs During the Cold War* (Montgomery, AL: Air University Press, 2006), 127–128. McLucas was Acting Secretary of the Air Force at the time.

13. Knaack, *Post–World War II Fighters*, 292; "Northrop F-5," *Wikipedia*, accessed July 18, 2011, at *http://en.wikipedia.org/wiki/Northrop_F-5*. This article is particularly well-sourced.

14. Knaack, *Post–World War II Fighters*, 297; Johnsen, *F-5/F-20/T-38*, 71–90, 99. For design of each component of the fire control and avionics systems, see Stuart, "Northrop F-5 Case Study," 171–202.

15. C.R. Anderegg, *Sierra Hotel: Flying Air Force Fighters in the Decade After Vietnam* (Washington, DC: Air Force History and Museums Program, 2001), 72–78.

16. McLucas, *Reflections of a Technocrat*, 128–129.

17. Johnsen, *F-5/T-38/F-20*, 90-91; "Aggressor Squadron," *Wikipedia*, accessed July 18, 2011, *http://en.wikipedia.org/wiki/Aggressor_squadron*. The Kfir was derived from the Dassault Mirage III. Enzo Angelucci, *The Rand McNally Encyclopedia of Military Aircraft, 1914–1980* (New York: The Military Press, 1983), 440.

18. "Naval Air Station Fallon," accessed July 11, 2011, *http://www.cnic.navy.mil/Fallon/About/index.htm*; Fred Krause, "Naval Air Station Fallon Adversaries, Part One: VFC-13 Saints," accessed July 21, 2011, *http://modelingmadness.com/scotts/features/krausevfc13g.htm*.

19. "Marine Fighter Training Squadron 401," accessed July 21, 2011, *http://www.yuma.usmc.mil/tenantcommands/vmft401.html*.

20. Northrop Grumman, "F-5 Tiger," accessed July 18, 2011, *http://www.as.northropgrumman.com/products/f5tiger/index.html*.

21. Pawlowski et al., "Origins and Overview of the SSBD Program," AIAA paper no. 2005-5, 7.

22. D.J. Maglieri and P.J. Bobbitt, "Overview of Flight Demonstrations to Demonstrate Persistence of Shaped Sonic Boom Signatures," presentation by Eagle Aeronautics to Dr. Richard Wlezien for Northrop Grumman, March 12, 2001, cited in Ibid., 7–8, 14.

23. Biographical Sketch, M.L. "Roy" Martin, Northrop Grumman Aerospace Systems, undated; Interview of Roy Martin by Lawrence R. Benson, Lancaster, CA, April 7, 2011.

24. Benson, Martin interview, April 7, 2011.

25. Pawlowski et al., "Origins and Overview of the SSBD Program," AIAA paper no. 2005-5, 5.

26. Northrop Grumman, "Demonstration of the Persistence of Shaped Booms Using a Modified F-5E Aircraft," April 2001, cited in

Ibid., 8, 14; David Graham to Lawrence Benson, "SSBD Design Questions," e-mail, August 4, 2011.

27. Northrop Grumman, "Northrop Grumman Awarded Contract To Demonstrate Less Intense Sonic Boom," news release, Aug. 3, 2001, accessed ca. July 15, 2011, *http://www.irconnect.com/noc/press/pages/news_releases.html?d=19142.*

28. Joe Pawlowski to Peter G. Coen et al., "QSP SBDWG Meeting Agenda," e-mail, August 16, 2001.

29. Andrea Brda to C.M. Darden and 28 others, "Quiet Supersonic Platform (QSP) Shaped Sonic Boom Workshop Announcement," e-mail, June 20, 2001; Charles Alcorn to C.M. Darden et al., "Shaped Boom Workshop," e-mail, July 3, 2001.

30. Joe Pawlowski to Peter G. Coen et al., "QSP Shaped Boom Demo Program Status," e-mail, August 1, 2001.

31. Joe Pawlowski to Peter Coen et al., SSBDWG meeting announcements, e-mail messages, August 16, 2001–April 9, 2002.

32. For a complete list of the SSBD Working Group, see Appendix A.

33. Pawlowski to Coen et al., "QSP SBD Program Status," August 1, 2001; Bruce Smith, "US Eyes Design to Lessen Sonic Boom, *Aviation Week* (August 13, 2001): 26; Guy Norris, "Quiet Supersonic Tests Set for 2002," *Flight International* (August 14, 2001), accessed ca. July 15, 2011, *http://www.flightglobal.com/articles/2001/08/14/134686/quiet-supersonic-tests-set-for-2002/html.*

34. Ed Haering to David H. Graham, cc: to Joseph Pawlowski et al., "F-5E QSP Data Needs and Accuracies," e-mail, June 28, 2001.

35. Interview, Edward A. Haering by Lawrence R. Benson, Dryden Flight Research Center, Edwards AFB, California, April 5, 2011; Ed Haering to David H. Graham et al., "Supersonic Probing of F-5E in 2001," e-mail, July 13, 2001.

36. M.L. "Roy" Martin (hereinafter, Roy Martin) to Ed Haering, "Re: Supersonic Probing of F-5E in 2001," e-mail, July 24, 2001; Roy Martin to Ed Haering, "Re: Supersonic Probing of F-5E in 2001," e-mail, August 17, 2001.

37. Joe Pawlowski to Peter G. Coen et al., "QSP SBDWG Meeting Agenda," e-mail, August 16, 2001.

38. Benson, Pawlowski/Graham interview, April 12, 2011.

39. Ed Haering to Joe Pawlowski et al., "New SBD name?," e-mail, February 22, 2002; Roy Martin to Ed Haering et al., "Re: New SBD name?," e-mail, February 22, 2002.

40. Joe Pawlowski to Peter Coen et al., "SSBDWG Meeting Reminder - 2/28/02," e-mail, February 22, 2002.

41. Benson, Pawlowski/Graham interview, April 12, 2011; Keith B. Meredith, John A. Dahlin, David H. Graham, Edward A. Haering, Juliet A. Page, and Kenneth J. Plotkin, "Computational Fluid Dynamic Comparison and Flight Test Measurement of F-5E Off-Body Pressures," 43rd AIAA Aerospace Sciences Meeting, Reno, NV, January 10–13, AIAA paper no. 2005-6, 2005, 2.
42. David Graham to Ed Haering, "F-5 Off-Body Pressures," e-mail, October 1, 2001; Ed Haering to David Graham, "Re: F-5 Off-Body Pressures," e-mail, October 5, 2001; Interview of Dana Purifoy by Lawrence Benson, Dryden Flight Research Center, April 8, 2011.
43. Ed Haering to David Graham, "Re: F-5E QSP Data Needs and Accuracies," e-mail, September 20, 2001; David Graham to Ed Haering, cc: Joe Pawlowski, "Re: F-5E QSP Data Needs and Accuracies," e-mail, September 20, 2001.
44. Ed Haering to Roy Martin, "Re: F-5E QSP Data Needs and Accuracies," e-mail, September 21, 2001.
45. Ed Haering to Roy Martin, "Re: F-5 flight and F-15B flight," e-mail, October 10, 2001.
46. Ed Haering to Roy Martin, "Re: F-5 Off-Body Pressures," e-mail, October 10, 2001.
47. Numerous e-mail messages between Roy Martin and Ed Haering et al., October 10–November 7, 2001.
48. Roy Martin to Ed Haering and Joe Pawlowski, "Re: Backup GPS," e-mail, November 12, 2001.
49. LCDR Darren Grove to Ed Haering et al., "Re: FW: F-5 GPS installation," e-mail, November 15, 2001.
50. Roy Martin to Jeff Ysells (NAVAIR) et al., "F-5E SBD Close Aboard Test Update," e-mail, November 21, 2001; Tommy White (NAVAIR) to Jeff Ysells, "F-5E SBD Close Aboard Test Update," e-mail, December 4, 2001.
51. Mark Wang to Ed Haering et al., "GPS Instrumentation Schematic," e-mail, December 18, 2001; Roy Martin to Ed Haering et al., "Re: F-5E drawing numbers," e-mail, December 21, 2001.
52. David Graham to Juliet Page, "Flightpath for F-5E ISSM," e-mail, February 9, 2002.
53. Ed Haering to David H. Graham, "Re: Urgent, F-5E probing priority," e-mail, January 2, 2002.
54. Roy Martin to Ed Haering, "Re: F-15B/QSP Review Kick-off," e-mail, January 2, 2002.
55. Ed Haering to Joe Pawlowski et al., "Re: F-5E status importance," e-mail, January 7, 2001.

56. Tommy White to Jeff Ysells et al., "Re: Flight Clearance Request, e-mail message," January 30, 2002; Scott White (NAVAIR) to Joe Pawlowski, "Re: Flight Clearance Request," e-mail, February 1, 2001.

57. Joe Pawlowski to Roy Martin et al., "FW: QSP Flight Clearance," e-mail, February 4, 2001.

58. Jeff Ysells to Joe Pawlowski et al., "Re: QSP Flight Clearance," e-mail, February 8, 2002; COMNAVAIRSYSCOM to COMAIRESFOR, "Interim Flight Clearance for F-5 Aircraft with Flight Test Instrumentation Installed...," teletype message, 0820006Z, February 2002.

59. Roy Martin to Jeff Ysells, "Re: QSP Mapcase," e-mail, January 23, 2002.

60. Edward A. Haering to Lawrence Benson, "Quick SSBD Question," e-mail, August 29, 2011.

61. Joe Pawlowski to John Cole et al., "F-5E Flight Test at NASA," e-mail, February 14, 2002.

62. Meredith et al., "CFD and Flight Test Measurement of F-5E," AIAA paper no. 2005-6 (January 2005), 3.

63. Ed Haering to Joe Pawlowski et al., "VERY preliminary ISSM data," e-mail, February 12, 2002.

64. Peter Coen to Dan Banks et al., "DFRC Funding Summary," e-mail, January 10, 2002; Peter Coen to Ed Haering et al., "Re: Telecon for Tuesday," e-mail, February 19, 2002.

65. Pawlowski to Cole et al., "F-5E Flight Test at NASA," e-mail, February 14, 2002.

66. Meredith et al., "CFD and Flight Test Measurement of F-5E," AIAA paper no. 2005-6 (January 2005), 1.

67. Ibid., 3–5.

68. Ibid., 5–9, David Graham and Roy Martin, "Aerodynamic Design and Validation of SSBD Aircraft," Northrop Grumman, PowerPoint briefing, August 17, 2004, slides 12–16. Figure 6-2 is copied from slide 14.

69. Meredith et al., "CFD and Flight Test Measurement of F-5E," AIAA paper no. 2005-6 (January 2005), 10.

70. David Graham to Ed Haering et al., "Re: GPS Antenna Position," e-mail, February 20, 2002, with previous message traffic; Roy Martin to Joe Pawlowski et al., "Time and Fuel Line," e-mail, February 22, 2002.

71. Ibid.

72. Graham and Martin, "Aerodynamic Design and Validation," slide no. 15.
73. Meredith et al., "CFD and Flight Test Measurement of F-5E," AIAA paper no. 2005-6, 10.

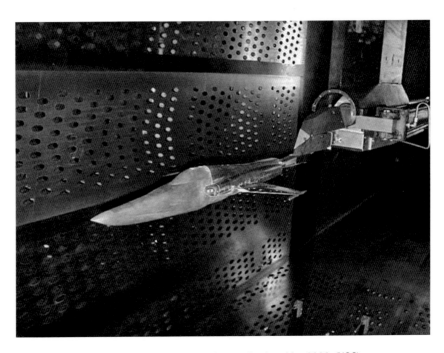

Improved SSBD wind tunnel model and mounting mechanism, May 2002. (NGC)

CHAPTER 7

Creating the Shaped Sonic Boom Demonstrator

In February 2001, 3 months after the QSP Phase I contract awards, some of the members of Northrop Grumman's integrated product team (IPT) for the Quiet Supersonic Platform—in collaboration with Eagle Aeronautics's sonic boom experts—began preliminary work on a new F-5E nose modification that would lower the strength of its sonic boom. As described in previous chapters, the basic principles and theories for reducing the pressure rise in the front half of an N-wave signature were fairly well understood. Even so, designing the exact geometrical contours (known aerodynamically as the loft) that would be certain to accomplish this goal while still retaining acceptable performance and handling qualities became a highly iterative process. Using a building block approach, the designers would eventually draw up almost two dozen basic configurations. The differences from one to the next were usually quite subtle, but the final configuration looked quite different than the original concept.

This design process combined high-order computational fluid dynamics, linear sonic boom–prediction models, wind tunnel evaluations, and some preliminary flight testing (the inlet spillage and shock measurements described in the previous chapter). The computer-generated image of F-5E shock waves in figure 7-1 hints at the value of continued advances in CFD to this endeavor not only for its powerful numbers-crunching and airflow-prediction capabilities but also for visualizing the nonlinear propagation of shock waves near an aircraft.[1] As implied by the experience described in this chapter, successfully creating an aerodynamically efficient low-boom aircraft would have been unlikely if not impossible before the advent of advanced CFD capabilities.

An Evolutionary Design Process

At the beginning of the SSBD project, the involved members of Northrop Grumman's QSP integrated product team assumed that almost all of the required geometrical changes to the F-5E would be limited to its nose assembly.

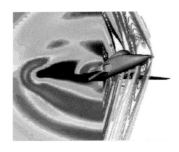

Figure 7-1. CFD-generated image of F-5E shock waves with pressure contours and expansion fields. (NGC)

They began by using relatively simple and speedy linear analysis tools to propose changes to the shape and volume of the nose that would achieve the desired pressure patterns for sonic boom reduction based on area-distribution principles.[2] Additional analyses refined the preliminary concept significantly. Using Euler codes, Eagle Aeronautics's Percy Bobbitt completed a detailed study in early March 2001 defining the shape of the equivalent area distribution for modifying the front portion of an F-5E (while staying within the overall length and width of an F-5F's forebody). CFD analysis indicated this loft would generate a flattop signature from 30,000 feet at Mach 1.4.[3]

Different flight conditions were also examined. Hideo Ikawa of NGC used a design optimization tool based on Christine Darden's computer program for the minimization of sonic boom parameters (which she had named SEEB in recognition of its descent from the minimization theories of Richard Seebass.)[4] Ikawa used this to calculate an area distribution that would create a flattop sonic boom at Mach 1.55. David Graham converted the results to a Mach 1.0 area distribution, which the NGC's configuration specialist, Jay Vadnais, adapted to create the initial geometry for the proposed shaped boom demonstrator. Graham ran this design through Peter Coen's PBOOM suite of linear tools to confirm it would generate a basically flattop sonic boom signature.[5] Keith Meredith of Northrop Grumman's Advanced Flight Sciences CFD group completed verification of the near-field effects of this design using Euler codes on April 18, 2001. The NGC team submitted its proposal to demonstrate sonic boom shaping using this F-5E modification to DARPA a couple of days later. The initial configuration with this nose glove (which included an underbody extension known as a fairing) was later designated SBD-01 (figure 7-2).[6]

After DARPA awarded the Shaped Sonic Boom Demonstration contract to Northrop Grumman Integrated Systems Sector based on this proposal in July 2001 and then provided funding in August, the design process resumed in earnest. SBD-02 was completed that same month, and SBD-04 was completed in September. Formation of the Sonic Boom Demonstration Working Group in August brought in more outside experts to help in the design process and evaluate the configurations. The steps required for each design iteration included (1) specifying an area distribution, (2) drawing up the aircraft lines, (3) projecting these surfaces onto a computational grid, (4) calculating a CFD solution, (5) validating the CFD solution, (6) evaluating its sonic boom signature based upon the shock wave pattern, and (7) correcting the area distribution based on these

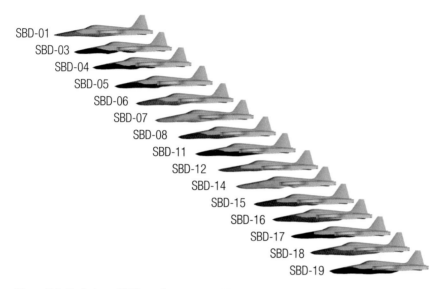

Figure 7-2. Preliminary SSBD configurations. (NGC)

evaluations. Initially, this cycle took about 2 weeks, but as the participants gained experience, they were consistently able to do this in about 3 days.[7]

Even so, the design process took considerably longer than initially expected. At first, the team expected that only about 6 major design iterations would be needed to reach a suitable configuration, but in the end they created 23 major configurations (most of which are shown in figures 7-2 and 7-3). Counting minor variations in many of these, the designers produced a total of 60 configurations.[8] From an aerodynamics perspective, it probably would have been easier to design a completely new forebody from scratch.[9]

The design team used a variety of linear methods, especially during the early months when assuring that the configurations would generate the properly shaped sonic boom signature was the primary concern. These included the aforementioned SEEB and PBOOM, both of which were developed at NASA Langley in the early 1990s, and VORLAX, a generalized vortex lattice methodology first developed for NASA Langley in 1977 to help determine both subsonic and supersonic airflows.[10] Other linear models included Wyle Lab's latest version of PCBoom[11] and NFBoom, a propagation code developed by Donald A. Durston of NASA Ames.[12] Darden's SEEB code was used throughout the design process, either directly or incorporated with other methods.[13]

As illustrated by figure 7-3, later configurations took on the characteristic pelican shape as the nose glove increased in size while the fairing under the fuselage grew longer and deeper.[14] This fairing created an area of expansion under the engine inlets, lowering the pressure of a shock wave that would normally coalesce with that from the bow.[15] Concerns about shock waves from the wings' leading edges and inlet spillage, which led to the ISSM tests, drove some of the design changes.[16] In its later configurations, the combined structure had doubled in size. Many of the changes reflected the increased application of the NGC's high-order GCNSfv computational fluid dynamics code (intro-

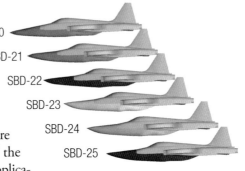

SBD-20
SBD-21
SBD-22
SBD-23
SBD-24
SBD-25

Figure 7-3. Later SSBD configurations. (NGC)

duced in chapter 6).[17] CFD specialists led by Keith Meredith used GCNSfv to account for nonlinear near-field and, later, mid-field shock and expansion effects. Previously, the GCNS code had normally been used to analyze onbody pressures, so it had to be revised to predict the offbody pressures as required for sonic boom purposes. Completing a single CFD prediction took 10 to 12 hours even on the powerful Silicon Graphics computer used for analyzing SSBD configurations. Fortunately, Northrop Grumman's CFD laboratory had quite a bit of spare computer time available during the SSBD design effort, and the project received management's support.[18]

The main purpose of the later design changes was to improve the aerodynamic performance and handling qualities of the modified F-5E while still maintaining its boom-shaping capability. Some configurations had to be rejected for not doing the latter. Regarding the predicted sonic boom signatures of the various configurations, "Everyone trusted Ken Plotkin's judgment."[19] Roy Martin played a similar role on the F-5E's aerodynamic performance. Early on, he recommended not making the nose glove any longer than the nose of a standard F-5F to avoid excessive lateral instability and being sure the width of the nose glove did not interfere with a smooth airflow into the engine inlets.[20] He also advised configuring the nose glove for optimum performance at 32,000 feet rather than the 30,000 feet used in some of the early models. In addition to allowing a higher Mach number because of lower temperatures, this would permit the chase plane probing the demonstrator's shock waves to also stay above 30,000 feet, which was the minimum height for Edwards AFB's high-altitude supersonic corridor.[21]

Continual feedback from other members of the SSBDWG, including parallel CFD analyses by Boeing and Lockheed Martin using their own proprietary codes, was also part of the process.[22] John Morgenstern and Tony Pilon of Lockheed Martin and Todd Magee of Boeing provided constructive help and advice. Morgenstern, for example, suggested a way to improve the NGC's method of calculating corrections to the area distributions that significantly reduced the turnaround time for design iterations. To help interpret inconsistencies in the CFD evaluations of the configurations by the three companies' computer codes, Juliet Page of Wyle Labs devised an easy and consistent format for comparing their data.[23] In addition, a recently developed three-dimensional propagation code by Eagle's Percy Bobbitt, with programming support from Old Dominion University's aerospace engineering department, was used to verify the results of these analyses.[24]

The SSBDWG's second Interim Design Review, the purpose of which was to present the working group's consensus on a technical solution to DARPA, was held on February 13, 2002—the day after the first ISSM flight tests at Edwards AFB (described in chapter 6). An e-mail sent that evening by Ed Haering at NASA Dryden to IDR attendees tentatively confirmed a positive correlation between near-field flight-test probe data and CFD predictions.[25] The working group was thus able to show DARPA that it was closing in on a design that could more accurately account for the effects of inlet spillage and associated shocks as well as the F-5E's less-than-ideal wing sweep. The configuration approved at the IDR had been designated SBD-24b (a minor alteration of SBD-24, shown in figure 7-3), the design of which initially was completed in January 2002.[26]

Wind Tunnel Testing at NASA's Glenn Research Center

The next big challenge was to confirm the pressure signatures of the approved configuration with wind tunnel testing, scheduled for March 2002 in the 8-foot-by-6-foot supersonic section of the Glenn Research Center's wind tunnel complex in Ohio. As described in chapter 4, computational fluid dynamics had become essential in designing aircraft, but wind tunnel testing was still required to help physically validate what the computer models predicted. So at the same time as the working group was planning, conducting, and analyzing the ISSM flight tests, it was also planning a wind tunnel experiment for the same basic purpose. After the successful correlation of February's ISSM flight-test data with CFD predictions, members of the group were hoping for a similar validation in the wind tunnel. However, when Keith Meredith sent the latest CFD results for the wind tunnel model to David Graham, who was already at Glenn, it was obvious that the model's support system would cause interference problems.[27]

The F-5E model chosen for the wind tunnel test was a 5-percent scale model (about 2½ feet long), which had been built in the 1970s (but apparently never used) for weapons integration and separation testing. The loft of configuration SBD-24b was scaled down to make a miniature nose glove that could easily

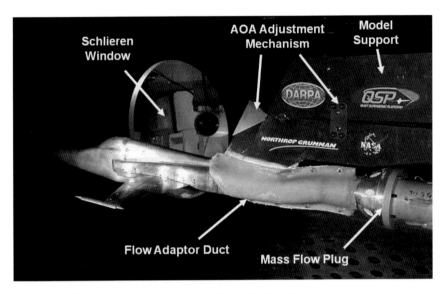

Figure 7-4. SSBD model in Glenn wind tunnel, March 2002. (NGC)

be exchanged with the model's regular nose. Because of the model's original purpose, it had a wind tunnel mounting strut where the tail section would have been. The goal of the SSBD was to modify only the shock waves affecting the front end of the sonic boom signature, so the lack of shock waves from tail surfaces was not critical to its design. The shock waves created by the mounting device, however, would cause problems.[28]

The model had several other limitations that would soon become painfully apparent. It had no internal-force balance, and its flow-through inlets lacked internal ducting, so the airflow passed straight through the internal cavity and exited in an uncontrolled manner. Because of the concerns about inlet spillage, a mass flow plug was placed just behind the model, connected by a fiberglass flow adaptor duct that covered the aft portion of the model's fuselage. This allowed inlet spillage to be controlled so that the testers could assess the effects of the airflow. To change the model's angle of attack (AOA), which was necessary for experimenting with different flight conditions, someone had to enter the tunnel and manually adjust the mounting bracket. This meant turning off and then restarting the 87,000-horsepower motors that drive the tunnel's high-speed compressor. Figure 7-4 shows the SSBD model's installation in the wind tunnel.[29]

The tunnel test plan called for examining the model in both its baseline and modified configurations at speeds of Mach 1.30, Mach 1.35, and Mach 1.40 using three angles of attack. In addition to sensors in the mass flow plug, there were survey probes at numerous stations along the tunnel walls to collect offbody pressure measurements. The test required use of the Glenn Wind

Tunnel for 80 hours; 20 of those with the air flowing. As these hours passed, the problems became ever more evident. To quote the definitive technical account of the test, "results were disappointing for both the baseline F-5E and the SSBD-24b configuration."[30]

The offbody pressures from the baseline F-5E showed unexplained and unrealistic sensitivities to the Mach numbers as well as significant changes in forebody, inlet, and wing shocks when only small changes in the latter two shocks were expected. The model's azimuth (directional) angles not only affected the wing shock as expected but also appeared to affect the pressure readings from the forebody. There were further discrepancies between baseline CFD predictions and wind tunnel data. Results with the SBD-24b configuration were somewhat different but no less troubling. For example, at both Mach 1.30 and Mach 1.40, there were unexpected pressure changes in the shocks from the modified forebody, which affected the plateau region of the pressure signature that would be critical for generating a flattop sonic boom N-wave.

Even allowing for expected interference from the model's support structure, the testers observed significant differences in shock location and strength from the CFD predictions. Figure 7-5 is an example of some of these discrepancies.

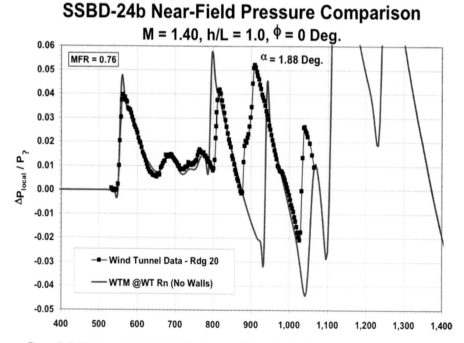

Figure 7-5. Wind tunnel data (dotted line) versus CFD prediction (solid line) at Mach 1.40, March 2002. (NGC) Key: h/L = height-to-length ratio; Φ = azimuth angle; MFR = inlet mass flow ratio; Δp = pressure increase in pounds per square foot; WTM = CFD wind tunnel model.

In this case, the wind tunnel data from the nose area matched CFD predictions fairly well (left third of the chart), departed from CFD predictions in the inlet and wing area (middle third), and failed as expected to provide any data from the rear of the model (right third). Even where the results followed anticipated trends, the data points were scattered and often not repeatable. Problems involving the wind tunnel itself included static pressure variations, off-centerline data, and questionable Mach and flow-angle calibrations.[31] "In conclusion," the results at Glenn "did not verify the shaped sonic boom design methodology; the near-field pressure did not support the CFD predicted signature, and significant model & tunnel data quality issues were identified."[32]

As would be expected, the Glenn wind tunnel test was a major topic at the next SSBDWG meeting on March 21, 2002. This was followed by the SSBD Preliminary Design Review (PDR) on March 26.[33] Disappointed but determined to overcome this serious setback, SSBD program management decided to attempt another wind tunnel test after correcting as many of the problems as possible.[34] (As discussed later in this chapter, this PDR also featured some good news about the availability of a Navy F-5E for the future demonstration.) Because of recent events and the need for more testing and CFD work, the critical design review (CDR), scheduled for mid-May, was postponed until the discrepancies could be resolved. As explained to the SSBDWG by Joe Pawlowski, "We are still in the process of understanding the issues with the previous test and will be generating additional test plans. NASA Glenn is investigating the data issues...."[35]

In preparation for this second try, Northrop Grumman specialists improved the model by adding an internal six-component force-and-moment balance to directly measure lift and other forces acting upon it. (In this context, "moment" refers to the torque that would tend to twist or pitch an airframe up or down.[36]) They also removed the mass flow plug to reduce the interference from extraneous shock waves. To help compensate for its absence, they inserted wedges in the duct and pressure instrumentation (known as rakes) at the aft exit to control mass flow. To move the model farther from the strut and its shock waves, they mounted it on a short sting. As shown in the accompanying photograph, the strut itself was reduced in size to alleviate blockage and lower the strength of the shocks it produced. After troubleshooting the problems encountered at Glenn, the project's CFD experts carefully analyzed the effects of proposed configuration changes prior to fabrication of the new model components. Meanwhile, specialists at Glenn conducted a flow survey and examined ways to improve the collection and validity of SSBD data in the wind tunnel.[37]

During this same period, the design team was making subtle revisions to SBD-24b. Based on continued analyses, the team digitally sculpted the aft section of the under-fuselage fairing to a narrower and more tapered shape to improve airflow and thereby reduce the possibility of viscous separation. Additionally, the design team realigned the angle of the nose assembly to minimize trim drag. CFD modeling assured that these changes did not adversely affect the desired sonic boom pressure distribution. This latest (and final) configuration was designated SSBD-24b4. Figure 7-6 shows the shape of its nose glove and fairing from the side, in cross sections, and under the fuselage (including two additional small

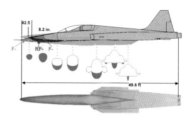

Figure 7-6. Final SSBD configuration 24b4 compared with other F-5s and showing its new underbody fairings. (NGC)

fairings extending back from the engine inlets). Although considerably higher and deeper than the nose of a standard F-5F, the SSBD nose glove matched its length and width. The additional weight of the added components, including flight-test instrumentation, would be more than offset by the absence of the F-5E's two 20 millimeter (mm) cannons, ammunition, radar, and other equipment. Extra ballast would therefore be required to keep its center of gravity within the limits of a normal F-5E or F-5F.[38]

With the 5-percent F-5E model's new mounting system and loft revised to match that of SSBD-24b4, the retest in the Glenn Center's 8-foot-by-6-foot wind tunnel section took place during the last week of May 2002. The tunnel was operated at Mach 1.367, which was as close as could be precisely calibrated to the Mach 1.40 speed planned for the eventual flight demonstration. Only 40 hours of tunnel time were required compared to 80 hours in March with 20 of the 40 hours being actual air-on time. This time, the results were encouraging. "Wind tunnel and CFD pressure distributions for the SSBD configuration showed a good correlation over most of the length of the model.... Wing shock strength and location were correctly predicted when matched with normal force."[39]

The results verified the ability of Northrop Grumman's GCNS code to predict offbody pressures of the SSBD configuration to at least 1.5 body lengths from both configurations of the model as well as sensitivity to lift, Mach number, and mass flow changes.[40] The data also confirmed CFD predictions that making moderate changes to Mach number, lift coefficient, and inlet mass flow did not significantly change near-field, offbody pressures. The only correlation that had fallen off since the March test involved pressure measurements near the inlet, which was not unexpected because of the removal of the well-instrumented mass flow plug. Although the specialists at the Glenn Center went out of their way to improve test procedures, there was some continued data scattering. Even so, "Results of the second test were significantly better than the first in nearly every way."[41]

The experience gained at NASA Glenn offered two major lessons learned for conducting sonic boom wind tunnel experiments. Because of the well-known influence of an airplane's lift coefficient on sonic boom overpressures, any models used need to be equipped to directly measure the forces acting upon them. And because sonic boom testing is less forgiving than regular aerodynamic testing for measuring onbody and offbody pressures, the wind tunnel used should be thoroughly analyzed in advance of model fabrication for characteristics such as the model-mounting arrangements, inlet-flow fields, Reynolds number, and test section flow characteristics.[42]

After a careful analysis of data from the successful second test, the Critical Design Review for the SSBD was held in El Segundo on July 18, 2002. Most importantly, the meeting approved the latest configuration and allowed Northrop Grumman to begin physical fabrication of the new nose and fairing.[43]

Although having encountered some unexpected challenges, the design process owed its ultimate success to several factors (some that could be considered lessons learned for similar projects in the future). Getting early test pilot involvement "proved invaluable for maximizing the use of similarity to previous aircraft modifications and for understanding the many capabilities and few important limitations of the F-5 aircraft."[44] The working group structure encouraged prompt contributions from a wide range of participants, which was especially useful in areas like sonic boom and wind tunnel testing, where NGC personnel had less experience. The relatively small design and analysis team dedicated to the project was able to react quickly to each new CFD result while QSP-SSBD program management assured access to the needed computer resources. Northrop Grumman's extensive database of past F-5 configurations, both produced and proposed, and its existing 6-DOF simulation that could be quickly modified, provided valuable data and saved time.[45]

Initial Fabrication and Final Wind Tunnel Testing

To prepare to manufacture the parts that would compose the nose glove and fairings, a 15-person vehicle-design team under Keith Applewhite used a computer-aided design (CAD) system to render the shape of the 24b4's surfaces into the tooling and precise specifications needed to make its outer panels and inner structures. The panels would be made from composite materials, while aluminum would be used for frames and stringers. One of the first steps for the CAD program was to smooth out minor imperfections in the fit and finish of the original CFD-generated loft for unobstructed airflow at supersonic speeds.

In coordination with the aerodynamic design team, other relatively minor tweaks continued during the coming months. CFD was used to investigate the offbody pressures of certain individual components (such as the nose-mounted pitot probe that measures airspeed), spillage from an inlet to the aircraft's environmental control system (ECS), wing lift, and horizontal tail angles. Some of the seemingly mundane but essential features that also required additional design work were nose-gear doors (which later proved problematic), arrangements for drainage, and exhaust ducting for the ECS. The functioning of this exhaust system would be critical because the fairing would be covered with a composite carbon-fiber skin known as LTM-45EL. This was a cost-effective material for making prototype and low-volume items partly because its resin was preimpregnated into the fabric and did not need curing at high temperatures and pressures.[46] One drawback, however, was that it would not tolerate direct heating of more than 250 °F. Dave Graham was more worried about this issue than the effects of the SSBD modifications on the F-5E's stability and control. The design teams devised a stainless steel ECS exhaust system to ensure dissipation of any heat buildup by using CFD to model its geometry, surface flow, and cooling capacity. They also decided to install thermal switches inside the fairing to alert the pilot in case of a dangerous hot-air leak.[47]

Even as some design work continued, a team of skilled craftsmen at Northrop Grumman's Advanced Composites Manufacturing Center in El Segundo were preparing to make the nose glove and fairing.[48] Leading this team was Mark Smith, a manufacturing engineer who had begun his career with Northrop in 1973 working on the F-5. To accomplish the fabrication as efficiently as possible, they followed rapid prototype procedures, which NGC had used when making small new vehicles in the past. The initial fabrication process began during July 2002 in NGC's Building 905. As shown in the two accompanying photos, some of the first steps were to create the foam tooling that served as templates for laminating (laying up) the surfaces to be used for shaping the outer skin panels. A total of 33 precisely engineered

Preparing tooling (top) to form the outer mold line of the nose glove (bottom) in August 2002. (NGC)

panels of various sizes eventually would be needed to cover the exterior of the nose glove, including the fairing extending under the fuselage. As with the aerodynamic design process, making all the required parts became more complicated (and time consuming) than originally planned.[49]

Although there was now confidence that SSBD configuration 24b4 could indeed lower the initial overpressure of its sonic boom signature, more testing and analysis were needed to verify its stability and control qualities. The value of CFD continued to be evident in this effort. Because of the small size of the high-speed wind tunnel model and its uncalibrated ducts, only CFD could be used to show the supersonic trimmed-drag effects of the nose glove.[50]

The CFD analyses indicated that the aerodynamics of SSBD-24b4 were within Northrop Grumman's existing (and very extensive) F-5 database but just barely for angles of attack at high speeds.[51] They also predicted it could meet the desired speed of Mach 1.4 at 32,000 feet on a standard day (i.e., one with the normal range of temperatures for that altitude) and would be statically stable about all its axes.[52] Even though preliminary CFD predictions of the aerodynamic coefficients showed acceptable handling qualities, two more wind tunnel tests were needed to validate the offbody pressures of the SSBD design and to update 6 degrees of freedom (6-DOF) motions used in computer simulations for the standard F-5E's flight maneuvers. These tests would help verify the plane's handling qualities for flight-readiness reviews to be conducted by both Northrop Grumman and the Navy. Fortunately, the SSBD F-5 would be flying mostly straight and level with relatively gradual turns, which did not require validating the entire F-5E flight envelope.

Wind tunnel testing of the SSBD 24b4's low-speed handling qualities required a model at least twice as large as the 5 percent scale model used for high-speed testing, one with both vertical and horizontal tail surfaces.[53] The SSBDWG originally had assumed that they could modify a legacy F-5E model from the 1970s, but despite inquiries both within and outside the company, David Graham was unable to locate one. So Northrop Grumman fabricated a 10-percent scale model using stereolithography, the CAD technique in which an ultraviolet laser device printed the three-dimensional loft of the model one layer at a time using a special resin-like material.[54] After getting an aluminum skin, the resulting 5-foot-long model (shown in the accompanying photograph) was tested in Northrop Grumman's 7-foot-by-10-foot low-speed wind tunnel for 25½ hours in August 2002 to validate CFD predictions for stability and control, including force-and-moment measurements throughout its reduced flight envelope. The existing 5-percent model was then tested for the same factors in a 4-foot supersonic wind tunnel section at the Air Force's Arnold Engineering Development Center (AEDC) in Tennessee during October 2002. Both tests started with the models having the standard

F-5E nose to establish a direct link to the existing database before moving on to the SSBD configuration. All the data collected were postprocessed to build increments between the standard and modified configurations for the stability and control analysis.[55]

Ten percent model of SSBD in Northrop Grumman's low-speed wind tunnel, October 2002. (NGC)

These two final wind tunnel tests revealed no significant changes in SSBD-24b4's longitudinal responses— except for the need to add 2 degrees of horizontal tail trim at most Mach numbers to compensate for a change in its pitching-moment coefficient. The lateral (directional) responses were comparable to that of a standard F-5E except for some reductions in yaw and roll stability. This made its handling qualities very comparable to an F-5F (with its longer forebody) when carrying a 275-gallon centerline fuel tank (which produced effects much like the underbody fairing). The controls worked well, with no significant change in power by the ailerons or horizontal tail surfaces and only a slight reduction in rudder power at high deflections and angles of attack. Previous concerns about sideslip effects on control power were deemed insignificant. Further aerodynamic and engineering analyses found that the effects of the SSBD modification on stability and control matched CFD predictions and would be suitable within its restricted flight envelope. In conclusion, both the subsonic and supersonic handling qualities of the modified aircraft were found to be satisfactory and not too different from other F-5 configurations with stability augmentation engaged.[56]

With the completion of wind tunnel testing and additional design refinements, the final configuration was confirmed at a DARPA-sponsored meeting in Huntington Beach, CA, in December 2002. Before the parts could be completed and installed, however, the Navy would have to approve this radical remodeling of one of its Tiger IIs and place it on loan to Northrop Grumman.

Obtaining the F-5E

The eventual modification of a stock F-5E would have to be done at a suitable facility with an adjacent runway, preferably located within easy flying distance to an FAA-approved supersonic corridor. Some members of the Northrop Grumman's QSP IPT gave some early consideration to doing it in Southern California at the Scaled Composites complex or another available facility at the Mojave Airport, in one of Dryden's hangars at nearby Edwards AFB, or in an NGC building at Air Force Plant 42 in Palmdale. Although these locations would be close to the manufacturing facilities in El Segundo and the test range at Edwards, they selected Northrop Grumman's maintenance depot in St. Augustine, FL (abbreviated as the NGSA), which is located at the St. Johns County Airport, as the most logical choice. The technicians and mechanics at the NGSA had unequalled experience in breaking down, refurbishing, and returning to flight status various Navy aircraft including F-5s. Doing the SSBD modifications would be similar to this work. Roy Martin later determined, however, that the 8,000-foot St. Johns runway, although suitable for a normal F-5, would not provide enough of a safety margin for the modified F-5E. As a solution, John Nevadomsky, the NGSA operations and safety officer, suggested flying all but the first Florida flight out of the nearby Cecil Commerce Center, which had been built up around a former military base named Cecil Field with a 12,500-foot runway. An Army reserve unit there later agreed to host the modified F-5 and its chase plane.[57]

Now all that was needed was to obtain a suitable aircraft. While hoping the Navy would agree to provide one of theirs, Northrop Grumman explored several other options. These included leasing an F-5E from the Swiss Air Force or from a couple of private companies that were considering the purchase of F-5Es from Switzerland or Taiwan.[58]

As explained in chapter 6, arranging for the U.S. Navy's aggressor squadron at NAS Fallon in Nevada to deploy one of its F-5Es to Edwards AFB for the 2-day ISSM flight tests had not been a simple matter. Even so, Roy Martin was happy to report from Fallon in late October 2001 that the staff of VFC-13, including its soon-to-be commander, W.J. Cole, had agreed to support the future SSBD (pending the approval of higher headquarters) with another similar deployment. Of course, they expected reimbursement by either NGC or NASA. Martin also informed fellow SSBD team members that VFC-13 would try to accommodate all the needed special equipment in the F-5E's cockpit.[59]

Getting the Navy's approval for a long-term loan of one of its F-5Es, which would have to be demilitarized and partially rebuilt into the SSBD configuration, was more complicated. The instrument for arranging this would be a Cooperative Research and Development Agreement (CRADA), which is basically a contract between a Federal and a non-Federal entity to facilitate joint R&D projects. As with the ISSM test, Mike Ingalls, NGSA's aircraft overhaul program manager who had excellent relationships with many Navy officials, was largely responsible for obtaining the F-5E. In January 2002, he informally approached Naval Air Systems Command officials about allowing Northrop Grumman to borrow and modify one for the SSBD. Receiving a favorable response, Joe Pawlowski drafted a formal request cosigned by Steven R. Briggs, a retired rear admiral who was vice president of the NGC Air Combat Systems division. Ingalls delivered this request to the NAVAIR program management office for multimission aircraft (code PMA-225) in late January while stopping at Patuxent River on his way to Switzerland with NAVAIR representatives.[60] There they discussed the purchase of surplus F-5s from the Swiss Air Force, which would be a key factor for obtaining use of one of the Navy's existing F-5Es.

In March 2002, after Joe Pawlowski had explained more about the program to Rear Adm. Timothy Heely of the Naval Air Warfare Center's Aircraft Division, Jim Sandberg, a former test pilot who was one of NGC's onsite personnel at Patuxent River, set up a full-scale briefing for Vice Adm. Joseph Dyer, NAVAIR commander. Dyer was very knowledgeable on the technical and safety aspects of experimental projects like the SSBD, having formerly been the Navy's chief test pilot and chief aviation engineer.[61] Attendees at this meeting, held on March 22, included Lisa Veitch representing DARPA; Peter Coen from NASA; and Charles Boccadoro, Steven Briggs, and Joe Pawlowski from NGC. After Veitch gave an introduction to the QSP program, Pawlowski gave a short briefing on the planned Shaped Sonic Boom Demonstration. Admiral Dyer responded favorably pending three conditions: having a Navy test pilot fly some of the sorties, meeting all Navy flight clearance criteria, and being given more involvement in the QSP program, especially if there was a follow-on to Phase II. The Navy's tentative approval to provide an aircraft allowed DARPA's Richard Wlezien to make a favorable go, no-go decision on the SSBD at the Preliminary Design Review held on March 26.[62]

Even with NAVAIR's tentative approval, negotiating terms of the CRADA became a drawn-out process that included coordination with the Navy's Type Command (TYCOM) in New Orleans, which was responsible for operating F-5 type aircraft. Because the planned purchase of the surplus F-5Es from Switzerland was not yet final, the Navy at first wanted the modified F-5E restored to its original condition after the demonstration. Negotiating terms of the CRADA required frequent discussions and exchanges of information with NAVAIR. Joe Pawlowski and other NGC personnel made three more trips to Patuxent River to negotiate terms of the CRADA and work with NAVAIR specialists regarding the flight clearance process.[63]

On December 10, 2002, 1 day after Admirals Dyer and Heely and Barbara Olsen from NGC's subcontracts office signed the CRADA, Pawlowski reported the "good news" to the membership of the SSBDWG. "The Navy has finally signed the CRADA and released the F-5E aircraft! We are now getting it ready to fly a baseline functional check flight next week to assess aircraft performance prior to modification."[64] The F-5E chosen for the modification, identified as BUNO 74-1519, came from the U.S. Marine Corps aggressor squadron at Yuma, AZ. (In Navy parlance, BUNO is an abbreviation for Bureau Number.)

As indicated by the first two digits of this number (inherited from the original Air Force–allocated tail number), it had been built in 1974. The plane was in need of a major overhaul, but NAVAIR authorized a lifetime extension of 50 flight hours given that the planned demo would put only minimal stresses on the airframe.

Roy Martin put the old F-5 through its paces during functional check flights (FCFs) on December 17 and December 18. The first flight revealed some minor handling abnormalities, but after the NGSA mechanics replaced some parts, he was able to make several runs along the supersonic corridor off the Florida coast. Instead of the expected Mach 1.55, however, the plane was able to achieve only Mach 1.4 at 32,000 feet. The temperature there was measured at –41 °F compared to –48 °F on a standard day. In view of the abnormally high temperature and having the option of using a pushover maneuver from a higher altitude to gain speed, Martin and Graham still hoped to be able to achieve the planned Mach 1.4 during the planned tests at Edwards despite the extra drag that the modification would impose. After further analyses, it was concluded that "a max Mach number of only 1.4 for the unmodified aircraft on a hotter than standard day is not a show stopper. There is no reason to delay installing the SSBD modifications to the...aircraft."[65] To begin preparing the Marine fighter plane for this work, it was towed into NGSA's Hangar 40 on January 9, 2003.[66] Much to the delight of all involved, in early February, NAVAIR notified Joe Pawlowski that the last of four congressional committees had approved the Navy's purchase of Swiss F-5s. This meant that after having disassembled and modified BUNO 74-1529, the NGSA would not have to restore the plane to its original condition after the demonstration as had been required in the CRADA.[67]

Final Fabrications and Modification

The modification team in St. Augustine, led by Dale Brownlow, began preparing the F-5E for its transformation during the second part of January by removing its existing nose section. At the same time, across the country in El Segundo, Mark Smith and his fabrication team began putting together the subassemblies. This work included constructing aluminum bulkheads and internal braces and aligning them in a nose-assembly tool, which (as shown in an accompanying photograph) served as a template. The final step was to attach the carefully shaped outer mold line (OML) composite skin panels to the nose-glove framework. The LTM45EL carbon fiber used for the panels was the same material to be used on Scaled Composites' SpaceShipOne that won the X-Prize for achieving the first private-sector suborbital flights in 2004.[68]

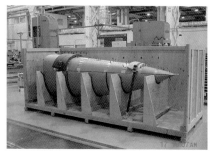

Nose bulkheads and framework being assembled and later crated for shipment with skin panels attached. (NGC)

The new nose and other recently delivered parts in a hangar at St. Augustine on April 3, 2003. (NGC)

Fabrication of the nose section was completed in mid-March, when it was carefully packaged as shown in the next photograph. The fairing was also completed for shipment to St. Augustine on March 17. As an extra precaution because of the skin's susceptibility to overheating, the designers provided a number of temperature-indicating tabs to be applied to the fairing's outer skin near the ECS exhaust outlet. During the future test program, postflight inspections of these "tell-tales" (which would change color if exposed to high temperatures) revealed no such problems.[69]

The next photograph shows the new nose being aligned with the stripped-down F-5E fuselage shortly after its arrival at NGSA Hangar 40 along with boxes and crates holding other components. To some extent, the rebirth of the Marine fighter as the F-5 SSBD was equivalent to handcrafting a new prototype. To help in this process, the NGSA workforce was augmented by Mark Smith and three technicians from El Segundo. Reflecting the high-level interest in completing the project, Charles Boccadoro and Joe Pawlowski spent the next several months taking turns overseeing the work in St. Augustine. As might be expected when putting a one-of-a-kind structure together for the first time, the modification team began encountering some problems.[70]

One of the more tedious issues involved securely mating the nose glove and fairing to the existing framework of the old F-5. A complex arrangement of bulkheads and aluminum stringers (shown in figure 7-7) served as a skeleton for the modified exterior. Based on NGC's file of F-5E plans and specifications, the fabrication design team had configured this aluminum framework for easy attachment to a production F-5E's existing structure— or so it was hoped. Unfortunately, over the past quarter century BUNO 74-1529 had undergone a lot of field maintenance and some overhauls with many minor changes not documented in available maintenance records. Because of these nonstandard specifications, about 90 percent of the shims and spacers shipped from El Segundo did not quite line up in the exact places they were supposed to. As a result, the NGSA technicians had to improvise and hand-fit new connectors.[71]

Not all of the project's troubles involved work on the aircraft. One late evening in April 2003, Charles Boccadoro was returning to his hotel on a narrow country road when he swerved to avoid a drunk driver in a black pickup truck. His rental car crashed into a drainage ditch, and when he regained consciousness, the engine compartment was on fire. Despite a badly injured kneecap, he was able to pull himself out the window and, with the help of some good Samaritans who came on the scene, get away before the passenger compartment was engulfed in flames. When emergency personnel arrived, his biggest concern was retrieving his laptop computer. Fortunately, a very good doctor in Jacksonville helped him begin rehabilitation from his injury, and even the data stored on the hard drive of his charred computer was recoverable. The accident may also have garnered him some sympathy at corporate headquarters, where unhappiness about the SSBD's delays and cost overruns had been mounting. In that regard, the SSBD project was fortunate to have continued receiving strong support from Scott Seymour, Boccadoro's superior at the Integrated Systems Sector.[72]

Figure 7-7. Framework for the SSBD nose glove and fairing. (NGC)

In addition to the F-5 SSBD's external shape being modified, much of its internal equipment was replaced. In effect, the former fighter plane was transformed into a flight--research platform. First, the NGSA team had removed the radar antenna, both 20 mm guns and ammunition boxes, the lead computing optical sight system, and the fire control computer. They also installed a sophisticated pitot static probe on the point of the new nose and relocated the battery, tactical air control antenna, and UHF identification friend or foe (IFF) antenna to other parts of the aircraft.[73]

Ed Haering and colleagues at NASA Dryden planned and obtained the special flight-test instrumentation needed in the SSBD. As during the ISSM test in February 2002, one of the key components was an Ashtech Z-12 carrier-phase differential GPS receiver.[74] In addition to recording data in its internal memory, it had a UHF modem for transmitting data to others. Once again, the NGC specialists in El Segundo integrated the receiver into a package that fit in the map case on the right side of the cockpit as shown in an accompanying photo. The GPS unit would determine the plane's exact position relative to the similarly equipped F-15B probe aircraft, the standard F-5E, and the array of microphones on the ground.[75]

Precisely measuring speed, acceleration heading, pitch, roll, yaw, and other aerodynamic factors required a suite of sensitive air-data acquisition instruments. This specialized equipment consisted of a calibrated Mach meter; temperature-controlled air-data transducers; an unheated outside air temperature probe with a signal conditioner and power supply to gauge true airspeed; a three-axis rate gyroscope; sensitive accelerometers; a signal-conditioning, power control module; an S-band telemetry transmitter; an Inter-Range Instrumentation Group (IRIG) format B time code generator; and three time recorders. To record pilot comments and the readings of cockpit instruments—such as air speed, altitude, and trim—two small "lipstick" video cameras were connected to a three-deck 8 mm airborne video and voice recorder mounted in the right-side gun bay. The telemetry system would allow real-time monitoring and recordings on the ground, while the onboard data-recording system would provide redundancy. Using UHF, the GPS data would be transmitted separately from the telemetry data stream on the S-band.[76] The accompanying photograph shows some of the GPS equipment and one of the little video cameras inside the cockpit.

"Lipstick" camera and GPS unit in cockpit. (NGC)

Since Northrop Grumman's St. Augustine facility conducted only routine flights of the aircraft it overhauled, it did not have the antennas or instrumentation needed to receive and record flight-test telemetry from the F-5 SSBD, as the former F-5E would be redesignated. To obtain these data during the required local check and envelope-expansion flights, Northrop Grumman had to install a special ground telemetry station as well as arrange advance clearances from the Federal Communications Commission (FCC) for the frequencies that would be used.[77] The equipment included a microwave antenna mounted on the roof of the NGSA hangar and an S-band telemetry receiver with related processors, displays, recorders, and printers.[78] Most of the instrumentation was installed and wired by late April.

On April 28, the modified F-5 with a big new olive drab nose and off-white underbody fairing attached to its original gray airframe was towed to the NGSA painting facility. One week later, it emerged having a bright white exterior with the sides of the fuselage featuring a red pinstripe shaped like a typical N-wave sonic boom signature and a parallel blue pinstripe representing the predicted flattop signature.[79] This clever design was the inspiration of a Northrop Grumman engineer named Joan Yazejian.[80] The transformation of the old F-5E fighter jet into the F-5 SSBD appeared to be complete.

All seemed to be going well until June 8, during preparations for the first high-speed taxi test scheduled for the following day. The inspectors noticed some unacceptable free play with the front doors for the nose landing gear. Unlike the aft portion, which was replaced with a newly designed wider door, the existing clamshell doors had been integrated into the new fairing using the same hinges and an extended actuation arm. This seemingly simple mechanical problem took the rest of the month to solve. Mark Smith, who was now leading the modification team in St. Augustine following Dale Brownlow's retirement in May, brought in three of the technicians who had fabricated the nose in El Segundo to help make the necessary fixes and get the landing gear flight worthy. As shown in figure 7-8, one of their ingenious solutions was securing the clamshell doors with segments of a piano hinge.[81]

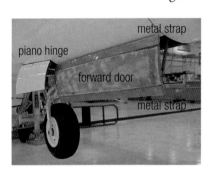

Figure 7-8. Changes made to nose gear door components. (NGC)

The delay this caused in starting flight operations almost made for an awkward situation with the Boeing Company. Regarding a chase plane and escort for the long cross-country flight to California, Roy Martin's preference had been to use another F-5-type aircraft, such as a T-38, which had the same basic performance. Charles Boccadoro had therefore arranged for Boeing to provide a T-38 chase plane as part of the arrangement for sharing in the SSBD's data. Martin went to Seattle in early June to help Boeing test pilot Mike Bryan bring the T-38 to Florida. (As practice for later flying the F-5 SSBD with its long nose, Martin frequently controlled the T-38 from the back seat on the flight to St. Augustine.) When it became obvious that the nose gear door would not soon be fixed, the time came to notify Boeing of the embarrassing delay. Just before that, however, Bryan told his hosts that his office at Boeing had just called to inform him that his T-38 was, in any case, needed back in Seattle.[82] He would return a month later when the F-5 SSBD was ready to fly.

Test Plan Approval and Flight Clearance

Planning for flight operations continued in parallel with the fabrication and modification work. These would begin with functional check flights and sub-sonic envelope-expansion maneuvers in Florida followed by a cross-country ferry flight to California, where additional FCFs and supersonic envelope-expansion profiles would be conducted in preparation for the actual sonic boom flight tests. Reflecting the plane's ownership, the Navy was responsible for flight safety, which helped assure that the planning process was very systematic and rigorous. In all, more than 50 NAVAIR specialists and officials would be involved in reviewing and supporting the SSBD project.[83]

Northrop Grumman's flight-test engineer for the SSBD was Steve Madison. With inputs from members of the working group and others involved in the project, he pulled together all the requirements needed to prepare for and conduct the Shaped Sonic Boom Demonstration. Briefings on these preparations were regularly presented, updated, and reviewed during 2002 and early 2003—both internally and by NAVAIR, which also assessed a detailed design analysis of the SSBD modification. This feedback helped in the drafting of a test plan, an early version of which was first posted on the SSBD Web site in October 2002. The briefings and planning documents covered a wide range of topics, including governing directives, inspection procedures, quality-assurance criteria, certification standards, personnel qualifications, and SSBD instrumentation. They also specified the sequence of actions to be followed for the aircraft's ground checks, envelope-expansion flights, functional check flights, and, finally, the sonic boom data-collection flights.[84] The process culminated in formal SSBD test plans for the data-collection flights as well as the preceding envelope-expansion flights. NASA Dryden published a related planning document to cover the in-flight near-field probes.[85]

Roy Martin getting into "the Pelican" for a ground test on July 23, 2003, just before its first flight. (NGC)

In preparing fight profiles, the planners followed a "predict, test, analyze"[86] procedure. They used data collected from the Northrop Grumman subsonic and AEDC supersonic wind tunnel tests combined with the existing F-5 performance database and CFD analyses to assure the modified F-5E would have enough stability (especially specific damping ratios) and other handling qualities in all of the planned flight conditions. Military Standard (MIL-STD) 1797, "Flying Qualities of Piloted Aircraft," defined these and other criteria in great detail. As chief SSBD test pilot, Roy Martin approved each of the planned maneuvers.[87]

Northrop Grumman conducted a series of flight-readiness reviews (FRRs). These covered the statuses of the aircraft configuration, engineering analysis, test plan, completion of modifications, safety-hazard analysis, and other readiness factors. On June 9, 2003, having been reassured by these reviews that the modified F-5 would be safe to fly, NAVAIR formally transmitted its eagerly awaited initial flight clearance (first announced on June 6), which closely followed the thresholds in NGC's previously published test plan. For the record, the message listed the various conditions and constraints that the aircraft would comply with, such as flying only in visual meteorological conditions, not exceeding Mach 1.45, keeping the most strenuous of several approved maneuvers under 3.0 equivalent gravitational forces (g's), turning with bank angles of no more than 60 degrees, and making no abrupt changes in pitch, roll, or yaw.[88]

Two days after the Navy's initial flight clearance notification, the nose gear problem was discovered.[89] Since the loose door would not affect nonflying activity, Roy Martin was able to take the F-5 SSBD out on the St. Johns runway for the first ground test as scheduled on June 9.[90] Keeping below the F-5's takeoff speed of 150 knots, this and two subsequent acceleration tests on July 23 were part of a systematic series of ground checkouts of all the aircraft's systems, including the revised environmental controls. The taxi runs tested the plane's brakes at various speeds and distances. On one run, the drag chute was deployed to simulate an abort at 130 knots. These operations also previewed functioning of the air-data instrumentation system, the Ashtech Z-12, and the telemetry equipment that would be so important in recording and transmitting results of the flight tests.[91]

After resolving the landing-gear problem and receiving another NAVAIR flight clearance on July 3,[92] the final flight-readiness reviews for the planned activities could begin. These were conducted in front of a nine-person board cochaired by representatives from NGC and NAVAIR. By late July, the final FRR for the envelope-expansion and ferry flights had assured NAVAIR that all pertinent quality assurance and safety issues for the new F-5 SSBD had been addressed successfully.[93] After a final ground test on July 23, the time had come to check it out in the air.

Endnotes

1. Source for figure 7-1: Peter G. Coen and Roy Martin, "Fixing the Sound Barrier: The DARPA/NASA/Northrop Grumman Shaped Sonic Boom Flight Demonstration," PowerPoint presentation, Experimental Aircraft Assoc. (EAA) Air Venture, Oshkosh, WI, July 2004, slide no. 4.

2. David H. Graham, John A. Dahlin, Keith B. Meredith, and Jay L. Vadnais, "Aerodynamic Design of Shaped Sonic Boom Demonstration Aircraft," 43rd AIAA Aerospace Sciences Meeting, Reno, NV, January 10–13, 2005, AIAA paper no. 2005-8, 3.

3. Domenic Maglieri to Lawrence Benson, "Comments on Chapter 7," e-mail, September 19, 2011.

4. C.M. Darden, "SEEB-Minimization of Sonic-Boom Parameters in Real and Isothermal Atmospheres," LAR-1179 (January 1994), NTRS abstract.

5. Peter G. Coen, "Development of a Computer Technique for Prediction of Transport Aircraft Flight Profile Sonic Boom Signatures," NASA CR 188117 (March 1991).

6. David Graham to Lawrence Benson, "SSBD Design Questions," e-mail, August 4, 2011. Figure 7-3 copied from Graham and Martin, "Aerodynamic Design and Validation of SSBD," slide no. 4.

7. Graham to Benson, August 4, 2011.

8. Robert Wall, "Noise Control," *Aviation Week* (August 4, 2003): 23–24, citing interview with Charles Boccadoro; John Croft, "Engineering through the Sound Barrier," *Aerospace America* 42, no. 9 (September 2004): 29. Figures 7-2 and 7-3 extracted from Graham and Martin, "Aerodynamic Design and Validation of SSBD," slide no. 4.

9. Croft, "Engineering through the Sound Barrier," 29, citing interview with Charles Boccadoro.

10. L.R. Miranda, R.D. Elliott, and W.M. Baker, "A Generalized Vortex Lattice Method for Subsonic and Supersonic Flow Applications," NASA CR 2895 (December 1977).

11. Kenneth J. Plotkin and Fabio Grandi, "Computer Models for Sonic Boom Analysis: PCBoom4, CABoom, BooMap, CORBoom," Wyle Report WR-02-11 (June 2002).

12. Graham and Martin, "Aerodynamic Design and Validation of SSBD" (August 17, 2004), slide no. 8. Daniel A. Durston's "NFBOOM User's Guide, Sonic Boom Extrapolation and Sound-Level Prediction," NASA Ames Research Center, was an unpublished document. His November 2000 edition is cited in Charles L. Carr et al., "Minimization of Sonic Boom on Supersonic Aircraft Using an Evolutionary Algorithm," 2003, accessed August 14, 2011, *http://portal.acm.org/citation.cfm?id=1756700*.

13. Graham to Benson, August 4, 2011.

14. Figure 7-3 copied from Graham and Martin, "Aerodynamic Design and Validation of SSBD," slide no. 6.

15. Michael Dornheim, "Will Low-Boom Fly?" *Aviation Week* (November 7, 2005): 69.

16. Ibid.; Graham et al., "Aerodynamic Design of SSBD," AIAA paper no. 2005-8, 3.

17. Meredith, "CFD Comparison and Flight Test Measurement," AIAA paper no. 2005-6, 5. Based on the Ames Research Center 3-Dimensional (ARC3D) code, GCNSfv was a multiple-block code that allowed both arbitrary face-matching and overlapping block interfaces. It used nodal finite volume and a diagonal beam-warming algorithm with viscous modifications. NGC had developed an extensive compilation of boundary conditions from various airframe components to use in airflow predictions. Boundary conditions and parametric linear interpolations were computed on multiple grids and a sequence of coarse and fine meshes using various techniques.

18. Benson, Pawlowski/Graham interview, April 12, 2011; Wall, "Noise Control," 24.

19. Benson, Pawlowski/Graham interview, April 12, 2011.

20. Interview of Roy Martin by Lawrence Benson, Lancaster, CA, April 7, 2011.

21. David Graham to Ed Haering et al., "FW: Release of F-5SBD Loft," e-mail, October 4, 2001.

22. Lockheed Martin's analysis in early 2002 used its SPLTFLOW-3D code. See John F. Morgenstern, Alan Arsian, Victor Lyman, and Joseph Vadyak, "F-5 Shaped Sonic Boom Demonstrator's Persistence of Boom Shaping Reduction through Turbulence,"

AIAA paper no. 2005-12, 43rd Aerospace Sciences Meeting, Reno, NV, January 10–13, 2005, 3–4.

23. Joe Pawlowski to Peter Coen et al., "SBDWG Meeting Reminder," e-mail, January 8, 2002; "CFD Used in Sonic Boom Test Program," *CFD Review* (September 23, 2003), accessed ca. August 30, 2011, *http://www.cfdreview.com/application/03/09/23/1327257. shtml.*

24. Osama A. Kandil, Z. Yang, and P.J. Bobbitt, "Prediction of Sonic Boom Signature Using Euler-Full-Potential CFD with Grid Adaptation and Shock Fitting," AIAA paper no. 2002-2543 (June 2002), cited in Pawlowski et al., "Origins and Overview of the SSBD," AIAA paper no. 2005-5, 8, 14. Kandil was a professor and Yang a research assistant.

25. Ed Haering to Joe Pawlowski et al., "VERY preliminary ISSM data," e-mail, February 12, 2002.

26. Joe Pawlowski to Peter Coen et al., "SBDWG Meeting Reminder," e-mail, January 22, 2002; Joe Pawlowski to Peter Coen et al., " IDR Rescheduled for Wednesday 2/13/02," e-mail, February 4, 2002; Graham to Benson, "SSBD Design Questions," August 4, 2011.

27. Graham to Benson, August 4, 2011.

28. The vertical tail surface of a typical fighter aircraft configuration such as the F-5 serves as a horizontal stabilizer and rudder. The horizontal surfaces serve as vertical stabilizers as well as elevators for climbing or descending. All of these components can create shock waves during transonic and supersonic flight.

29. David H. Graham, John A. Dahlin, Judith A. Page, Kenneth J. Plotkin, and Peter G. Coen, "Wind Tunnel Validation of Shaped Sonic Boom Demonstration Aircraft Design," 43rd Aerospace Sciences Meeting, Reno, NV, January 10–13, 2005, AIAA paper no. 2005-7, 2–3; Graham and Martin, "Aerodynamic Design and Validation of SSBD," slides nos. 17–19, with figure 7-4 copied from slide no. 20.

30. Graham et al., "Wind Tunnel Validation of SSBD," AIAA paper no. 2005-7, 3.

31. David Graham, "Wind Tunnel Boom Test," PowerPoint presentation, July 18, 2002, with figure 7-5 copied from slide no. 7.

32. Graham et al., "Wind Tunnel Validation of SSBD," AIAA paper no. 2005-7, 4.
33. Joe Pawlowski to Peter Coen et al., "SSBDWG Meeting Reminder," e-mail, March 19, 2002; Joe Pawlowski to Richard Wlezien et al., "SSBD PDR 3/26/0," e-mail, March 20, 2002.
34. Joe Pawlowski to Peter Coen et al., "SSBDWG Meeting - 4/18/02!!," e-mail, April 9, 2002.
35. Joe Pawlowski to Peter Coen et al., "SSBD CDR Status," e-mail, May 1, 2002.
36. Steven A. Brandt et al., *Introduction to Aeronautics: A Design Perspective* (Reston, VA: AIAA, 2004), 87–88.
37. Graham et al., "Wind Tunnel Validation of SSBD," 4–5; Graham, "Wind Tunnel Boom Test," July 18, 2002.
38. Graham and Martin, "Aerodynamic Design and Validation of SSBD," figure 7-6 extracted from slide no. 5.
39. AIAA paper no. 2005-7, 5.
40. Graham and Martin, "Aerodynamic Design and Validation of SSBD," slide no. 26.
41. Graham et al., "Wind Tunnel Validation of SSBD," AIAA 2005-7, 6.
42. Ibid.
43. Joe Pawlowski to Peter Coen et al., "SSBD CDR Postponed," e-mail, June 17, 2002; Joe Pawlowski to Lisa Veitch, "Re: SSBD CDR Meeting Reminder - 7/18/02," e-mail, July 17, 2002; Peter Coen to Ed Haering, "SSBD Status," e-mail, August 7, 2002.
44. Graham et al., "Aerodynamic Design of SSBD," AIAA paper no. 2005-8, 6.
45. Ibid., 6–7.
46. R. Francombe, "LTM—A Flexible Processing Technology for Polymer Composite Structures," paper presented at NATO Research and Technology Organization (RTO) Applied Vehicle Technology (AVT) specialists meeting on low-cost composite structures, Loen, Norway, May 7–11, 2001, accessed August 28, 2011, *http://ftp.rta.nato.int/public//PubFullText/RTO/MP/RTO-MP-069-II///MP-069(II)-$$TOC.pdf*.
47. Benson, Pawlowski-Graham interview, April 12, 2011; Graham et al., "Aerodynamic Design of SSBD," AIAA paper no. 2005-8, 3–4.
48. Joe Pawlowski to Lawrence Benson, "EXT: Re: More SSBD Questions," e-mail, August 17, 2011.

49. Benson, Graham-Pawlowski interview, April 12, 2011; Croft, "Engineering through the Sound Barrier," 30.

50. Benson, Pawlowski-Graham interview, April 12, 2001; Graham et al., "Aerodynamic Design of SSBD," AIAA paper no. 2005-8, 3–4.

51. Graham and Martin, "Aerodynamic Design and Validation of SSBD," slide no. 28.

52. Ibid., slide no. 11.

53. The latter, which serve as both a stabilizer and elevator, is sometimes referred to as a stabilator.

54. Graham et al., "Aerodynamic Design of SSBD," AIAA paper no. 2005-8, 5; Marshall Brain, "How Stereolithography 3-D Layering Works," accessed July 30, 2011, *http://computer.howstuffworks.com/stereolith.htm.*

55. Graham to Benson, "SSBD Design Question," April 4, 2011.

56. Graham and Martin, "Aerodynamic Design and Validation of SSBD," slide nos. 31–33.

57. Benson, Martin Interview, April 7, 2011; Roy Martin to G. Allen West, "Preliminary QSP SOW Definition," e-mail, June 25, 2001; Roy Martin to Joe Pawlowski, "Factors to be resolved concerning F-5E mod for QSP," e-mail, March 29, 2002.

58. Benson, Martin Interview, April 7, 2011; Martin to West, June 25, 2001.

59. Roy Martin to Joe Pawlowski et al., "Trip Report Fallon NAS," e-mail, October 26, 2001.

60. Benson, Graham-Pawlowski interview, April 12, 2011; Pawlowski to Benson, "Re: More SSBD Questions," e-mail, August 31, 2011.

61. Biography of Joseph Dyer (who later became an executive at IRobot Corp. and chair of NASA's Aerospace Safety Advisory Panel), accessed July 16, 2011, *http://investor.irobot.com/phoenix.zhtml?c=193096&p=irol-govBio&ID=195535.*

62. Pawlowski to Benson, August 31, 2011; Pawlowski to Peter Coen, "Presentation Material for SSBD PDR," e-mail, March 25, 2011.

63. Benson, Pawlowski-Graham interview, April 12, 2001; Pawlowski to Benson, August 31, 2011.

64. Joe Pawlowski to Peter Coen et al., " Good News! & SSBDWG Meeting Notice," e-mail, December 10, 2001.

65. David Graham to Peter Coen et al., "Baseline F-5E Supersonic Functional Check Flight," e-mail, January 23, 2002, quoted;

Roy Martin to Ed Haering et al., " Florida Temperature," e-mail, January 24, 2002.

66. Northrop Grumman, "SSBD Aircraft Assembly Photo-Log," April 17, 2003, slide no. 9.

67. Pawlowski to Benson, "Re: EXT: More SSBD Questions," e-mail, September 3, 2011.

68. Advanced Composites Group, "SpaceShipOne Makes Historic Space Fight," news release, October 2004, accessed ca. August 10, 2011, *http://www.advanced-composites.co.uk/aerospace_archived_news_index_pre2007.html.*

69. Joe Pawlowski to Lawrence Benson, "Re: SSBD Chapter 7 for Review," e-mail, September 29, 2011.

70. Benson, Graham-Pawlowski interview, April 12, 2011; Benson, Boccadoro interview, August 12, 2011. The three aircraft mechanics from El Segundo were Malcolm Croxton, Jack Allan, and Darrell Norwood: Boccadoro to Benson, e-mail, "Names," August 20, 2011.

71. Benson, Graham-Pawlowski interview, April 12, 2011. Source for figure 7-7: Northrop Grumman, "F-5 SSBD Aircraft Description," September 2, 2004, slide no. 4.

72. Benson, Boccadoro interview, August 20, 2011.

73. Steve Madison, Ken Ferguson, and Roy Martin, "Test Plan: Quiet Supersonic Platform Shaped Sonic Boom Demonstrator Data Collection," Revision A, March 31, 2003, Appendix C, confirmed during review of this chapter.

74. Founded in 1987 in Santa Clara, CA, Ashtech in 1997 became part of Magellan Corp., which was in turn a subsidiary of Orbital Sciences Corp. In 2001, Magellan Corp. was acquired by the Thales Group, which later sold its GPS business to an investment company that renamed it Magellan Navigation—with the Ashtech brand continuing to focus on high-precision global satellite navigation system equipment.

75. Edward A. Haering, James E. Murray, Dana D. Purifoy, David H. Graham, Keith B. Meredith, Christopher E. Asburn, and Lt. Col. Mark Stucky, "Airborne Shaped Sonic Boom Demonstration Pressure Measurements with Computational Fluid Dynamics," 43rd AIAA Aerospace Sciences Meeting, Reno, NV, January 10–13, 2005, AIAA paper no. 2005-9, 5.

76. Madison et al., "Test Plan: QSP SSBD," 5–6, Appendices B and C; Haering et al., "Airborne Sonic Boom Demonstration Pressure Measurements," AIAA paper no. 2005-9, 5; Haering to Benson, "Re: Quick SSBD Question," e-mail, August 29, 2011.

77. Benson, Pawlowski-Graham interview; Bruce King to Ed Haering et al., "Re: Urgent? Florida TM freq scheduling," e-mail, March 12, 2003, with previous message traffic inserted.

78. Steve Madison, "Shaped Sonic Boom Demo (SSBD) Test Plan Review," PowerPoint presentation, as of February 2011, slide no. 37.

79. Date stamps on digital photos showing the work at NGSA.

80. Pawlowski to Benson, e-mail, August 31, 2011.

81. Benson, Graham-Pawlowski interview, April 12, 2011; Benson, Boccadoro interview, August 20, 2011. Source for figure 7-8: "Shaped Sonic Boom Demonstration Overview," AIAA Southern California Aerospace Systems and Technology Conference, Buena Park, CA, May 21, 2005, slide no. 12.

82. Statement of Work, "Boeing T-38 Chase Aircraft Support for F-5SSBD Flight Test Program," April 17, 2003; Benson, Martin interview, April 7, 2011.

83. Northrop Grumman Integrated Systems, "SSBD/SSBE Team Roster," September 16, 2004, slide nos. 13 and 14.

84. Joe Pawlowski to Peter Coen et al., "Draft SSBD Test Plan Available for Review," e-mail, October 15, 2002; Madison, "SSBD Test Plan Review."

85. Steve Madison, Ken Ferguson, and Roy Martin, "Test Plan: Quiet Supersonic Platform Shaped Sonic Boom Demonstrator Data Collection," Revision A, March 31, 2003; Steve Madison, Roy Martin, Keith Applewhite, and Eric Vartio, "Test Plan: Quiet Supersonic Program Flight Envelope Clearance," Revision 6, May 5, 2003.

86. Benson-Martin interview; MIL-STD-1797A, "Flying Qualities of Piloted Aircraft," December 19, 1997, accessed ca. September 1, 2011, *http://www.mechanics.iei.liu.se/edu_ug/tmme50/MIL-HDBK-1797.PDF.* (The next edition, MIL-STD-1797B, was issued on February 15, 2006.)

87. Ibid.

88. COMNAVAIRSYSCOM Patuxent River MD/4.0P, to AIRTEVRON Three One China Lake CA et al., "F-5

Interim Flight Clearance for TYCOM Designated Aircraft," teletype message, 162007Z, July 2003, referencing COMNAVAIRSYSCOM 092003Z, June 2003, which was the initial flight clearance.

89. Pawlowski to Benson, e-mail, August 31, 2011.
90. Coen and Martin, "Fixing the Sound Barrier," slide no. 14.
91. Northrop Grumman, "F-5 SSBD Test Plan Review," undated, slide nos. 7–9.
92. Pawlowski to Benson, e-mail, August 31, 2011.
93. Madison, "SSBD Test Plan Review," slide no. 18; Pawlowski et al., "Origins and Overview of the SSBD," AIAA 2005-5, 10.

The Navy's F-5E, Northrop Grumman's F-5 SSBD, and Dryden's F-15B on August 29, 2003. (NASA)

CHAPTER 8

Proof at Last

Making and Measuring Softer Sonic Booms

The weather appeared favorable in the Jacksonville area on the morning of July 24, 2003, as the Northrop Grumman team prepared for the maiden flight of the F-5 SSBD. The former F-5E had not flown since the previous December, before its transformation into a sonic boom test bed. Temperatures were in the 70s, climbing toward a high in the upper 80s. Relative humidity hovered at almost 100 percent as the day began, but visibility was satisfactory, and seasonal thunderstorms were not forecast until the afternoon.[1] Before beginning this flight (and all subsequent flights), the pilot and ground personnel complied with a safety checklist and stringent "go, no-go" criteria based on an extensive hazard analysis.[2] With everything in order, the time had finally arrived for the rebuilt F-5, now unofficially nicknamed the Pelican, to take to the air.

Functional Check Flights

Roy Martin climbed into the F-5 SSBD's cockpit and gave a thumbs up to Charles Boccadoro as his unique aircraft began its taxi from Northrop Grumman's St. Augustine facility. With many of his colleagues watching intently, Martin's modified F-5 soon cleared the St. Johns County Airport's 8,000-foot runway. Serving as a chase plane, Mike Bryan's T-38 accompanied it into the air, with John Nevadomsky in the back seat. After a busy but uneventful functional check flight into a restricted flying area (W-158A) off the Atlantic coast, both planes flew inland to the western edge of Jacksonville's extended city limits. There they landed on the 12,500-foot runway of Cecil Field, which would serve as their temporary base of operations for the next several days. As predicted by the preflight simulations and analyses, the F-5 SSBD displayed slightly reduced directional stability but nothing unexpected. From Cecil Field they flew an additional FCF and envelope expansion on July 27 and another on the morning of July 28.[3] For record-keeping purposes,

Flying over Air Force Plant 42, showing Northrop Grumman's Site 3, with arrow pointed toward Building 307, and Site 4, with Building 401 in lower right corner. (NASA)

these and all subsequent flight checks and flight tests by the F-5 SSBD were assigned a sequential number starting with QSP-1. (See appendix B for a table listing all the flights.)

As with all the future flight tests, the three local sorties lasted less than an hour. The main purpose of the Florida FCFs was to confirm the plane's subsonic airworthiness, the calibration of its air-data measurements, the telemetry from its new instruments to the temporary ground station during orbits over St. Augustine, and that all systems were working properly for the upcoming cross-country trip to California. Martin also verified that the modified F-5 could still exceed the speed of sound despite the drag of its larger body by pushing it to Mach 1.1.[4] On July 28, having successfully completed the final local sortie earlier in the morning, the two planes headed west on the long ferry flight to Palmdale, CA, making four stops along their way (a journey described previously in the Introduction).[5]

Avoiding a rare summer thunderstorm building over Lancaster, the two aircraft gracefully touched down on the runway at Air Force Plant 42 in Palmdale on Tuesday afternoon, July 29. Belying its name, Plant 42 is more than a single facility; it is actually a 5,800-acre installation offering spacious and secure locations for a number of separate hangars, industrial facilities, shops, offices, and miscellaneous buildings used by Northrop Grumman, Lockheed Martin (including its Skunk Works), Boeing, and other aerospace companies and Government organizations. Northrop Grumman occupied facilities in Sites 3 and 4 (shown in the accompanying photo). The latter includes Building 401, the huge structure where Northrop had completed assembling its B-2 stealth bombers—originally, in total secrecy. The SSBD team would operate out of a section of Building 307 in Site 3.[6]

After landing, the F-5 SSBD and its T-38 escort taxied to Site 3, where they were welcomed by a maintenance crew as well as some of the Northrop Grumman engineers from El Segundo who had designed the plane and were anxious to see the finished product. Because of the threatening weather, the F-5 was quickly moved into the shelter of Building 307; Section 4 of which was reserved for the SSBD team. When Roy Martin got back to his home in Lancaster that evening, he found that the storm had uprooted large trees onto two of his nearby neighbors' homes. Such unseasonal storms were not much of a concern for the Shaped Sonic Boom Demonstration but the unusually warm summer was. Weather balloons had been measuring much higher-than-normal atmospheric temperatures, which, because of the Mach number parameters used in designing the F-5 modifications, could pose a serious threat to the SSBD's success.

Functional flight checks resumed on Saturday, August 2, with two FCFs (QSP-4 and QSP-5) flown mainly over Edwards AFB's restricted airspace. For the remainder of the SSBD, Roy Martin and Cmdr. Darryl "Spike" Long, the Navy's chief test pilot at Naval Air Weapons Station (NAWS) China Lake, CA, alternated as pilots of the F-5 SSBD. With Mike Bryan having taken Boeing's T-38 back to Seattle, a NASA Dryden F/A-18 piloted by Jim Smolka flew chase. The focus of the Palmdale functional check flights was supersonic envelope expansion. Another sortie was conducted on August 4, and after a long hiatus to address the performance problems described below, the final FCF was flown on August 15.[7]

The subsonic and supersonic envelope expansions in Florida and California validated handling qualities for 15 flight conditions with a series of exercises called maneuver blocks. Each maneuver block included pitch stick and rudder doublets (two opposing movements of the controls in quick succession), 30-degree bank-to-bank rolls, steady heading sideslips, and close-loop turns. The subsonic maneuver blocks were flown both with and without the stability augmentation system, which automatically positions the horizontal tail and rudder to dampen pitch and yaw oscillations. The system was always turned on during the supersonic maneuver blocks.[8]

In addition to their intended purpose, these flights had other benefits. As pointed out in the published summary of the SSBD, "Although the supersonic envelope expansion flights were not designed to support shaped sonic boom data collection, they proved to be invaluable as trial runs for the flight crews, the flight test ground controllers, and for the ground data crews."[9] (As described below, the sensors and other data-collection equipment were already in place.)

Unfortunately, however, concerns expressed earlier about having to delay the flight testing until summer proved well-founded. Continuing weather balloon measurements, including those from GPS radiosondes, showed ambient temperatures at the planned flight-test altitude to be about 17 °C (30.6 °F) higher than the standard atmosphere temperature for which the demonstration had been planned. (The standard temperature at 32,000 feet is about −55 °C or −48.5 °F.) The high temperature would not only hamper the F-5's ability to reach the Mach number of 1.4 for which its modifications had been designed, but it would limit its endurance as well.[10] Indeed, Roy Martin could only get the F-5 SSBD up to about Mach 1.2 on his first supersonic envelope-expansion flight on August 2.[11]

Because of this unfortunate weather pattern, members of the SSBD Working Group and others involved in the project recommended postponing the flight tests until November, when upper atmosphere temperatures would surely be more favorable. Although this made sense from a purely technical standpoint, Charles Boccadoro had to consider other issues, such as the response of corporate management. So he vetoed any further delay, telling team members, "Time is our enemy." As he later explained to the author, "To be behind schedule and over budget is true hell for a program manager."[12] In a bold move to allow the demonstration to proceed, he reached out to NAVAIR's propulsion directorate and General Electric Aviation's engine maintenance division to get permission for Northrop Grumman technicians—led by NGSA's flight-test engineer, William "W.D." Thorne—to do a compressor wash and uptrim the F-5 SSBD's jet engines for maximum thrust.[13]

Finally, at 1030 Pacific daylight time (PDT) on the morning of August 25, 2003, with no wind and the surface temperature already at 91 °F, Spike Long entered the cockpit of the Northrop Grumman F-5 SSBD for its first sonic boom flight test (QSP-8). Turning on only the left engine to conserve fuel, he taxied to Runway 25. After about two minutes of GPS data logging to acquire and confirm satellite signals, he turned on the right engine and took off at 1102. After reaching 4,400 feet, he pushed the engines into military power (maximum thrust without the use of afterburners) and quickly climbed to 25,000 feet. On the way, he was joined by NASA F/A-18B number 846 piloted by Jim Smolka as a photo chase aircraft and NASA F-15B number 836 piloted by Dana Purifoy, who was there to perform the first in-flight probes of the F-5 SSBD's shock waves.[14] Mike Thomson was in the back seat.[15] NASA Dryden had improved the F-15B's nose boom since the ISSM tests, when its Sonix digital absolute pressure transducer (which converted physical shock wave measurements to electrical signals) could take only 17 samples per second. For the SSBD, NASA Dryden added a Druck analog differential transducer that was capable of 400 samples per second.[16]

After finishing the photo opportunity, the tuneup of the F-5's engines soon proved to have been successful. Lighting his afterburners, Long climbed to 47,000 feet and, closely followed by Purifoy's F-15B, turned toward the supersonic corridor. Nine miles west of Mojave, the F-5 SSBD began accelerating in a 12-degree dive, reaching a gravity-assisted maximum speed of Mach 1.41 before leveling off at 31,500 feet. It led the F-15B through the probing run at Mach 1.38 despite an outside air temperature of only −38.2 °F (−39 °C).[17] This speed was within the margins required for adequate flight tests. Charles Boccadoro was in an office with Tom Weir, NGC's director of advanced design, when they received word that the plane had "hit its mark."[18] He immediately called Steve Walker at DARPA to report the good news.

The high-fidelity Ashtech GPS tracking system also proved its capabilities during the flight test with relative positioning between the F-5 and the F-15 being displayed to within mere centimeters both in the two aircraft and down at the Dryden control room. Long's postflight report commented somewhat sardonically that "the painfully precise NASA engineers did request several redo's. I figure Dana was off by a foot or two from time to time."[19] Sixty-five minutes after climbing into the cockpit, Spike Long's feet were back on the ground. Later in the day another Navy pilot, Lt. Cmdr. Dwight "Tricky" Dick of Composite Fighter Squadron (VFC)-13, arrived at Palmdale from NAS Fallon with his camouflaged F-5E. All was in place for SSBD's next flight test—the one that would compare the two F-5s' sonic booms as they reached the ground.

Preparations at Edwards Air Force Base

During the months that Northrop Grumman was fabricating the new parts in El Segundo, then modifying and preparing the Shaped Sonic Boom Demonstrator in St. Augustine, members of the working group and additional specialists at NASA's Dryden and Langley Centers were planning and preparing for the Shaped Sonic Boom Demonstration at Edwards AFB. The basic plan was to measure the sonic boom of the standard F-5E from Fallon followed less than a minute thereafter by that from the F-5 SSBD to compare the initial overpressures and loudness of the two sonic booms in the same basic atmospheric conditions. Since this would be the first ever opportunity to show the effects of sonic boom shaping, the participants wanted to employ the latest in pressure measuring and acoustic recording technology both on the ground and in the air. To take full advantage of the opportunity presented by the demonstration flights, NASA's Dryden and Langley Centers as well as Wyle Laboratories made plans for an extensive number of ground sensors.[20] Implementing this proved significantly more complicated than simply going out into the desert and setting up the equipment.

With Ed Haering as the lead planner, the working group's sonic boom specialists determined the best area under the Edwards high-altitude supersonic corridor to place the array of sensors. This would in turn determine the exact route to be flown by the aircraft. They chose to use Cords Road, which ran east to west under the base's R-2515 restricted airspace, as the centerline for the aircraft's flightpath (shown in figure 8-1).[21] The land along Cords Road is divided into a checkerboard pattern, either under the jurisdiction of the Bureau of Land Management (BLM) or privately owned. Because the private owners were difficult to contact, the planners limited their sensor sites to BLM sections. Based on maps and aerial photographs, they selected the specific locations on which to distribute sensors in the desired configuration while keeping them within a reasonable walking distance of existing roads. The sites selected were on either side of a 6-mile stretch of Cords Road just north of Harper Dry Lake, which is located about 30 miles north of California Highway 58 between Barstow and Boron.[22] It was expected that this road, the lakebed, and other local landmarks would be readily visible to the pilots flying 32,000 feet above.[23]

Arranging to use the area for the SSBD proved to be a long and complicated coordination process for Ed Haering and some colleagues at NASA Dryden. Mostly unoccupied by humans, the area is the habitat of several threatened or endangered species, most notably the desert tortoise and Mojave ground squirrel. Starting in the fall of 2002, Haering and Mike Beck, the environmental specialist at NASA Dryden, spent a great deal of time working with the BLM office in Barstow and the Edwards AFB environmental management division to get permits and meet other requirements to use the land. For example, they had to arrange tortoise protection training for all the personnel who would be working on the sensor array, which was eventually accomplished with the help of a video recording of the training. A BLM-approved biologist (paid for out of project funds) was required to oversee the site selections as well as the setup and removal of

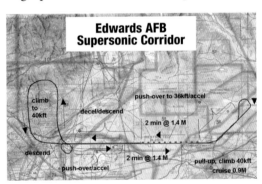

Figure 8-1. Map of SSBD flight plan. (NASA)

equipment. Vehicles were prohibited from driving off the existing dirt roads, and when desert tortoises were encountered on the roads (as happened on several occasions), they could not be picked up and moved out of the way.[24]

To provide spatial sampling along the center of the flightpath, plus bracketing to the side, the SSBD ground sensor team set up their equipment in and around three major sites adjacent to Cords Road—designated West, Center, and East—and at Sites North and South, 2 and 3 miles from the road. The sensors around the central site were arranged in a cruciform pattern. To gain an extra boost in speed from prevailing winds and avoid having the sonic booms reach Boron and communities along U.S. Highway 395, the working group chose to have the planes fly from west to east for the measurements.[25] (This is the reverse of the direction flown on a similar route by Chuck Yeager during the XS-1's first supersonic flight.)

In addition to the pressure sensors, microphones, and associated electronic devices, the equipment positioned out in the desert included specially equipped vehicles, generators, battery packs, cables, and portable toilets. Since these items would be left unattended in the isolated area for much of the time, the testers had to hire armed guards to watch over them. As NASA Dryden's security contractor explained, "equipment left alone for a few days would probably be stripped clean."[26]

By the time the F-5 SSBD landed at Palmdale, the ground array had been laid out and its equipment made ready for use. NASA Langley had deployed an acoustic instrumentation van with ultrasensitive Brül and Kjaer (B&K) 4193 low-frequency condenser microphones and supporting equipment. Among the personnel from Langley was Christine Darden, whose minimization research had helped pave the way for designing the SSBD. Wyle Labs provided similar sensor capabilities using identical microphones feeding into advanced TEAC and National Instruments recording systems. NASA Dryden contributed two Boom Amplitude and Direction Sensor (BADS) systems and other instruments. Designed by James Murray, David Berger, and Ed Haering, each BADS system (shown in an accompanying photo) included six Sensym differential pressure transducers attached about 6 feet apart to an octahedron-shaped framework. The separation of the sensors would help determine the angles of incoming acoustic rays. With its sophisticated pneumatic and electronic instrumentation, the BADS system could measure pressure changes to within plus or minus .003 psf at more than 8,000 times per second.[27]

In addition to the members of the three organizations mentioned above, Northrop Grumman, Lockheed Martin, Boeing, and Gulfstream personnel also helped set up and monitor the stations. Because of the delayed start of the demonstration and another commitment, the Langley Acoustics Division's van had to depart after the initial flight test of August 25, which left Site North

Boom Amplitude and Direction Sensor near Cords Road, August 2003. (NASA)

unused for the remainder of the SSBD.[28] Other than that, the ground array network functioned largely as planned for the rest of the flight tests.

The same cannot be said for another highly anticipated means of measuring the shock waves. In addition to the near-field probing by the F-15B and the far-field measurements by the ground sensors, the SSBD test plan called for two midfield probes. To expand on prior in-flight sonic boom experiments, Ed Haering had got the working group to include provisions for recording the F-5 SSBD's shock wave signature as it evolved through the atmosphere. In August 2002, he began working with Gulfstream and Raytheon to obtain executive-type jets for this purpose.[29] The companies agreed to supply a Gulfstream G-V and Raytheon Premier I. NASA Dryden equipped each of them with one of its Small Airborne Boom Event Recorder (SABER) devices, which involved installation of two static pressure ports on top of their fuselages. Dryden would also loan an Ashtech Z-12 to Raytheon to provide the Premier I with accurate position and velocity data. (The Gulfstream G-V was already equipped with this GPS system.)

Including these two civilian aircraft in the SSBD flight tests required many months of coordination and logistical preparations. The test plan called for each jet to fly below the F-5 SSBD at subsonic speeds, one at levels between 3,000 feet and 10,000 feet and the other at levels between 15,000 feet and 12,000 feet. Unfortunately, acoustic analyses eventually confirmed that aerodynamic noise from the two planes could adversely affect the microphones. Problems were also encountered with mounting the instrumentation. As a result, participation by the two corporate jets had to be canceled before the demonstration.[30]

Making Aviation History

After Spike Long's successful probing flight on August 25, the SSBD team was anxious to make the long-awaited ground measurements of the shaped and unshaped sonic booms, but the potential for thunderstorms on August 26 forced postponement of the flight. The weather forecast was better for the next day, so in the wee hours of August 27, the members of the ground recording team made their way through the desert to the remote equipment sites north of Harper Dry Lake.[31] With only the Milky Way lighting the dark sky overhead, the ground crews looked for their headlights to illuminate reflectors on sticks marking sensor locations and watched out for any desert tortoises on the roads. David Graham and three passengers had to stop and wait patiently for about 15 minutes until a tortoise got out of their way.[32] The crews needed to set up and check their equipment before dawn, since the flight test was scheduled to occur shortly thereafter when the layer of air over the desert floor was usually at its calmest. In the NGC control center at Plant 42, Ken Plotkin was examining the latest readings from weather balloons and reports from the field. When he was sure that atmospheric conditions were acceptable for undistorted sonic boom signatures to reach the sensors—with no turbulence and a slight thermal inversion near the surface and acceptable winds aloft—Plotkin gave the go-ahead for the launch.[33] At Edwards AFB, the NASA Dryden control room got ready to monitor the flights.

At 0626 local time, just as the Sun was rising above the distant horizon,[34] Roy Martin in the F-5 SSBD (call sign Leahi 05) and Lt. Cmdr. Dick in the Navy F-5E (Leahi 06) took off from the long runway of Air Force Plant 42 on flight number QSP-9. They flew into a clear sky with only a few clouds at 12,000 feet and compared the readings on their altimeters and Mach meters as they flew north. Twenty minutes later, when they reached 45,000 feet, Dick did a gradual S maneuver to get the required 45 seconds behind Martin for the sonic boom run. The original test plan had called for the F-5 SSBD to follow the standard F-5B, but because of the difficulty the enlarged F-5 could have in keeping up with the sleeker F-5E in the higher-than-normal temperatures, their places had been reversed.[35] Cruising at Mach 0.9 over the Mojave Airport at 0644, Martin lit his afterburners, began a 12-degree dive, and accelerated to a maximum speed of Mach 1.38. Because a thick surface haze prevented Martin from seeing Cords Road or other landmarks while flying into the rising sun, he had to rely on his hand-held GPS unit to navigate through the supersonic corridor on the prescribed course. After leveling off at 32,000 feet, the F-5 SSBD was able to maintain a steady Mach 1.36 (about 920 mph) from 14 miles west of the recording array until after passing over it at about 0646. Flying at the same speed, Dick followed by the desired 45 seconds. Returning

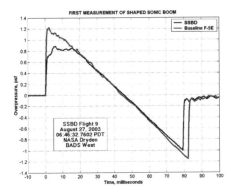

Figure 8-2. F-5E and F-5 SSBD sonic boom signatures. (NASA)

to Palmdale, Martin touched down on Plant 42's runway at 0703 and Dick 1 minute later. After taxiing back to NGC Site 3 and shutting down their engines at 0722, the two pilots joined other team personnel to learn more about the measurements of their sonic booms.[36]

Working at one of the sensor sites, David Graham thought the second boom sounded distinctly louder than the first.[37] Data displayed on the screen of a laptop computer quickly confirmed the difference. It showed the sonic boom generated by Dick's standard F-5E had an initial pressure rise of 1.2 psf while that from Martin's modified aircraft registered only 0.82 psf. Of equal significance, a blue line plotting the F-5 SSBD's pressure measurements showed that its signature reached the ground with the predicted flattened shape. Overlaying the blue line was a red one showing the F-5E's typical N-wave. These lines closely matched the red and blue pinstripes that so hopefully decorated the Pelican's fuselage. Although reproduced in only black and white on the printed page, figure 8-2 shows this historic sonic boom exactly as recorded by the BADS unit at Site West—the first sensor ever to measure a shaped sonic boom.[38]

When these results reached the control center at Plant 42, Ken Plotkin did a little victory dance. He then called Albert George at Cornell University, interrupting a counseling session with one his students, to inform him that the flight test had finally confirmed his and Seebass's theory beyond all doubt. "I knew it would," the professor replied. When the image had been received at the NASA Dryden control room, Peter Coen jumped up from his seat and announced, "It's a classic flattop!"[39]

The original SSBD test plan had called for eight sonic boom data-collection sorties with a break for in-depth data analysis after the first four.[40] The delays in starting the flight tests, the cancellation of the Gulfstream and Raytheon probe missions, deteriorating weather conditions, and other factors reduced the number of data-collection flights to five. After a quick turnaround of the F-5 SSBD, Spike Long took off on the third of these (QSP-10) at 0905 PDT along with Tricky Dick in the F-5E. After reaching the vicinity of Lake Isabella, Long climbed to 47,000 feet while Dick, reaching 45,000 feet, maneuvered his plane to follow by the required interval. Long hit a maximum speed of Mach 1.41 as he descended through a milky haze into the flightpath—the fastest flight yet. He was able to maintain a steady Mach 1.38 at 31,800 feet for the sonic boom run at 0930. Although he flew about one-half mile north of Cords Road because of overcorrecting for high-altitude crosswinds, the sensor array recorded the sonic booms satisfactorily. Surface winds, however, were now too high for signatures as good as on the early morning flight. Dick's F-5E followed by 45 seconds with similar results. Long landed at Plant 42 at 0946, 2 minutes before Dick touched down. In his postflight report, Long wrote that "Tricky from VFC-13 has been an excellent addition to the test team" and that "we should make this guy a test pilot now!"[41]

Roy Martin and Dwight Dick flew a very similar sortie (QSP-11) from 0620 to 0653 on the following morning, which provided the ground observers with more sonic boom data.[42] The plots of the sonic boom signatures remained fairly consistent, confirming that the first recordings on the morning of August 27 were no fluke. However, the quality of the ground data continued to decline after that flight because of less favorable atmospheric conditions.[43]

The last data-collection flight on August 29 (QSP-12) focused on in-flight probing combined with more ground data collection. Spike Long in the F-5 SSBD and Tricky Dick in the F-5E took off from Palmdale shortly after 0830 local time. They joined up with Dana Purifoy and Mike Thomson in NASA F-15B number 836 and NASA F/A-18B number 846 piloted by Dick Ewers. As the aircraft climbed to altitude, Carla Thomas in the back seat of the F/A-18B took photos of the other three flying in formation (including the one in front of this chapter). Dick's F-5E then broke away and headed home to Fallon while Long and Purifoy got into position northeast of Lake Isabella before turning and descending for their second near-field probing run through the supersonic corridor. Because the temperature at 32,000 feet was even 10 degrees warmer (at −29 °F) than earlier in the week, Long could reach only Mach 1.38 during his shallow dive. He had to keep gradually descending just to maintain Mach 1.34 during the rest of the more than 80-mile probing run, which included the segment over the ground array. After completing his mission at 0920, Long reiterated one of the major lessons learned during the flight tests: "The sun angle was the biggest detriment during data collection and course management was more luck than skill. If future events are planned we will reverse the run-in course to avoid this scenario."[44] After Long's departure for Palmdale, Ewers's F/A-18B rejoined Purifoy's F-15B in the supersonic corridor for some NASA Dryden midfield probing unrelated to the SSBD project. Because of the less-than-favorable weather, QSP-12 was the final SSBD flight test.

After Northrop Grumman posted a combined news release with DARPA and NASA on August 28 announcing the previous day's successful demonstration, what had once been a rather obscure project, mainly of interest to those involved in aeronautics or the aviation industry, quickly received wider publicity.[45] On September 3, the NGC, DARPA, and NASA officials hosted a briefing at the National Press Club in Washington, DC, with representatives from the general media as well as the aviation press attending.[46] Charles Boccadoro presented a graphically rich slide show on Northrop Grumman's participation in the QSP and the SSBD.[47] He also relayed an anecdote about first learning of the result. "About 20 minutes [after takeoff], we were all listening to our chief test pilot Roy Martin over the radio when the guys in the desert... called back and said they could hear the difference. We weren't expecting that. We knew we had something at that point."[48] With the attendees having seen the graphic depiction of the sonic boom signatures in Boccadoro's presentation, Richard Wlezien, now NASA's vehicle systems program manager in the Office of Aerospace Technology, stated that "the results were about as unambiguous as you could get."[49] Added Peter Coen: "The team was confident that the SSBD design would work, but field measurements of sonic booms are notoriously difficult.... [So] we were all blown away by the clarity of what we measured." To put the achievement in its historical context, Wlezien also noted, "This demonstration is the culmination of 40 years of work by visionary engineers."[50]

In addition to numerous accounts of the achievement in aeronautics and aviation publications, a flurry of articles and reports began appearing in the general media, ranging from a short wire story by the Associated Press published in newspapers nationwide to two televised reports several minutes in length on CNBC showing video of the aircraft interspersed with interviews of Roy Martin, David Graham, and Joe Pawlowski.[51] Northrop Grumman was now reaping a harvest of favorable publicity as an intangible return on its investment in sonic boom research. Some of the corporation's major contributors to the SSBD are shown in the accompanying photo, taken during a company party in El Segundo on October 14, 2003, while celebrating the project's success.

The timing of the SSBD project's success was fortuitous. In May 2003, the FAA—citing the findings by the National Research Council that there were no insurmountable obstacles to building a quiet small supersonic aircraft—had begun seeking comments on its noise standards in advance of a technical workshop on the issue.[52] The noise standards included Part 91 of Title 14 of the Code of Federal Regulations (CFR), which prohibited supersonic flight over the United States. On November 30, 2003, in Arlington, VA, the FAA held the Civil Supersonic Aircraft Technical Workshop. It allowed subject matter experts to submit comments on recent supersonic research data and present their findings on the mitigation of environmental impacts, as well as to inform the public. In response, the Aerospace Industries Association, the General Aviation Manufactures Association, and most aircraft companies reported that the FAA's sonic boom prohibition was still the most serious impediment to creating the market for a supersonic business jet.[53]

Key members of Northrop Grumman's QSP/SSBD team. Front row: Charles Boccadoro, Joe Pawlowski, David Graham, and Roy Martin. Rear row: Steve Komadina, Mark Smith, and Paul Meyer, who was vice president of advanced systems at the time. (NGC)

The major aircraft and engine companies participating in the QSP all made presentations on addressing the sonic boom and jet-noise problems. Among them, Steve Komadina and David Graham discussed the promising results of their QSP research and SSBD flight tests.[54] Complementing cases made by Gulfstream and other manufacturers for a supersonic business jet, Richard G. Smith III of Berkshire Hathaway's NetJets provided an analysis of the potential SSBJ market and suggested the need for a comprehensive public-private risk-sharing consortium.[55] On the all-important issue of human response to sonic booms, Peter Coen and Brenda Sullivan updated attendees on NASA Langley's latest analyses and its reconditioned boom-simulator booth. Building on the recent progress in taming the sonic boom, Coen outlined planned initiatives in NASA's Supersonic Vehicles Technology program. In addition to leveraging the results of QSP research, NASA hoped to engage industry partners in planning follow-on projects involving critical supersonic technologies. Of special relevance to the FAA workshop, NASA was actively considering options for an experimental low-boom aircraft that could fly over populated areas for the sake of definitive surveys on the public's response to reduced sonic signatures both outside and indoors.[56] Meanwhile, NASA was preparing to use the only existing reduced-boom airplane for more research into the physics of sonic booms and how to control them.

Preparing for the Shaped Sonic Boom Experiment

Although the Shaped Sonic Boom Demonstration had been extremely successful as a proof of concept—more than meeting the objectives of DARPA's contract with Northrop Grumman—the unfavorable high-altitude temperatures, limited number and scope of the flight tests, relatively quiet conditions in the lower atmosphere, and less-than-ideal functioning of some of the instrumentation left many of the participants hungry for more sorties and more data. Systematically gathering and analyzing flight-test data had always been a hallmark of NASA's aeronautics mission and, before that, of the NACA's. Among an extensive list of SSBD lessons learned compiled by one of the NASA researchers was this rather tongue-in-cheek entry: "If at first you succeed (8/27/03), you might be cancelled. Hide your success until you get a reasonable amount of data."[57]

Another lesson learned from the SSBD is that a new technology program "will cost twice your initial estimate."[58] With the SSBD venture having lasted almost a year longer than first anticipated and the QSP program having cost Northrop Grumman more than $4 million in discretionary funds, its corporate management was not eager to spend even more money to extend the project purely for the sake of scientific research.[59] And although the modified F-5E was a unique platform for performing such research, it was existing on borrowed time because of the requirement to return the plane to the Navy. So even as the demonstration was underway, Peter Coen was working to get NASA to sponsor about 20 more flight tests in cooler weather, when the F-5 SSBD could reach Mach 1.4 and stronger surface winds would be likely for gathering a wider range of data.[60] He was successful in obtaining approximately $1 million from NASA Aeronautics's Vehicle Systems Division to reimburse contractors (especially Northrop Grumman) and cover some additional expenses for a follow-on test program. Using NASA funding, the Air Force Research Laboratory managed an Air Vehicles Technology Integration Program (AVTIP) contract with Northrop Grumman to conduct SSBE flight operations. Meanwhile, NAVAIR agreed to amend the CRADA to allow the additional flight tests and extend its loan of the F-5 SSBD to Northrop Grumman until the end of January 2004.[61]

To reflect the follow-on testing's scientific purpose while indicating continuity with the SSBD, NASA named it the Shaped Sonic Boom Experiment. NASA also continued the Government-industry management structure successfully used for the SSBD by retaining most of its working group. The 26 members of the SSBE Working Group (listed in appendix A) included representatives from the Langley and Dryden Centers, the AFRL, Northrop Grumman, Wyle Laboratories, Eagle Aeronautics, Gulfstream Aerospace, Raytheon Aircraft, Boeing, and Lockheed Martin. As SSBE program manager, Peter Coen served as the group's chairman, with Joe Pawlowski managing Northrop Grumman's lead role in the flight operations.[62]

The SSBE working group convened in person in Arlington, VA, on November 13, 2003—just after the FAA civil supersonic workshop earlier in the day—to establish data requirements and review a preliminary test plan. Once again, it called for back-to-back flights with the F-5 SSBD and a standard F-5E as well as near-field probing by the F-15B. Based on lessons learned in the previous tests and the desire for more data, the planners also made several changes. In view of the poor visibility experienced by the pilots in August, the primary flightpath in the new test plan would run from east to west—away from the morning Sun. In addition to measuring signatures at the SSBD's original design speed of Mach 1.4 at 32,000 feet above sea level, the 20 planned flight tests afforded the opportunity to fly some of the missions at both higher and lower speeds and altitudes for evaluating sonic boom characteristics under off-design conditions. The planners added two entirely new flight-test profiles to the schedule: having the F-5 SSBD generate a focused boom and allowing the two F-5s to fly over the ground array in close formation. In case of poor weather or unforeseen problems, the working group ranked the 20 planned data missions in their order of priority in case some of them had to be canceled because of weather or other problems. Yet for some missions, the planners hoped to measure the effects of turbulence on the sonic boom signatures.[63]

Reflecting other lessons learned, there were also some major changes in data-collection arrangements and capabilities. Because of the long drives required for the ground crews to reach the area near Harper Lake and the need to guard the equipment left there during off-duty hours, the working group agreed to set up ground instrumentation at a less remote location. They decided to use a quiet section within the perimeter of Edwards AFB known as North Base—located across Rogers Dry Lake from the major Air Force and NASA facilities—where security would not be a significant issue. Based on findings during the SSBD in August, the planners also agreed to concentrate most of the monitoring equipment into a smaller area. And to compensate for the inability to use corporate jets for in-flight measurements, the working group approved a Dryden proposal to have a Czech-made Blanik sailplane from the USAF Test Pilot School (TPS)—an almost perfectly quiet aircraft—measure midfield shock waves above the ground turbulence level.[64] Its role would be similar to that of the YO-3A light aircraft used during the SR-71's probing by the F-16XL in 1995 (described in chapter 4) and the Goodyear blimp during the National Sonic Boom Evaluation in 1966 and 1967 (chapter 1).

Preparations to implement this ambitious agenda continued at a fast pace during the remainder of 2003, with NAVAIR reviewing the test plan (including the new pushover maneuver needed to focus a sonic boom) and VFC-13 formally agreeing to once again deploy one of its F-5Es to Palmdale if at all possible. Representatives from NASA Dryden, Northrop Grumman, and Lockheed Martin conducted an initial flight-readiness review on December 6. Because the F-5 SSBD had not flown since August, Northrop Grumman technicians had to perform multiple inspections and maintenance actions to make sure it would be ready for another round of flight tests. By Friday, January 9, 2004, preparations were complete—with the test plan having been approved, the F-5 SSBD certified for flight, the ground array being put in place, and recording instrumentation installed on the glider. The 20 planned flight tests would need to be conducted within the next 2 weeks to allow time for flying the F-5 back to St. Augustine and preparing it for return to the Navy by January 31.[65]

As shown in figure 8-3, the ground recording array included 26 sites (conveniently designated Alpha through Zulu) placed 500 feet apart in a straight line through desert terrain paralleling the North Base runway at a heading of 240 degrees magnetic. Two supplemental sites (M2 and Q2) were located 100 feet from Sites Mike and Quebec for fine-scale sampling, and two lateral sites (Mike Left and Mike Right) were placed 500 feet on either side of Mike to ensure the measurement of maximum overpressures in case the aircraft flew slightly off track. There were also Far North and Far South sites equipped with automatic sensors 2 miles from the main array to determine whether the F-5 SSBD's flattop signature would also persist off to the sides of the sonic boom carpet. To monitor surface conditions, a portable weather station was deployed between sites M and M2.[66]

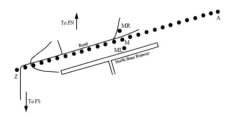

Figure 8-3. SSBE ground array (NASA). Key: MR=Mike Right, ML=Mike Left, FN=Far North, FS=Far South.

With the approval of the Edwards AFB Environmental Management Division, the ground crew set up a total of 42 different sensors under and to the side of the designated flightpath. Using postprocessed carrier-phase Differential GPS (DGPS) measurements, the sensor location for each site was surveyed to within just 0.03 meters.[67] Sites I through X, ML, MR, and Q2 were equipped with B&K Type 4193 low-frequency condenser microphones and associated amplification, power supplies, and processing equipment. On sites I through L, the microphones fed into a TEAC RD-145T digital audio tape (DAT) recorder. On sites M, ML, MR, and N, the microphones were connected to a laptop computer, with a special National Instruments data acquisition (DAQ) card installed, which offered immediate viewing of sonic boom signatures. The microphones on sites O through X used two DAT PC208AX

 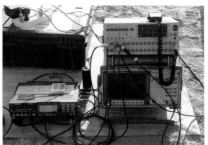

B&K microphone and recording equipment. (NASA)

DAT recorders. (A microphone and recording equipment are shown in accompanying photos.)[68]

The sites at either end of the array (A through H and Y through Z) as well as M2, FN, and FS were equipped with Dryden-built automatic sonic boom recording systems. These consisted of the previously described BADS used in the SSBD, the SABER system designed for mounting on aircraft, and another Dryden device called the Boom Amplitude and Shape Sensor (the BASS-o-matic), a completely autonomous single-transducer version of the BADS. Dryden produced 10 of the latter just in time for the SSBE. All three systems used Sensym SCXL004DN pressure sensors as their basic transducer element with those for the smaller BASS and SABER systems affixed to a flat plate placed on the ground. After processing with filters, preamplifiers, buffers, and computer software, the data collected was saved on flash memory cards. The three NASA systems tagged the exact times of the sonic boom measurements with built-in GPS receivers, which were later used to interpolate the exact times of the data captured by the B&K microphones in the TEAC, Sony, and National Instruments recordings.[69]

Personnel from NASA Dryden, NASA Langley, Northrop Grumman, Wyle Labs, Eagle Aeronautics, Gulfstream Aerospace, Boeing, and the FAA (all with desert tortoise protection training) set up and operated the ground array. Although its location was more convenient than the remote array along Cords Road, the ground crews once again had to arrive before dawn to set up and check out the equipment for morning flights. They then monitored radio communications from the approaching aircraft to learn when the supersonic sprints toward their ground array would start.

NASA Dryden contracted with the USAF Test Pilot School to install B&K 4193 microphones and a SABER recording system on the two-seat L-23 Super Blanik sailplane (shown in an accompanying photo) and make midfield measurements of the sonic booms over the ground array as the F-5s flew overhead. They fastened the microphone on the left wingtip with a bullet-nose

Blanik L-23 sailplane with microphone on left wingtip and installed recording equipment. (NASA)

attachment and windscreen to minimize wind noise during flight. A B&K Nexus amplifier raised the voltage level to improve the signal-to-noise ratio for recording on the SABER, which automatically sensed the arrival of the shock waves. The testers also installed a Thales Navigation Z-Xtreme carrier-phase differential GPS receiver for position and velocity data. A handheld GPS unit with a moving map display was also provided to help the pilot fly a precise route under the F-5s' flightpath.[70]

The in-flight probing system on NASA F-15B Number 836 (depicted in figure 8-4) was further improved from that used during the SSBD in August. NASA Dryden installed a new data recorder that lowered transducer noise and more than doubled the analog data-sampling rate to 977 times per second. The pneumatic reference tank (also called a lag tank), which was part of a complicated system used to ensure the accuracy of the pressure measurements made by the nose boom, was also replaced. During the SSBD, Dryden had

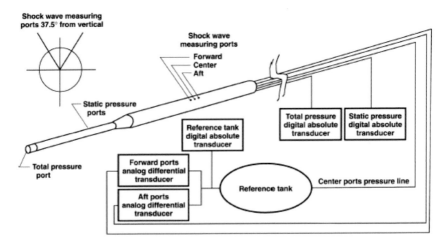

Figure 8-4. Schematic of F-15B nose boom probing system. (NASA)

relied on a spare 252-cubic-inch SR-71 nitrogen tank to perform this function. However, when the F-15B dived from over 45,000 feet down to below 32,000 feet for the supersonic runs, the reference tank saturated the analog differential transducer readings for almost all the probes. This left only the digital transducer at 17 samples per second to provide good data. To correct this deficiency, Dryden technicians fabricated a new 100-cubic-inch reference tank, which would work very well during the SSBE.[71]

Conducting the SSBE: Twenty-One Flights in 10 Days

With all preparations in place and everyone involved eager for the tests to get under way, flight operations began on Monday morning, January 12, 2004. As during the SSBD, Cmdr. Darryl "Spike" Long from China Lake had been scheduled to alternate with NGC's Roy Martin as the F-5 SSBD's pilot. Unfortunately, Long came down with a stubborn head cold just before test start and had to remove himself from flight status for the entire span of the SSBE. This meant Martin would get in a lot of flying time during the next 2 weeks, performing all 21 SSBE missions plus 4 more sorties to return the plane to St. Augustine. Once again, Lt. Cmdr. Dwight "Tricky" Dick from Fallon would fly the baseline F-5E, completing eight missions from January 13 to January 15 during its deployment to Palmdale. Dana Purifoy would pilot NASA F-15B number 836, performing four successful probe missions on January 21 and January 22. Jim Smolka and Dick Ewers would fly chase in NASA F/A-18 number 846 during three missions on January 12 and January 14. The Blanik L-23 sailplane would be flown by three Air Force officers from the USAF Test Pilot School: Lt. Col. Mark "Forger" Stucky, Lt. Col. Robert "Critter" Malacrida, and Maj. Vince "Opus" Sei. Civilian employee Gary Aldrich towed them to altitude in a Pawnee Pa-25, releasing the glider at about 10,000 feet. The two TPS aircraft flew on each of the first 11 SSBE missions from January 12 to January 16 and three more times on January 19.[72] To obtain the best possible data, they would need to launch and navigate the glider to be over the ground array and under the sonic boom shock waves at just the right times and locations.

After getting into the F-5 SSBD's cockpit at 0900 Pacific Standard Time (PST) on January 12 and carrying out all the preflight chores, Roy Martin (still using the call sign Leahi 05) took off 40 minutes later. On this and most subsequent SSBE missions, Martin flew a fairly standard route and profile to get to the ground array. He went north through Edwards AFB restricted airspace to the vicinity of Koehn Lake and then far enough east (often near Goldstone on later missions) to make a right turn toward the southwest and descend from 40,000 feet to begin a run through the supersonic corridor about 20 miles north of Barstow. He continued supersonically on a west-southwest heading over Edwards North Base and the northern edge of Rogers Dry Lake then slowed to a subsonic speed before reaching Mojave.[73]

Although the initial sortie's primary purpose was to serve as a functional check flight, it began with NASA F/A-18 number 846 joining up with the F-5 SSBD at 1055 for an attempt to take some schlieren photography of its shock waves during the supersonic segment of the flight test. As Martin's F-5 approached the ground array after reaching the desired Mach 1.4 speed at 32,000 feet, he did a 0.5-g pushover maneuver for 3 seconds to practice creating a focused sonic boom before recovering at 30,800 feet and Mach 1.38. Meanwhile, the L-23 (call sign Cobra 77) glided quietly below. After passing over Mojave, Martin turned back toward Edwards and did a low-speed run by its control tower at only 100 above ground level to help confirm the accuracy of the F-5 SSBD's air-data system. Proceeding south, the Pelican was back on the runway at Palmdale at 1031 local time after a busy 50 minutes in the air.

Wasting no time, Martin was back in the F-5's cockpit at 1230 local time to begin preparing for the next sortie, which was supposed to include in-flight probing by the NASA F-15B 836. After waiting for the launch of Dana Purifoy's F-15 and an airborne pickup (establishing the communications link between the two aircraft) at 1316, Martin took off to join up with the F-15. Just as the two aircraft began their supersonic run toward the ground array at 1331 PST, one of the F-15's engines suffered a compressor stall, and Purifoy had to drop out of the formation. Martin continued over the ground array at Mach 1.4 and 32,000 feet with the L-23 again on station below. He then began a subsonic descent over California City, followed by another low-level flyby of the Edwards tower to check air-data calibrations. Meanwhile, Purifoy had seen indications of a landing gear problem with his F-15B, so Martin—with sufficient jet fuel still remaining—flew around to perform a visual inspection to ensure its wheels were down. After the F-15 was safely on the ground, Martin returned to Palmdale, landing at 1401 PST.[74] Fortunately, the SSBE's schedule afforded plenty of opportunities to perform in-flight probing toward the end of the flight-test schedule, and the old F-15's engines would be in better condition by then.

The ambitious schedule for January 13 called for three flight tests (QSP-15–17). With Lt. Cmdr. Dwight Dick having arrived from Fallon on the previous day, all of them featured back-to-back sonic boom measurements of the F-5 SSBD and standard F-5E (Leahi-06) at Mach 1.4 and 32,000 feet. He and Roy Martin took off on the day's first flight test at 0656 PST, just as the Sun was rising. All went according to plan as they flew over the ground array and the glider 45 seconds apart. After landing at Palmdale at 0730, however, they learned that field personnel had reported a malfunction in Martin's GPS telemetry. At 0930, after the Ashtech Z-12's software had been rebooted, Martin got back in the cockpit for another sonic boom run with Dick. They flew this one almost identically with the previous test with the L-23 again participating. This time Martin's GPS transmissions seemed okay, but those from Dick's F-5E had experienced the same problem as the F-5 SSBD's signals had on the previous flight, and therefore, it too had to be reprogrammed. The third flight test of the day, which lasted from 1300 to 1333, flew the same scenario as the first two, except this time both GPS systems behaved themselves, much to the relief of the data collectors.[75]

The same pilots flew the same basic course twice again on January 14 (QSP-18 and 19), but this time to generate sonic booms at off-design speeds for data-analysis purposes. After a 0957 takeoff on the first flight, they entered the supersonic corridor from 45,000 feet and accelerated to Mach 1.43 for a data-collection run at 32,000 feet over the ground array and the glider. The day's second flight, which began at 1327, followed the same profile but crossed the ground array at only Mach 1.35. Both planes were back on the Palmdale runway by 1400.[76]

To take full advantage of the F-5E's final day at Palmdale, the F-5 SSBD flew three more back-to-back missions with it on January 15. Martin and Dick took off on the first flight at 0655. At 0709, just 2 minutes before beginning the pushover from 45,000 feet toward the ground array, the control center informed Martin that his GPS modem was once again failing to transmit. He continued on the flightpath, crossing over the ground array and the glider, flying Mach 1.43 at 32,000 feet. He was followed by Dick's F-5E maintaining the same speed, with both planes back on Plant 43's runway by 0727. The Ashtech Z-12 was reprogrammed in time for the next flight, which took off at 0957. Martin and Dick flew almost the same profile as during the early morning flight except for doing the sonic boom run at only Mach 1.35.[77]

A published paper on the SSBE later commented, "While onboard-recorded data tends to be cleaner than telemetered data, failure in the onboard recording system could result in loss of mission. Having both onboard and telemetered data gives redundancy for data collection, which saved several flights in this program."[78]

The test scenario for QSP-22, which took off at 1257 on the afternoon of January 15, was quite different from any of the previous missions involving both F-5s. The primary purpose of this test was to eliminate any effects of changing winds during the interval between the two signatures by measuring them almost simultaneously. To do this, the F-5E was supposed to fly in the F-5 SSBD's Mach cone, within 580 feet behind and about 19 degrees below, as they passed over the ground array at Mach 1.35 and 32,000 feet. When the pilots returned to Palmdale at 1329, they learned that the F-5E's GPS modem stopped transmitting from the time it began its pushover until just after passing Site Mike.[79] In this case, the Z-12's internal memory preserved the GPS data. The sonic boom measurements on the ground (described in the next section) would prove to be somewhat disappointing for another reason. The F-5E had arrived with its centerline pylon still attached. Because of the extra drag, Dick's airplane was out-accelerated by the F-5 SSBD at the start of the supersonic run and was unable to catch up and follow as closely as planned.[80]

With the F-5E having returned home to Fallon, flight-test operations on January 16 were the lightest of the SSBE with only one sortie in the afternoon. After a low-speed taxi to make sure that a problem discovered with the nose-wheel actuator had been fixed, Roy Martin took off in the F-5 SSBD at 1503. He then made an uneventful solo run at Mach 1.35 and 32,000 feet over the ground array and the glider, landing back at Palmdale at 1538. Although this was Friday, the test team would not be taking the weekend off.

Roy Martin took off in the F-5 SSBD three times on Saturday morning, January 17, at 0703, 0945, and 1138. All of these were solo supersonic runs planned to measure sonic booms generated at off-design flight conditions. On the first mission, Martin did a pushover maneuver from 32,000 feet and Mach 1.375 to create a focused sonic boom. On the second mission, he flew over the ground array at a steady-state speed of Mach 1.375 but this time at 36,000 feet. He also flew at 36,000 feet on the third mission (after noting turbulence at between 30,000 feet and 32,500 feet), but this time he sped overhead at Mach 1.45—the highest speed permitted by the NAVAIR flight clearance.[81]

The fast pace of flight tests resumed on Monday, January 19, with three more solo sorties by Roy Martin.[82] Taking off at 0659, he performed a 3-second 0.1-g pushover down to 30,000 feet that focused a sonic boom toward the L-23 then recovered at 31,000 feet to cross over the ground array at Mach 1.375. On all the remaining six flight tests, the F-5 SSBD would make passes over the ground array in both directions to get the most out of each sortie. Martin took off on the first of these at 0954. As on many of the previous flights, he made his first pass from east to west at Mach 1.375 and 32,000 feet with the L-23 again below. He then made a supersonic U-turn to the right and, still at 32,000 feet, returned over the array from west to east at Mach 1.3. Even with the extra afterburner time, the F-5 SSBD still had 1,100 pounds of fuel left after landing at 1026. Following the postflight inspection and refueling, it was back in the air at 1159 for another dual run. This time Martin made the first pass at Mach 1.4 and 32,000 feet and the second at Mach 1.31 and 31,200 feet.[83]

This flight test (QSP-29) was the last in which the L-23 sailplane partici-pated. During its 14 flights, Gary Aldrich took off in the Pawnee Pa-25 with the

L-23 in tow about 15 to 20 minutes before the F-5 began its sonic boom run. The sailplane usually released at about 10,000 feet and glided over the ground array at between 6,000 feet and 8,000 feet.[84] The lower altitudes helped it record the weak signatures of booms echoing off the ground.[85] Skillful piloting and the plane's slow speed kept it on station during the supersonic runs. The data recorded by the L-23 sailplane would help enable analysts to determine the distortions in sonic boom signatures caused by turbulence in the lower atmosphere and recorded by ground sensors. Approximately 16 seconds after each of the F-5s flew overhead, their sonic booms were heard by the L-23 pilot and automatically recorded by the SABER system. About 6 seconds later, it would record a second weaker boom that had been reflected off the ground.[86] In this interval, the glider flew about 400 to 500 feet. Figure 8-5 depicts how one of the acoustic rays in a sonic boom would pass the L-23 on its way to the sensor at Site N as well as how a reflected ray from the sonic boom tat had reached the ground earlier would also cross its path.[87]

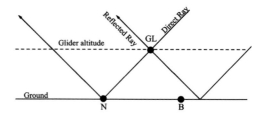

Figure 8-5. Direct and reflected acoustic rays passing L-23 glider (GL) over the ground array. (NASA)

The last 2 days of the SSBE flight tests, January 21 and January 22, were devoted to near-field measurements of the F-5 SSBD's shock wave patterns, while the ground array continued to collect far-field data. Dryden's F-15B number 836 piloted by Dana Purifoy successfully performed probe after probe as Roy Martin flew the F-5 at several different speeds while maintaining the optimum attitude of 32,000 feet. On each flight, the two aircraft flew side-by-side in a close formation from Palmdale until reaching the start point for the supersonic runs. As Martin pushed over and accelerated, Purifoy slid his F-15 Eagle below and behind the F-5. When Martin leveled out at 32,000 feet and radioed that he had reached the desired Mach number, Purifoy pushed his left stick with the throttles forward just enough to move slowly through the F-5's shock waves. He had to keep his eyes on the F-5 while also keeping a special vector symbol in his heads-up display (HUD) on the horizon to remain level. Since the pressure changes from F-5's shock waves disrupted the altimeter, the crewmembers in the back seat helped provide situational awareness and relayed other instrument readings, such as the indicated Mach number. Engineers in Dryden's control center—which had a precise display of the aircraft's relative positions as well as telemetry of the nose boom's pressure measurements showing when it had crossed the bow shock—advised Purifoy when to begin slowing down and back out of the shock waves. Unless nearing the end of the supersonic run, the F-15 would then quickly follow up with another probe.[88]

On January 21, Martin took off for the first probing mission (QSP-30) at 0702, joined up with Purifoy's F-15B with Mike Thomson in the backseat, and led them on a westbound Mach 1.4 run over the ground array. After beginning a right turn over Site Mike, they did an eastbound run at Mach

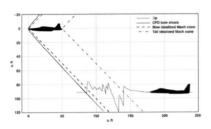

Figure 8-6. Illustration of F-15B probing F-5 SSBD shock waves. (NASA)

1.35. During these first two runs, the F-15 completed 12 successful probes. The pressure signature recorded on the sixth of these (identified as probe 30-6) is depicted in figure 8-6. Both planes were turned around in time for Martin to make a 1124 takeoff for another similar mission. This time, they made both their supersonic runs at Mach 1.375, completing 10 successful probes.[89]

The last day of SSBE testing on Thursday, January 22, provided six more two-way supersonic probing runs. Roy Martin's F-5 SSBD took off at 1124 on its first sortie, 1342 on its second, and 1534 on its third with Dana Purifoy's F-15B already in the air before each launch. Craig Bomben flew in the back seat of the F-15 on the first and third of these flights and Frank Batteas sat in for the second flight. The F-5 flew at 32,000 feet for all of the probing runs but at various speeds: twice at Mach 1.375 on the first flight; at Mach 1.4 and Mach 1.35 on the second flight; and at Mach 1.375 and Mach 1.4 on the third and last of the flight tests (QSP-33). The two pilots conducted 45 successful near-field probes during the four missions.[90]

In all, the F-15B completed 68 near-field probes at distances of 60 to 720 feet from the F-5 SSBD. The majority of probes were made directly below at an average distance of about 100 feet, but others were flown off to one side or the other. Although all the probes provided valuable shock wave measurements, there were some lessons learned. The F-15 collected the most detailed data when it moved through the shock wave pattern at the slowest possible speed relative to the F-5.[91] It was also noted that the F-15 tended to be pushed around more by the shock waves when probing off to the side than when directly under the F-5.[92]

In planning the SSBE, historical January weather statistics for the area around Edwards AFB had indicated that only early morning flights would meet the stringent atmospheric criteria required to collect useful sonic boom data.[93] As it turned out, nearly perfect weather conditions as well as the overall reliability of the participating aircraft allowed the functional check flight and all 20 planned flight tests to be completed in only 11 days. The SSBD project had gone through its share of problems and delays over the past 2½ years, but as Joe Pawlowski remarked when looking back years later, it also experienced "a lot of small miracles."[94]

Even though the flight tests ended in what seemed like plenty of time to return the F-5 SSBD to the Navy, Roy Martin did not get much rest after his final postflight debriefing. He was back at Northrop Site 3 the next day to fly the F-5 back to St. Augustine, this time accompanied by an F/A-18 from Air Test and Evaluation Squadron 31 (VX-31) at China Lake. On the third leg of their journey, the F/A-18 had an in-flight emergency and had to divert to Memphis, TN, while Martin landed at Birmingham, AL. He was stuck there for 3 nights awaiting another escort and sitting out inclement weather in Alabama and northern Florida. On Tuesday, January 27, accompanied by a Northrop Grumman Citation XL corporate jet, Martin finally made it to the St. Johns County Airport, where the Pelican's journey had begun 6 months before.[95]

Collecting and Analyzing the Data

The ground array crews recorded an abundance of sonic boom data from the 21 SSBE flights. NASA Dryden provided the sensors (BADS, BASS, and SABER) for 13 of the sites: A through H, M2, Y through Z, F through N, and F through S (see figure 8-3). Wyle Laboratories equipped Sites I through L and along with Langley provided the equipment for Sites M, MR, ML, and N. Gulfstream Aerospace, supported by Northrop Grumman, equipped the remaining 11 locations—sites O through X. FAA personnel from the U.S. Department of Transportation's John Volpe Center in Cambridge, MA, also helped. As shown in an accompanying photograph taken on January 13, some of the men and women manning the SSBE ground array (whose names also appear in several previous chapters) had been involved in sonic boom research for many years. Every morning in advance of the first flight, team members would set out much of the recording equipment and check the settings of the B&K 4231 systems with sound calibrators. Monitoring radio calls from the aircraft, they would turn on the recorders when the F-5 SSBD began accelerating toward the array and turn them off about 30 seconds after the final boom. Dryden's BADS, BASS, and SABER systems activated automatically upon sensing the pressure waves. Immediately after each flight, personnel at Sites M, ML, MR, and N would review data on the quick-look DAQ card systems positioned in the center of the array, so as to make any adjustments for the next flight. Personnel at the BADS and BASS sites could also review their data before the next flight.[96]

Fourteen members of the SSBE's data-collection team. Front row, left to right: Edward Haering (Dryden), Brenda Sullivan (Langley), David Graham and David Schein (NGC), James Murray (Dryden), Mark Stucky (USAF TPS, in flight suit), Domenic Maglieri (Eagle), and Gary Aldrich (USAF TPS, in flight suit). Rear row: Kenneth Plotkin (Wyle), David Read and Judy Rochat (FAA), Peter Coen (Langley), John Swain (Wyle, without hat), and Joe Salamone (Gulfstream). (Photo by Plotkin)

The ground array sensors made over 1,300 sonic boom recordings, the F-15B captured near-field shock wave signatures on 45 probes, and the L-23 recorded 29 midfield signatures plus secondary booms reflected off the desert floor. All of this data was supplemented by records of atmospheric conditions, including the weather balloons launched at Edwards AFB, Vandenberg AFB (CA), China Lake, and other military installations in the region.[97]

The redundancy of recordings and the analysts' ability to cross-reference and interpolate data from the various sources helped compensate for certain equipment limitations. For example, the atmospheric measurements made by highly accurate DGPS rawinsonde-equipped weather balloons was used to compensate for pneumatic lag and related inaccuracies in the F-5 SSBD's pitot-static air-data system, which was later discovered to have a small leak in the plumbing used to apply pressure to the nose boom. The SSBE's comprehensive data-collection strategy also helped compensate for a few major malfunctions. For example, on six flights, the raw GPS data from the F-5 SSBD or baseline F-5E were lost when being downloaded, but their previous GPS transmissions to the ground array had been saved. A record of GPS readings is essential for precise Mach number calibration and sonic boom data reduction. Fortunately, SSBE analysts were able to correlate the somewhat less accurate data from the ground instruments with those recorded from other flights by both the onboard and ground equipment. This allowed them to recalibrate time and position data for the signatures on the six affected flights for greater precision.[98]

Almost as soon as the flight tests had been completed, sonic boom researchers began the task of cleaning, processing, organizing, analyzing, comparing, and interpreting the vast amount of data collected. Over the next several months, key participants also presented briefings on the SSBD and SSBE at conferences and symposiums from coast to coast.[99] On August 17, 2004, NASA's Langley Research Center hosted a review of preliminary findings at a Shaped Sonic Boom Experiment Closeout Workshop. Members of the SSBE Working Group presented a number of reports followed by panel discussions on the significance of the results, issues raised, and next steps. One of the next steps was to prepare and publish a series of formal scientific and technical papers on most aspects of the SSBD and SSBE.[100] The first of these had been presented at a joint conference of the American Institute of Aeronautics and Astronautics and the Council of European Aerospace Societies (CEAS) in Manchester, England, in May 2004. The AIAA then devoted an entire session to the SSBD-SSBE at its 43rd Aerospace Sciences Meeting and Exhibit in Reno, NV, held from January 10 through January 13, 2005. An additional paper on the SSBE was presented at the 26th AIAA Aeroacoustics Conference in Monterey, CA, May 23 through May 25, 2005. (See appendix D for a list of these reports.)

The more than 1,300 sonic boom measurements made by the 42 ground sensors displayed variations caused by their location along the array, changing atmospheric conditions, differences in aircraft speed and altitude, and some random anomalies. The lines plotted for the signatures from the F-5 SSBD usually had a jagged shape at the plateau region of its flattop signatures presumably caused by intrusions from wing and inlet shock waves as well as atmospheric turbulence. Even so, reviews of the signatures confirmed consistently lower overpressure in the flattop signatures of the F-5 SSBD versus the stronger N-wave signatures of the standard F-5E. The SSBD's signatures also appeared to be consistently shorter than those of the F-5E.[101]

One of NASA's reasons for sponsoring the SSBE was the desire to learn about the effects of turbulence on shaped sonic boom signatures. In this regard, the January weather and numerous flights later in the day, although not distorting the sonic boom signatures as much as expected, did not disappoint the analysts. In general, however, the data recorded proved the persistence of the shaped sonic boom signature through turbulence.[102] An analysis by John Morgenstern and colleagues at Lockheed Martin using CFD modeling confirmed a consistent 18-percent reduction in the initial pressure impulse as well as a reduction in perceived loudness of 6.4 decibels in the F-5 SSBD's sonic

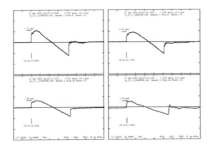

Figure 8-7. One shaped sonic boom signature as measured by Site B (top left), the L-23 glider (bottom left), Site E (top right), and after being reflected back to the glider (bottom left), all within 1.3 seconds. (NASA)

booms as compared to those from the standard F-5E.[103]

The data collected by the Air Force L-23 sailplane allowed analysts to compare the sonic boom signatures recorded at ground level with clean signatures measured above by the L-23 and signatures from the shock waves reflected back into the atmosphere. At first, the L-23's flight plan called for it to be directly over the center of the ground array (near Site M) when the sonic boom arrived, but soon the researchers had it begin up boom of the center point so that the shock waves measured along the ground array could literally travel past the L-23. Approximately 6 seconds after each of the F-5s flew overhead, the sonic boom was heard by the L-23 pilot and automatically recorded by its SABER system.[104]

Ken Plotkin, Ed Haering, Domenic Maglieri, Brenda Sullivan, and Gulfstream's Joseph Salamone used these data to determine exactly how the surface signatures were distorted by boundary-layer conditions. Figure 8-7, for example, shows how one of the sonic booms generated by the F-5 SSBD on January 15 (QSP-21) was affected by atmospheric conditions. The four graphs plot the signatures of one sonic boom captured by B&K 4193 microphones as it evolved behind and through the L-23 glider to ground sensors and again back up to the glider—all within about 1.2 seconds.[105]

Using such comparative data, the researchers were able to develop a turbulence-subtraction algorithm. Because turbulence causes almost identical perturbations in the pressure patterns of both the front and rear shocks, a mathematical template based on the rear shock distortion pattern could be used to subtract the effects of turbulence and smooth out the front of the signature to show what it would be in a standard atmosphere with minimal surface winds—in the case of the two F-5s, cleaner N-waves or flattops. Figure 8-8 shows the application of this technique to

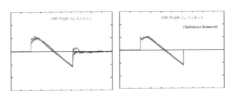

Figure 8-8. Removal of effects of turbulence from five signatures. (NASA)

the SSBD signatures recorded at several sites during the same Mach 1.35 flyover (QSP-21) on January 15.[106]

Although the rapid changes in atmospheric pressure were the most common metric used for measuring the strength of sonic booms, how loud these abrupt changes sounded to human ears would be a key factor in determining what might be acceptable to the public. Brenda Sullivan of Langley's Structural Acoustics Branch found that the shaped sonic booms were even quieter than their reduced pressure rise would indicate. The reduction of average overpressure from 1.2 psf in the sonic booms of the F-5E to 0.9 psf in those from the F-5 SSBD would equate to an expected reduction of 2.5 decibels in perceived loudness. Based on an advanced acoustical analysis of 132 booms recorded by 19 microphones on seven of the flight tests, she found the sonic booms made by the F-5 SSBD actually averaged 4.7 decibels quieter than those made by the F-5E. She concluded that the combined effects of both slower rise time and lower peak pressure of the shaped sonic boom signatures helped cause this result.[107] This reduction in loudness was found to persist through turbulence.[108]

All previous flight tests measuring focused booms involved aircraft that normally generated N-wave signatures. The F-5 SSBD presented the first chance to do so from an aircraft designed to create shaped sonic booms. This opportunity especially interested Ken Plotkin and Domenic Maglieri, who had collected and analyzed focused boom data in the past and were currently doing related research for NASA.[109] As discussed in earlier chapters, even a civil aircraft designed to generate an acceptably quiet sonic boom carpet when cruising supersonically could still create more powerful focused booms when accelerating, turning, or beginning to descend with a pushover maneuver. Based on the F-5 SSBD's design parameters and performance limitations, such as not transitioning from subsonic to supersonic in level flight, it was decided to use a pushover maneuver for the Shaped Sonic Boom Experiment. The three push-overs flown by Martin on January 12, 17, and 19 successfully placed focused booms on ground array sensors (Site Q on QSP-13 and Site E on QSP flights 24 and 27). On these sorties, Martin would quickly push his stick forward to attain unloaded forces of 0.1 g to 0.5 g and hold the pushovers for about 3 seconds until making an approximate 1.75-g recovery into level flight. Figure 8-9 illustrates how

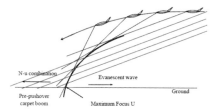

Figure 8-9. Diagram of acoustic rays during a pushover maneuver to create a focused (U-wave) sonic boom. (Wyle)

this kind of maneuver projects multiple acoustic rays to form a caustic (the dark curved line) that intersects the surface with a narrow sonic footprint.[110]

The coverage of the focused boom footprints, projected forward more than 6 miles from the start of the pushover maneuver, closely matched that predicted using PCBoom4. The actual focused boom signatures, however, revealed new data. On January 17, for example, the signature of the focused boom was recorded at Site E as a sharp U-wave with an overpressure of 3 psf as predicted, and then it evolved into a shape corresponding to an N-wave as it crossed additional sensors while also being overlapped by the premaneuver flattop signatures. The focused boom weakened and eventually dissipated as it passed Site W. The signatures of the postfocus carpet boom, however, did not display the flattop shaping of the sonic booms created by the F-5 SSBD when flying straight and level. Instead, they were recorded as conventional N-waves. The researchers concluded that this resulted from a flight condition in which the lift load was only one quarter of that for which the F-5 SSBD was designed. "The focusing maneuver was successful, but a lesson learned was that a focusing maneuver will be at a flight condition that does not correspond to a minimum boom cruise condition."[111] This unexpected finding was "a useful result" that "led to a realization that a focus maneuver of a low-boom shaped aircraft will generally correspond to an off-design condition. That result can be exploited by making an off-design condition a complex wave whose focus factor is less than that of a simple boom with two shocks."[112]

The only event that failed to provide hoped-for data was the close-behind flight of the two F-5s at Mach 1.35 on January 15 (QSP-22). Instead of the usual separation of 45 seconds (or about 63,000 feet), the F-5E was supposed to follow no more than 700 feet (or 0.3 seconds) behind and about 200 feet below the F-5 SSBD. This would subject their sonic boom signatures to the same amount of turbulence. Unfortunately (and somewhat ironically), the sleeker baseline F-5E could not quite keep up with the modified F-5, so its sonic boom signature arrived 2 seconds later. Because turbulence in the lower atmosphere was higher than during any of their other flights together, the effects of turbulence scatter on the sonic booms could not be precisely determined.[113]

The in-flight sonic boom measurements made during the F-15B probes, which measured the F-5 SSBD's shock waves shortly after coming off its airframe, would be especially valuable for use in computational fluid dynamics. Figure 8-10 depicts the near-field pressure signature as measured on the sixth probe made at Mach 1.414 during the early morning flight test on January 21 (QSP-30).[114]

The superimposed photo of the F-5 SSBD clearly shows the propagation of its shock waves, starting with a strong bow shock. This prevents the next three forebody shocks from moving forward to reinforce

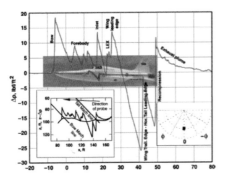

Figure 8-10. Near-field probing data, with photo showing sources of shock waves. (NASA)

the bow shock farther down in the midfield and far field. After another strong shock from the wings and their leading edge extension (abbreviated as LEX on the chart), there is a strong expansion (a type of decompression) before another shock from the trailing edge of the wings and the horizontal stabilizer. After a final recompression shock wave, the pressure drops as the plume from the engines' exhaust dissipates. The box on the lower left shows the movement of the probe relative to the angle of the shock cone, and the box at the lower right shows the orientation of the F-15B below the F-5. On some probes, it was learned that the bow shock could advance 10 feet or more from the ideal Mach cone.[115]

A technical paper published on the airborne portion of the SSBE emphasized the importance of precision in gathering this data; among the lessons learned: "For all measurements, accurate global positioning system–based timing is essential for data correlation with multiple aircraft.... Shock measuring plumbing design require[s] careful sizing.... [and] ....Real-time monitoring of measured shock waves is essential for efficient shock wave probing."[116] A delay in communications from the Dryden control room to the F-15B on the relative positions of the aircraft during probes allowed only gross corrections for the next probe. It would have been desirable to upgrade the F-15B's cockpit instrumentation, especially with a relative position display, but the compressed schedule for completing the SSBE did not allow enough time.[117]

After applying various corrections, recalibrations, and adjustments to the raw data and inputting flight conditions (e.g., Mach number, lift coefficient, angle of attack), meteorological data, and other variables, the researchers were able to begin validation of computational fluid dynamics codes. At the Langley workshop in August 2004, NGC's Keith Meredith showed how the data collected could be incorporated into CFD analyses. First, Ed Haering selected 6 of the 68 probes based on such factors as the number of data points in each recorded signature, the constancy of the F-5 SSBD's Mach numbers and flightpath angles (which implied the plane's lift, AOA, and trim), and how close to exactly parallel it and the F-15B flew during the probes. He reprocessed the GPS flightpath data recorded during these probes in five ways to select the process that would provide the best CFD comparisons. Meredith then incorporated data from the

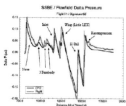

Figure 8-11. A CFD image (left) and graph of the CFD-generated shock wave signature, with the actual signature recorded during the flight test also shown on the graph. The squiggly horizontal line under the F-5 SSBD depicts the path flown by the F-15B during this probe. (NGC)

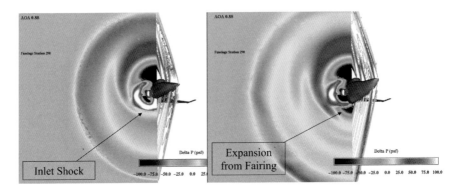

Figure 8-12. CFD comparison of normal F-5E (left) and F-5 SSBD (right) at Mach 1.4 with shock waves shown propagating from forebody and pressure contours from the engine inlet. (NGC)

six selected signatures with NGC's GCNS model using the offset grid with 14.2 million data points. Some of the CFD runs required minor adjustments to compensate for slight deviations in the probing measurements, but in general, the CFD results were in excellent agreement with the flight-test data. Figure 8-11 depicts the second probe on January 22 (QSP-31) flying at Mach 1.38 with an overlay of the recorded and CFD-generated sonic boom signatures.[118]

A collaborative overview of the SSBD project prepared by Joe Pawlowski, David Graham, Charles Boccadoro, Peter Coen, and Domenic Maglieri assessed the significance of what the Shaped Sonic Boom Demonstration and Experiment had recorded. "The vast amount of data collected during these tests will be invaluable to future supersonic aircraft designs in that it will allow designers to go forward with confidence in the ability to predict, and thereby control, sonic booms."[119] Using computational fluid dynamics, the data revealed the effects of sonic boom shaping in practice as well as in theory. In what might seem counterintuitive to a layperson, the stronger bow shock generated by the F-5 SSBD's nose glove actually resulted in a weaker and more slowly spiking bow wave in the far-field (e.g., on the ground) because trailing shock waves would be less prone to merge with it. Furthermore, the SSBD's carefully designed underbody fairing produced an area of expansion (decompression) that decreased the shock wave from the engine inlets. What in decades past could only be imagined in the mind's eye of sonic boom researchers could now be imaged in vivid colors based on real-world flight-test data. Figure 8-12 shows how the F-5 SSBD's modifications changed the normal F-5E shock waves and pressure dips into a pattern that reduced the strength of its sonic boom. The pressure in the dark areas above both aircraft and below the F-5 SSBD equates to at least −75 psf (below ambient air pressure), while the stronger bow shock extending down from the front of the

F-5 SSBD equates to more than 50 psf.[120] Although the original colors are not reproduced on paper, the results are clearly evident even in tones of gray and black.

The achievements of the Shaped Sonic Boom Demonstration and Experiment did not go unrecognized. Among the many tributes and honors given to the participants, *Aviation Week and Space Technology* bestowed one of its "Laurels" for 2003 on Charles Boccadoro, Richard Wlezien, and Steven Walker; the AIAA awarded its Aircraft Design Award for 2004 to Charles Boccadoro, Joseph Pawlowski, and David Graham; and NASA recognized the entire SSBD team with a 2004 Turning Goals into Reality Partnership Award.[121] The Navy, which had previously planned to use Bureau Number (BUNO) 74-1519 for spare parts, recognized the significance of this old but now unique F-5 by allowing it to be preserved. The disassembled aircraft was delivered in August 2004 to the Valiant Air Command Museum in Titusville, FL, where, after being reassembled and restored by volunteers (as shown in a photo at the end of the last chapter), the Pelican has been on display with the museum's collection of "warbirds" since June 2006.[122]

Endnotes

1. Weather Warehouse, "Jacksonville Cecil Field Airport, 7/24/03," accessed August 11, 2011, *https://weather-warehouse.com/WxHubP/ WxSPM531874661_174.28.158.213/1_Jacksonville*.
2. Steve Madison, Roy Martin, Keith Applewhite, and Eric Vario, "Test Plan: Quiet Supersonic Program Flight Envelope Clearance," Appendices H and I, Revision 6 (May 5, 2003).
3. Benson-Martin interview, April 7, 2011; Peter G. Coen, David H. Graham, Domenic J. Maglieri, and Joseph W. Pawlowski, "Origins and Overview of the Demonstration Program," PowerPoint presentation, January 10, 2005, slide no. 17, SSBD Flight Test Summary; Northrop Grumman, "Northrop Grumman F-5E Modified for Sonic Boom Demonstration Completes First Flight," news release, July 29, 2003.
4. Steve Madison, Roy Martin, Keith Applewhite, and Eric Vario, "Test Plan: Quiet Supersonic Program Flight Envelope Clearance," Appendices H and I, Revision 6 (May 5, 2003); Graham et al., "Aerodynamic Design of SSBD," AIAA paper no. 2005-8, 6; Northrop Grumman, "Shaped Sonic Boom Demonstration Program Overview," undated, 5.
5. The Introduction provides a detailed description of the cross-country ferry flight, based primarily on a telephone interview of Roy Martin by Lawrence Benson, May 31, 2011.
6. "Air Force Plant 42," accessed August 10, 2011, *http://www. dreamlandresort.com/black_projects/plant42.htm*; Madison, "SSBD Test Plan Review," slide no. 51.
7. Northrop Grumman, "Shaped Sonic Boom Demo Flight Test Program," PowerPoint presentation, August 17, 2004, slide no. 5, "SSBD/SSBE Team Roster: Test Pilots (SSBD)."
8. Graham et al., "Aerodynamic Design of SSBD," AIAA paper no. 2005-8, 5.
9. Pawlowski et al., "Overview of the SSBD," AIAA paper no. 2005-5, 10.
10. Ibid., 12. Source for temperatures at 32,000 feet: Aerographer/ Meteorology, Table 1-6, U.S. Standard Atmosphere Heights and Temperatures, accessed August 27, 2011, *http://www.tpub.com/ content/aerographer/14269/css/14269_75.htm*.
11. Benson-Martin interview, May 31, 2011.
12. Benson-Boccadoro interview, August 12, 2011.
13. Ibid.; Croft, "Engineering through the Sound Barrier," 30.

14. NGC, Flight Report, QSP-8, August 25, 2003, with attached pilot report by Cmdr. "Spike" Long.
15. Dana D. Purifoy to Lawrence Benson, "Re: Name(s) of F-15B Backseater during SSBD/SSBE," e-mail, October 6, 2011.
16. Edward A. Haering and James E. Murray, "Shaped Sonic Boom Demonstration/Experiment Airborne Data SSBD Final Review," PowerPoint briefing, August 17, 2004, slide no. 27; Edward A Haering, James E. Murray, Dana D. Purifoy, David H. Graham, Keith B. Meredith, Christopher E. Ashburn, and Lt. Col. Mark Stucky, "Airborne Shaped Sonic Boom Demonstration Measurements with Computational Fluid Dynamics Comparisons," AIAA paper no. 2005-9, 43rd Aerospace Sciences Meeting, Reno, NV, January 10–13, 2005, 6.
17. NGC, Flight Report, QSP-8, August 25, 2003.
18. Benson-Boccadoro interview, August 12, 2011.
19. Cmdr. "Spike" Long, QSP Pilot Report, August 25, 2003.
20. Madison et al., "Test Plan QSP SSBD Data Collection," 11–12.
21. Source for figure 8-1: Coen and Martin, "Fixing the Sound Barrier," slide no. 19.
22. The lake used to have year-round water and still has a marsh with a wildlife viewing area. Bureau of Land Management, "Harper Dry Lake," accessed August 20, 2011, *http://www.blm.gov/ca/st/en/fo/barstow/harper.html,* as of August 2011.
23. Kenneth J. Plotkin, Edward A. Haering, Domenic J. Maglieri, Joseph Salamone, and Brenda M. Sullivan, "Ground Data Collection of Shaped Sonic Boom Experiment Aircraft Pressure Signatures," AIAA paper no. 2005-10, 43rd Aerospace Sciences Meeting, Reno, NV, January 10–13, 2005, 2.
24. Ibid., 3; Numerous e-mail messages from and to Ed Haering and Mike Beck on meeting desert tortoise protection and other land-use requirements.
25. Madison et al., "Test Plan QSP SSBD Data Collection," 2, 11–12; Plotkin et al., "Ground Data Collection of SSBE," AIAA paper no. 2005-10, 2–3.
26. Ed Haering to David McCurdy, "Security and toilets," e- mail, January 16, 2003. As it turned out, many of the sticks at Harper Lake used to position the sensors and all but one of the reflector stakes used to find locations in the dark were ripped up or stolen. Ed Haering, "Draft SSBD Lessons Learned," June 6, 2004, 5.
27. Ken Plotkin, "SSBD Ground Boom Data Wyle and Northrop," PowerPoint briefing, September 19, 2003, slide nos. 2–4; Plotkin et

al., "Ground Data Collection of SSBE," AIAA paper no. 2005-10, 2–5; Ed Haering to Lawrence Benson, "Re: SSBD Chapter 8 for Review," e-mail, October 26, 2011.

28. Plotkin, "SSBD Ground Boom Data," September 19, 2003, slide nos. 2–4; Plotkin et al., "Ground Data Collection of SSBE," AIAA paper no. 2005-10, 2.

29. Ed Haering to Joe Pawlowski et al., "Re: SSBDWG Meeting Notice - Flight Test Planning," e-mail, August 19, 2002.

30. Madison et al., "Test Plan: QSP SSBD Data Collection," as of April 17, 2003, 10; Benson-Haering interview, April 5, 2011; Graham Warwick, "F-5E Shapes up to Change Sonic Boom," *Flight International*, August 5, 2003, 30.

31. Sweetman, "Whooshhh!" *Popular Science* (July 2004); Croft, "Engineering through the Sound Barrier," *Aerospace America* (September 2004): 30.

32. Sweetman, article cited above. A new Moon had set the previous evening and would not rise until after 0600: Time&Date.com, "Moonrise and Moonset in Los Angeles," August 2003, accessed August 26, 2011, *http://www.timeanddate.com/worldclock/astronomy. html?n=137&month=8&year=2003&obj=moon&afl=-11&day=1.*

33. "CFD Used in Sonic Boom Test Program," *CFD Review*, September 3, 2003; Kenneth Plotkin et al., "Ground Measurements of a Shaped Sonic Boom," AIAA paper no. 2004-2923 (May 2004), 3–4.

34. Time&Date.com, "Sunrise and Sunset in Los Angeles."

35. Croft, "Engineering through the Sound Barrier," 30.

36. NGC, Flight Report, QSP-9, August 27, 2003.

37. As quoted by Sweetman in "Whooshh!" *http://www.popsci.com.*

38. Figure 8-2 copied from Coen and Martin, "Fixing the Sound Barrier," slide 21.

39. Benson-Plotkin interview, May 2, 2011; Croft, "Engineering through the Sound Barrier," 24, 31.

40. Madison et al., "Test Plan QSP SSBD Data Collection," as of April 17, 2003, 9–10.

41. NGC, Flight Report, QSP-10, August 27, 2003, with attached pilot report by Cmdr. Spike Long attached; Plotkin et al., "Ground Measurements," AIAA paper no. 2004-2923, 4.

42. NGC, Flight Report, QSP-11, August 28, 2003.

43. Plotkin, "SSBD Ground Boom Data," slide no. 7.

44. NGC, Flight Report, QSP-12, August 29, 2003, with attached pilot report by Cmdr. Spike Long (quoted); Plotkin et al., "Ground Measurements," AIAA paper no. 2004-2923, 4.

45. Jim Hart (NGC), Jan Walker (DARPA), Kathy Barnstorff (NASA Langley), and Gray Creech (NASA Dryden), "Northrop Grumman/ Government Team Shapes Aviation History with Sonic Boom Tests," news release, August 28, 2003, accessed ca. September 1, 2011, *http://www.irconnect.com/noc/press/pages/news_releases.html?d=44396*.

46. Science Blog, news advisory, "DARPA, Northrop Grumman, NASA to Brief Media on Results Achieved in Supersonic Boom Reduction Demo," August 28, 2003, accessed August 23, 2011, *http://www. scienceblog.com/community/older/archives/K/1/pub1157.html*.

47. Northrop Grumman, "Quiet Supersonic Platform (QSP) Shaped Sonic Boom Demonstrator (SSBD) Program," PowerPoint presentation, Washington Press Club, September 3, 2003.

48. "NG, NASA, DARPA Announce Successful Sonic Boom Reduction," *Aerospace Daily* (September 4, 2003): 6.

49. Ibid.

50. Kathy Barnstorff, Jan Walker, and Jim Hart, "NASA Opens New Chapter in Supersonic Flight," news release, September 4, 2003, accessed ca. September 1, 2011, *http://www.nasa.gov/centers/langley/ news/releases/2003/03-060.html*; "F-5SSBD Press Conference," *http:// sonicbooms.org/News/F5SSBD_PressConf.html*.

51. Andrew Bridges (AP Science Writer), "Changing Jet's Shape Takes the Bang out of Sonic Booms" (September 4, 2004); Videos of CNBC's Tech Watch with Jane Wells, September 10, 2003, provided to author by Joe Pawlowski.

52. National Research Council, *Commercial Supersonic Technology: The Way Ahead* (Washington, DC: National Academies Press, 2001), 41–43.

53. James R. Asker, "FAA Seeks Information on Sonic Boom Research," *Aviation Week* (June 2, 2003): 21; David Bond, "The Time is Right," *Aviation Week* (October 20, 2003): 57–58.

54. Steve Komadina, "Quiet Supersonic Platform (QSP)," PowerPoint presentation, FAA Civil Supersonic Aircraft Technical Workshop, Arlington, VA, November 13, 2003; David Graham, "Shaped Sonic Boom Demonstrator Program," same venue, November 13, 2003. Presentations accessed September 5, 2011, *http://www.faa.gov/about/ office_org/headquarters_offices/apl/noise_emissions/supersonic_aircraft_ noise/*. Spokespersons for Boeing, Lockheed Martin, Gulfstream, Raytheon, General Electric, and Pratt & Whitney also made one or more presentations.

55. Richard G. Smith III, "NETJETS," PowerPoint presentation, FAA Civil Supersonic Aircraft Technical Workshop, Arlington, VA, November 13, 2003.

56. Peter G. Coen (for Kevin P. Shepherd), "Human Response to Sonic Booms" (November 13, 2003); Brenda M. Sullivan, "Metrics for Human Response to Sonic Booms" (November. 13, 2003); Peter G. Coen, "Supersonic Vehicles Technology: Sonic Boom Technology Development and Demonstration" November 13, 2003 (all presented at FAA Civil Supersonic Aircraft Workshop, November 13, 2003); Aimee Cunningham, "Sonic Booms and Human Ears: How Much Can the Public Tolerate," *Popular Science* (July 30, 2004) accessed ca. September 10, 2011, *http://www.popsci.com/ military-aviation-space/article/2004-07/sonic-booms-and-human-ears.*

57. Haering, Draft SSBD Lessons Learned, 10.

58. Martin and Coen, "Fixing the Sound Barrier," slide no. 22, "Lessons Learned from Initial Tests."

59. Benson-Boccadoro interview, August 20, 2011.

60. Irene Mona Klotz, "Test Shows Shape Sheds Sonic Boom's Bang" (September 26, 2003), accessed ca. September 10, 2011, *http://www. upi.com/Science_News/203/09/26/UPI-32491064603528/.*

61. NGC-NASA, "Shaped Sonic Boom Experiment Program Overview," ca. September 2004, 1.

62. SSBD/SSBE Team Roster.

63. SSBE Overview, 1–2; "Shaped Sonic Boom Demonstrator Data Collection Test Plan Addendum, Quiet Supersonic Platform Shaped Sonic Boom Experiment" (ca. December 2003), 3–5, 7.

64. SSBE Overview, 1–2; Test Plan Addendum SSBE, 5–6.

65. SSBE Overview, 3; Tom McCoy, "Flight Readiness Review for Shaped Sonic Boom Experiment" (January 6, 2004).

66. SSBE Overview, 5; Plotkin et al., "Ground Data Collection SSBD," AIAA paper no. 2005-10, 3–4; Edward A. Haering, James E. Murray, Kenneth J. Plotkin, and Brenda Sullivan, "NASA Shaped Sonic Boom Experiment, Ground Sensors and Data Final Data Review," PowerPoint presentation, August 17, 2004, figure 8-3 copied from slide no. 9.

67. Haering et al., "NASA SSBE Ground Sensors," slide nos. 10–11.

68. Plotkin et al., "Ground Data Collection of SSBE," AIAA paper no. 2005-10, 4–5; Haering et al., "NASA SSBE Ground Sensors," slide nos. 15–17.

69. Haering et al., "NASA SSBE Ground Sensors," slide nos. 10–14.

70. Haering et al., "Airborne SSBD Pressure Measurements," AIAA paper no. 2005-9, 7; Haering and Murray, "SSBD/E Airborne Data," slide nos. 15–16.

71. Haering et al., "Airborne SSBD Pressure Measurements," AIAA paper no. 2005-9, 6; Haering and Murray, "SSBD/E Airborne Data," slide nos. 15–16.

72. Benson-Martin interview, April 7, 2011; Benson-Purifoy interview, April 8, 2011; QSP Flight Reports, January 12–22, 2004; "SSBD/SSBE Team Roster," slide no. 6, "Test Pilots (SSBE)."

73. Martin and Coen, "Fixing the Sound Barrier," slide no. 25, "Revised Ground Track"; Author's review of Flight Reports QSP-13 through QSP-33, January 12–22, 2004.

74. NGC, Flight Report, QSP-14, January 12, 2004.

75. NGC, Flight Reports, QSP-15, QSP-16, and QSP-17, January 13, 2004.

76. NGC, Flight Reports, QSP 18 and QSP 19, January 14, 2004.

77. NGC, Flight Reports, QSP 20 and QSP 21, January 15, 2004.

78. Haering et al., Airborne SSBD Pressure Measurements," AIAA paper no. 2005-9, 33.

79. NGC, Flight Report QSP 21, January 15, 2004.

80. NASA-NGC, "SSBD Program Overview," 5.

81. NGC, Flight Reports, QSP-24, 25, and 26, January 17, 2004. The last report noted that the data run was made at Mach 1.43, but the speed was later revised to Mach 1.45 in the SSBE Flight Test Summary matrix used in NGC-NASA briefings during 2004. This would equate to a true airspeed of over 950 mph, based on *http://www.hochwarth.com/misc/AviationCalculator.html*.

82. As this was the Martin Luther King Federal holiday, the team had to get special permission to conduct operations.

83. NGC, Flight Reports, QSP-27, 28, and 29, January 19, 2001.

84. Plotkin et al., Ground Data Collection of SSBE," AIAA paper no. 2005-10, 5; Haering et al., "NASA SSBD Ground Sensors," August 17, 2004, slide 84.

85. Haering, Draft SSBD Lessons Learned, 3.

86. Haering et al., "Airborne SSBD Pressure Measurements," AIAA paper no. 2005-9, 13.

87. Haering et al., "NASA SSBE Ground Sensors," August 17, 2004, slide no. 87 used for figure 8-5.

88. Edward A. Haering, James E. Murray, Dana Purifoy, David H. Graham, Keith B. Meredith, Christopher E. Ashburn, and Lt. Col. Mark Stucky, "Airborne Shaped Sonic Boom Demonstration

Pressure Measurements with Computational Fluid Dynamics Comparisons," AIAA paper no. 2005-9, 43rd Aerospace Sciences Meeting, Reno, NV, January 10–13, 2005, 10–11 (also source for figure 8-6).

89. NGC Flight Reports, QSP-30 and 31, January 21, 2001; SSBE Flight Test Summary.

90. NGC Flight Reports, QSP-32 and 33, January 22, 2001; SSBE Flight Test Summary; Purifoy to Benson, October 6, 2011.

91. Haering et al., Airborne SSBD Pressure Measurements with CFD," AIAA paper no. 2005-9, 3, 15.

92. Haering, Draft SSBD Lessons Learned, 1.

93. NASA/NGC, "SSBE Program Overview

94. Benson-Graham-Pawlowski interview, April 12, 2011.

95. SSBE Flight Test Summary; NASA-NGC, "SSBE Program Overview," 5; Benson-Martin interview, April 7, 2011.

96. Plotkin et al., Ground Data Collection of SSBE," AIAA paper no. 2005-10, 5.

97. NASA-NGC, "SSBE Program Overview," 3; Haering et al., "Airborne SSBD Pressure Measurements," AIAA paper no. 2005-9, 14–15.

98. Haering et al., "Airborne SSBD Pressure Measurements," AIAA paper no. 2005-9, 8–10.

99. These included the University of Southern California on March 17; the Society of Experimental Test Pilots (SETP) in San Diego, CA, on March 26; the Society of Automotive Engineers in Wichita, KS, on April 21; the SETP in Annapolis, MD, on April 22; the Confederation of European Aerospace Societies in Manchester, England, on May 11; and the Experimental Aircraft Association in Oshkosh, WI, in July.

100. Agenda, "Shaped Sonic Boom Experiment Closeout Workshop," Pearl Young Theater, NASA Langley Research Center, Hampton, VA, August 17, 2004. As evident in the footnotes, briefings from this workshop and the papers published by the AIAA have been essential sources for writing this and the three previous chapters.

101. Haering et al., "NASA SSBE Ground Sensors" (August 17, 2004), slide nos. 25–56; Plotkin et al., Ground Data Collection of SSBE," AIAA paper no. 2005-10, 1, 5, 8.

102. Plotkin et al., "Ground Data Collection of SSBE," AIAA paper no. 2005-10, 5, 8–9; Haering et al., "NASA SSBE Ground Sensors" (August 17, 2004), slide nos. 88–89.

103. John M. Morgenstern, Alan Arslan, Victor Lyman, and Joseph Vadyak, "F-5 Shaped Sonic Boom Demonstrator's Persistence of Boom Shaping Reduction Through Turbulence," AIAA paper no. 2005-12, 43rd Aerospace Sciences Meeting, Reno, NV, January 10–13, 2005, 1, 7–9, 14.

104. Haering et al., "Airborne SSBD Pressure Measurements," AIAA paper no. 2005-9, 13.

105. Plotkin et al., "Ground Data Collection of SSBE," AIAA paper no. 205-10, 5–10; Haering et al., "NASA SSBE Ground Sensors" (August 17, 2004), slide nos. 58–89, with figure 8-7 copied from slide no. 88.

106. Kenneth J. Plotkin, Domenic J. Maglieri, and Brenda M. Sullivan, "Measured Effects of Turbulence on the Loudness and Waveforms of Conventional and Shaped Minimized Sonic Booms," AIAA paper no. 2005-2949, 26th AIAA Aeroacoustics Conference, Monterey, CA, May 23–25, 2005, 5–6. Figure 8-8 copied from Haering et al., "NASA SSBE Ground Sensors" (August 17, 2004), slide nos. 76 and 77.

107. Brenda M. Sullivan, "Sonic Boom Loudness Results for Shaped Sonic Boom Experiment," SSBE Closeout Workshop, August 17, 2004.

108. Plotkin, Maglieri, and Sullivan, "Measured Effects of Turbulence on Loudness," AIAA paper no. 2005-2949, 7-8.

109. Domenic Maglieri to Lawrence Benson, "Re: SSBD Chapter 8 for Review," e-mail, October 16, 2011.

110. Kenneth J. Plotkin and Roy Martin, "NASA Shaped Sonic Boom Experiment: Pushover Focus Maneuver, Final Data Review," PowerPoint presentation, NASA Langley Research Center, Hampton, VA, August 17, 2004, with figure 8-9 copied from slide no. 6; Kenneth J. Plotkin, Roy Martin, Domenic J. Maglieri, Edward A. Haering, and James E. Murray, "Pushover Focus Booms from the Shaped Sonic Boom Demonstrator," AIAA paper no. 2005-11, 43rd Aerospace Sciences Meeting, Reno, NV, January 10–13, 2005, 1–3.

111. Plotkin et al., "Pushover Focus Booms from the SSBD," AIAA paper no. 2005-11, 4–14, 4–5 quoted.

112. Ibid.

113. Morgenstern et al, "F-5 SSBD's Persistence Through Turbulence," AIAA paper no. 2005-12, 10–11.

114. Haering et al., "Airborne SSBD Pressure Measurements," AIAA paper no. 2005-9, figure 8-10 extracted from 16.

115. Ibid.; Haering, Draft SSBD Lessons Learned, 1.
116. Haering et al., "Airborne SSBD Pressure Measurements," AIAA paper no. 2005-9, 33 quoted.
117. Haering, Draft SSBD Lessons Learned, 2.
118. Keith Meredith, "SSBE—Flight Test to CFD Comparison," PowerPoint presentation, Langley Research Center, Hampton, VA, August 17, 2004, figure 8-11 extracted from slide nos. 14 and 16.
119. Pawlowski et al., "Overview of the SSBD Program," AIAA paper no. 2005-5, 13.
120. Meredith, "SSBE—CFD Comparison," slide no. 9 used for figure 8-12.
121. "2004 Aircraft Design Award," *Aviation Week* (October 4, 2004): 11; NASA Turning Goals into Reality 2004 Awards, accessed August 25, 2011, *http://www.aeronautics.nasa.gov/events/tgir/2004/2004_award_winners.pdf*.
122. Valiant Air Command Warbird Museum, *Unscramble* [bulletin], August/September 2004 and July 2006; Museum Gallery, accessed October 18, 2011, *http://www.vacwarbirds.org/*.

F/A-18B no. 852, used to perfect Low Boom/No Boom flight maneuvers. (NASA)

CHAPTER 9

Continuing Research; Postponing Development

Exactly 2 weeks after the Federal Aviation Administration held its civil supersonic technical workshop in Arlington, a Concorde airliner flew for the last time. It touched down in Bristol, England, on November 26, 2003—3 weeks before the 100th anniversary of the Wright brothers' first flight.[1] The Concorde's retirement after more than 30 years of Mach 2 service marked the first time in modern history that the trend toward ever-faster modes of transportation had gone into reverse. Although the causes of the Concorde's demise were primarily economic, its inherently loud sonic boom was the main reason it had been unable to offer airlines a suitable route structure. Its absence now left the market for high-speed travel open solely to smaller and intrinsically quieter supersonic business jets for the foreseeable future.

Losing the Momentum for a Low-Boom Demonstrator

After the generally promising studies on the feasibility of a Quiet Supersonic Platform and the success of the Shaped Sonic Boom Demonstrator, the stars seemed to be aligning in favor of taking the next steps toward developing a small supersonic civilian airplane. Various market analyses projected a viable market for supersonic business jets, with the most optimistic predicting the potential for up to 500 SSBJs over a 20-year period. In a colorful portrayal of the aviation industry's growing interest, a June 2004 article in *Fortune* magazine stated that companies "are starting to get excited about a new generation of hot little jets that would warp the very fabric of space-time while meeting noise and environmental regulations."[2]

During the next several years, major American and European aircraft manufacturers and a few individual investors pursued assorted SSBJ concepts with varying degrees of cooperation, competition, and commitment. Some of these and other aviation-related companies also worked together on supersonic strategies through two major consortiums: Europe's High-Speed Aircraft Industrial

Project (known by its French acronym, HISAC)—comprised of more than 30 companies such as EADS (parent company of Airbus), Dassault, and Sukhoi as well as universities and other organizations; and the Supersonic Cruise Industry Alliance (SCIA)—referred to as the Super Ten. The SCIA included airframe manufacturers Boeing, Cessna, Gulfstream, Lockheed Martin, Northrop Grumman, and Raytheon; engine builders Rolls-Royce, GE, and Pratt & Whitney; and the fractional ownership company NetJets. This group's ambitious goal was supersonic civilian flight within 10 years.[3] To make this possible, the SCIA told NASA that its top priority was to solve the issue of sonic boom noise over populated areas as soon as possible, something that would almost certainly require flight tests with an experimental low-boom aircraft.[4]

Meanwhile, acoustics specialists at NASA Langley including Kevin Shepherd and Brenda Sullivan had resumed an active program of studies and experiments on the human response to sonic booms. They upgraded the HSR-era simulator booth with an improved computer-controlled playback system, new loudspeakers, and other equipment to more accurately replicate the sound of various boom signatures, such as those recorded at Edwards (described later in this chapter). In 2005, they also added predicted boom shapes from several low-boom aircraft designs.[5] At the same time, Gulfstream was creating its own new mobile sonic boom simulator to help demonstrate the difference between traditional and shaped sonic booms to a wider audience. Although Gulfstream's folded-horn design could not reproduce the extremely low frequencies of Langley's simulator booth (with its rack of large subwoofers), it created a "traveling" pressure wave that moved past the listener and resonated with postboom noises, features that were judged more realistic than other simulators.[6]

In September 2003, 2 months before holding its civil supersonics workshop, the FAA had started the Partnership for Air Transportation Noise and Emissions Reduction (PARTNER) Center of Excellence along with NASA and a number of universities. One of the center's purposes was to conduct and share research into sonic boom acceptability. After having begun the process for considering a new American metric on acceptable sonic booms, the FAA then helped prompt the International Civil Aviation Organization (ICAO) and its Committee for Aviation Environmental Protection to put the issue on its agenda in the interest of global consistency. Addressing existing SST-era noise restrictions for a new generation of smaller supersonic airplanes would require international agreements and probable action by the U.S. Congress. Doing this would necessarily be time consuming. Carl Burleson, the director of the FAA's Office of Environment and Energy, warned, "It's one thing to develop a new scientific metric. It's another to develop a public consensus."[7]

In addition to the major aircraft companies, sometimes referred to as original equipment manufacturers (OEMs), two new privately held companies were also making significant progress on SSBJ designs. These were Supersonic Aerospace International (SAI), led by J. Michael Paulson, son of Gulfstream founder Allen E. Paulson, and Aerion Corporation, led by former Learjet President Brian Barents and chaired by its chief benefactor, billionaire investor Robert M. Bass. In October 2004, both companies revealed their SSBJ concepts at a conference of the National Business Aviation Association (NBAA).[8]

In 2000, just before his death, Allen Paulson bequeathed $25 million to his son to form SAI and contract with Lockheed Martin's Skunk Works to design a low-boom SSBJ. The resulting design, refined over the next several years, was a 132-foot-long, two-engine, 12-passenger aircraft featuring canards and an inverted V-tail capable of flying up to Mach 1.8. SAI called it the Quiet Small Supersonic Transport (QSST). Based on extensive CFD and wind tunnel testing, its sonic boom was estimated to be only 1 percent as loud as the Concorde's.[9]

Aerion Corporation was formed in 2002. Its 8-to-12-passenger design, which the company continued to refine in future years, featured a 136-foot fuselage and tapered biconvex wings (similar to those of the F-104 fighter) with natural laminar flow. Richard Tracy, the company's chief technology officer, had owned Reno Aeronautical, which worked on the wing's design as one of DARPA's contractors during the QSP. Not being a low-boom concept, Aerion's SSBJ was optimized for cruising both supersonically at about Mach 1.6 and transonically at up to Mach 1.15.[10] Because of the Mach cutoff effect, the latter option was intended to allow it to fly as fast as possible over land while costing less than competing designs.[11] Although flying at transonic speeds might be permitted under the ICAO's rule, which prohibits "the creation of a disturbing sonic boom,"[12] the FAA's blanket ban on civilian aircraft flying more than Mach 1 would still have to be relaxed to meet this goal.

Despite their progress in sonic boom research and efficient low-boom designs, both SAI and Aerion would have to negotiate joint ventures with major aircraft corporations before they could begin any serious development work. On their part, the OEMs—some of which were pursuing their own SSBJ design efforts—continued to await the kind of sonic boom research and testing by NASA that would lead to relaxation of the onerous national and international noise regulations. They wanted such assurance before making the multibillion dollar commitment needed to develop and produce an SSBJ.

In this context of both renewed enthusiasm and continued uncertainty within the aviation industry, as well as finite Government funding, NASA's aeronautics organizations hoped to sustain the momentum in developing supersonic technologies fostered by DARPA's Quiet Supersonic Platform program and the SSBD-SSBE. The next step most desired by both the aviation industry and NASA proponents was an X-plane, preferably one designed from nose to tail for generating sonic booms quiet enough to satisfy the public and thereby help lead to changes in the Federal Aviation Administration's supersonic rule. Despite a reduction in overall aeronautical research, NASA's FY 2006 budget included projects in four areas that focused narrowly on breakthrough technology demonstrations of benefit to the public, including sonic boom mitigation. NASA Vehicle Systems Program Manager Richard Wlezien (referring to the SSBD) explained the objective as follows: "The F-5 boom was shaped, but not mitigated.... The next step is to show [an] acceptable sonic boom."[13]

Although it was a relatively modest proposal compared to many NASA programs, sustaining funds would not be easy. The Agency's budget requests for aeronautics declined steadily: $959 million for FY 2004, $919 million for FY 2005, and $852 million for FY 2006.[14] The aeronautics budget request was only $593.8 million for FY 2007, but this largely reflected a change in accounting procedures for operations at NASA's research centers.[15] Based perhaps on only the raw data, an article in *Aviation Week & Space Technology* lamented that "NASA is attempting to absorb a 40% cut in its aircraft technology development programs without a clear national aeronautics policy to guide it."[16] In any case, it was clear that new programs needed to pursue President George W. Bush's goal of establishing a lunar base as a stair step for an eventual piloted mission to Mars, in addition to the ongoing demands of the Space Shuttle program and International Space Station, were forcing NASA to make some hard choices. Even so, the prospects for pursuing a new sonic boom demonstrator continued to move forward during the first half of 2005 with strong backing from the Supersonic Cruise Industry Alliance.[17]

In July 2005, NASA announced the Sonic Boom Mitigation Project. It inherited recently awarded contracts of approximately $1 million each for 5-month concept explorations on the feasibility of either modifying another existing aircraft or (more likely) designing a new demonstrator. The participating companies were Boeing Phantom Works, Raytheon Aircraft, Northrop Grumman teamed with Gulfstream, and Lockheed Martin teamed with Cessna. The best of the concepts would provide the basis for the experimental low-boom aircraft that, if successful, could be used for human response surveys. Robert E. Meyer, Dryden's associate director for programs, was named as mitigation project manager.[18] As summarized by Peter Coen, NASA's supersonic vehicle sector manager, "these studies will determine whether a low sonic boom demonstrator can be built at an affordable cost in a reasonable amount of time."[19]

With the support of most of the aerospace industry, the Sonic Boom Mitigation Project appeared to be on a fast track. NASA specialists were already reviewing the companies' existing research to help draft an RFP for building the demonstrator. Preston "Pres" Henne, a Gulfstream senior vice president who had long been a strong SSBJ advocate, expected a selection to be made by early 2006 and the experimental airplane to be flying before the end of 2008.[20] His forecast was corroborated by Dryden's Robert Meyer. "It will probably be an X-plane, although we don't have a designation for it yet.... We're approaching this fairly aggressively. We hope to award the contract to the winning company early next year and perform flight tests in 2008."[21]

These predictions soon proved to be premature. On August 30, 2005, Lisa Porter, NASA's newly appointed associate administrator for aeronautics, informed participants that NASA could no longer fund the new demonstrator. After less than 2 months of gestation, the Sonic Boom Mitigation Project was terminated while still in its first trimester.[22] In retrospect, just as the technological capabilities and business case for an experimental low-boom airplane seemed to be reaching a critical mass, its cancellation postponed any chance of resolving the ban on civilian supersonic flight for at least another decade. Although NASA would explore cheaper alternatives while continuing other avenues of sonic boom research, the demise of the mitigation project marked a major detour in the quest for quiet supersonic flight.

Despite this setback, there was still one significant boom-lowering experiment in the making. Gulfstream Aerospace Corporation, which had been teamed with Northrop-Grumman in one of the stillborn mitigation studies, had already patented a new sonic boom mitigation technique.[23] Testing this invention—a retractable lance-shaped device to extend the effective length of an aircraft's nose section—would be the next major sonic boom flight demonstration at Edwards AFB.

Inventing the Low Boom/No Boom Maneuver

In the meantime, NASA Dryden was doing some relatively modest (i.e., low cost) but very innovative sonic boom flight experimentation intended mainly to improve low-boom simulation capabilities. In a joint project with the FAA and Transport Canada in July 2005, researchers from Pennsylvania State University (a key contributor to the PARTNER Center of Excellence's noise-reduction effort) strung an array of advanced microphones at Edwards AFB to record sonic booms created by Dryden F/A-18s far above and miles away. Eighteen volunteers, who sat on lawn chairs alongside the row of microphones during the flyovers to experience the real thing, later gauged the fidelity of the recordings. These could then be used to help improve the accuracy of the booms replicated in simulators, such as the advanced mobile simulation system developed by Gulfstream.[24]

This experiment (and more to come in future years) was made possible by a clever new flight profile, featuring an inverted push-over maneuver, called "Low Boom/No Boom." The initial inspiration for developing this technique was the measurement of a low sonic boom from a sounding rocket returning to Earth at a steep angle and low Mach number. NASA Dryden's Ed Haering used PCBoom4 to model a flight profile that could emulate this result with an F/A-18 Hornet. In essence, the profile applied Frank Walkden's 1958 theory on how an airplane's lift affects the strength of its sonic boom, which had first been measured by NASA's in-flight probes made above and below a B-58 in 1963 (figure 1-4). This meant that the weaker shock waves from an aircraft's upper surfaces could propagate a quieter sonic boom signature when flying upside down.

Jim Smolka, applying his piloting skills, then refined Haering's modeling into a flyable maneuver in a series of flight tests using NASA F/A-18B number 852 equipped with an Ashtech Z-12 differential GPS unit and a Research Quick Data System (RQDS) that converted normal air data into pulse-code modulated data for transmission to the ground stations. With its precise telemetry and an extensive ground array of BASS and BADS pressure sensors and microphones, these flight tests were able to determine exactly how to create controlled sonic booms. The new technique allowed F/A-18s to generate shaped ("low boom") signatures. It also could produce the evanescent sound waves ("no boom") that remain after the refraction and absorption of shock waves generated at low Mach speeds before they reach the surface.[25]

The basic Low Boom/No Boom technique (depicted later in figure 9-6) involves cruising just below Mach 1 at about 50,000 feet, rolling into an inverted position, diving at a 53-degree angle, keeping the aircraft's speed at Mach 1.1 during a portion of the dive, and pulling out to recover at about 32,000 feet. This flight profile took advantage of four attributes that contribute to reduced overpressures: a long propagation distance (the relatively high altitude of the dive), the weaker shock waves generated from the upper surfaces of an aircraft (by beginning the dive while inverted), low airframe weight and volume (the relatively small size of an F/A-18), and a low Mach number. This technique allowed Dryden's F/A-18s, which normally generate overpressures of 1.5 psf in level flight, to produce overpressures under 0.1 psf. Using these maneuvers, Dryden's test pilots could place these focused quiet booms precisely on specific locations, such as those with observers and sensors. Not only were their overpressures low, but they also had a longer rise time than the typical N-shaped signature. High-fidelity recordings of these reduced booms would be used in the new generation of acoustic simulators.[26]

Silencing the Bow Shock with Quiet Spike

Quiet Spike was the name that Gulfstream gave to the telescoping nose-boom concept, which it began developing in 2001. Based on CFD modeling and results from Langley's 4-foot-by-4-foot supersonic wind tunnel in 2002, company experts were convinced that its patented Quiet Spike device could mitigate a sonic boom greatly by creating only mild nose shock from its narrow tip followed by weak shocks from the cross-section transitions between adjacent telescoping sections, asymmet-

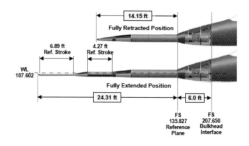

Figure 9-1. Specifications of Quiet Spike F-15 nose extension. (Gulfstream)

rically shaped to propagate less powerful pressure waves in parallel to the ground.[27] However, the company needed a way to test the structural and aerodynamic suitability of the device and also obtain supersonic flight data on its shock-scattering abilities.

NASA's Dryden Flight Research Center had all the capabilities to accomplish these tasks. Under this latest public-private partnership, Gulfstream fabricated a telescoping 470-pound nose boom (made of molded graphite epoxy over an aluminum frame) to attach to the radar bulkhead of Dryden's frequently modified F-15B number 836. As shown in figure 9-1, a motorized cable and pulley system could extend the spike up to 24 feet and retract it back to 14 feet. After extensive static testing at Gulfstream's Savannah, GA, facility, Gulfstream and NASA technicians at Dryden attached the specially instrumented spike to the radar bulkhead of the F-15B in April 2006 and began conducting further ground tests (see photo). Michael Toberman was Dryden's project manager. Key engineers included Dryden's Leslie Molzahn and Thomas Grindle and Gulfstream's Frank Simmons III, Donald D. Freund, and Robert A. "Robbie" Cowart.[28]

After safety reviews, aerodynamic assessments, and six baseline flights to measure the F-15B's flight data with its standard air-data nose boom, Dryden conducted 32 Quiet Spike flight tests from August 10, 2006, to February 14, 2007.[29] After carefully verifying the Quiet Spike's behavior during several subsonic envelope-expansion flights completed on October 3, veteran NASA test pilot Jim Smolka took it to Mach 1.2 on October 20 to begin incrementally expanding its supersonic flight envelope up to Mach 1.4 and 45,000 feet. On December 13, NASA Dryden's F-15B number 837 began in-flight pressure-measurement probes of its spike-equipped counterpart. Aerial refueling by AFFTC's KC-135 Stratotanker—with the Quiet Spike fully extended—allowed a longer mission within the R2508 restricted area and along an extended high-altitude supersonic corridor in coordination with the FAA's Los Angeles Center. The chase F-15B, flown by Thomas Hill of AFFTC's

F-15B no. 836 during vibration testing of Gulfstream's Quiet Spike at the Dryden Flight Research Center. (NASA)

F-15B number 836 in flight with Quiet Spike, September 2006. (NASA)

46th Test Group, collected data at distances of 100 feet to 700 feet from the Quiet Spike during 31 successful probes at Mach 1.4. On January 19, 2007, Smolka finished expanding the Quiet Spike's supersonic envelope by testing it at Mach 1.8. Except during supersonic sideslip maneuvers, the only discrepancy between the F-15B with and without Quiet Spike was reduced directional stability at speeds above Mach 1.4. The spike itself proved to be structurally sound in all flight conditions, and it even went beyond expectations by being extended to its full length while flying supersonically.[30]

It was known from the outset that the weak shock waves generated by the Quiet Spike would rather quickly coalesce with the more powerful shock waves generated farther back on the F-15's unmodified high-boom airframe. Therefore, the in-flight probes collected pressure signatures from less than 1,000 feet away using similar techniques as during the SSBD-SSBE tests. Figure 9-2 shows one of these signatures, made from 95 feet directly below the Quiet Spike F-15B flying at Mach 1.4, compared with a CFD prediction.

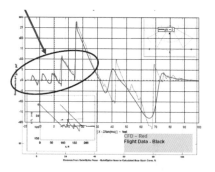

The flight test confirmed the Quiet Spike's ability to generate a relatively weak saw-tooth pattern. Also as anticipated, the powerful fifth shock wave (generated from the F-15B's inlets and wings) resulted in a sonic boom at ground level similar to that from a standard F-15B.[31] As Pres Henne put it, "Frankly, the F-15, compared to what we would have in [our] airplane, is a flying brick. It has strong shocks coming off of it, and there is no way we can stay ahead of that."[32] Analyses, however, indicated that by themselves, the weak shock waves from the front of

Figure 9-2. CFD prediction and in-flight pressure measurement with shock waves from Quiet Spike nose boom and F-15B radome circled. (NASA)

the aircraft would not have coalesced, and only a muffled sonic boom would have been heard from them on the ground. As with the SSBE, the data collected from the Quiet Spike tests would be of continuing value for developing and refining CFD capabilities.[33]

On February 13 and February 14, 2007, with all the major test objectives having been accomplished, the Quiet Spike F-15B flew to the former Kelly AFB, TX, and then on to Savannah, where Gulfstream and NASA technicians restored the aircraft to its normal configuration. A report to the Society of Experimental Test Pilots in September, prepared by 10 of the NASA Dryden and AFFTC personnel involved in the project as well as Gulfstream's Cowart, concluded, "The Gulfstream and NASA experience working on [this] joint flight research project was very good. The project took longer than expected, but once the flying started, it progressed rapidly and achieved all test objectives."[34]

For this successful test of an innovative design concept for a future SSBJ, James Smolka and Leslie Molzahn of NASA Dryden and Robbie Cowart, Donald Howe, and Frank Simmons of Gulfstream subsequently received *Aviation Week & Space Technology*'s Laureate Award in Aeronautics and Propulsion in March 2008. (Just 1 month after celebrating this honor, both Gulfstream and Dryden were saddened by the death in an aircraft accident of Gerard Schkolnik, Gulfstream's director of supersonic technology programs since 2006, who before that had been a Dryden engineer for 15 years.)[35]

Focusing on Fundamentals: The Supersonics Project

In January 2006, after several months of internal deliberations, Headquarters NASA announced its restructured Aeronautics Research Mission Directorate (ARMD) and a new aeronautics strategy. As explained by Lisa Porter, the associate administrator for the ARMD, "NASA is returning to long-term investments in cutting-edge fundamental research in traditional aeronautical disciplines...appropriate to NASA's unique capabilities."[36] One of ARMD's four new program areas was Fundamental Aeronautics with Rich Wlezien as program director. Likewise, Fundamental Aeronautics had four research areas: rotary wing, subsonic fixed wing, supersonics, and hypersonics.

In May 2006, NASA released more information on Fundamental Aeronautics, including a detailed 5-year plan for what was named the Supersonics Project with Peter Coen as its principal investigator. No longer calling for any single objective as tangible (or costly) as a low-boom demonstrator, NASA confirmed the benefits of sustaining research into sonic boom reduction technology as one of its goals. "[I]t is only through NASA investment in new technologies and improved design methods that the benefits of increased cruise speed will become a reality for the general public. These benefits include...reduced travel time for business and pleasure, rapid delivery of high value cargo including time critical medical items, and rapid response of disaster first responders" as well as potential military missions.[37]

The 60th anniversary, on October 14, 2007, of the XS-1 breaking the sound barrier brought renewed attention to the era of supersonic flight, including the continued barriers to developing any more civilian aircraft that could do the same.[38] At the end of the month, NASA convened its first conference on Fundamental Aeronautics in New Orleans with more than 400 specialists from the aviation industry, universities, and Government agencies attending.[39] The session on the Supersonics Project covered numerous challenges requiring innovative multidiscipline solutions within the areas of efficiency, performance, systems integration, and environmental impact—including the sonic boom. Milestones in each area were projected over 5 years. To serve as longer range goals on which to focus this research, the Supersonics Project projected capabilities for the next three generations of supersonic civil aircraft. The first was an SSBJ (designated N+1) that could carry 6 to 20 passengers 4,000 nautical miles at Mach 1.6—Mach 1.8 by 2015. The next generation, projected for 2020, was a small airliner (N+2) that could carry up to 70 passengers at the same speed and range. Furthest in the future, between 2030 and 2035, was an efficient multi-Mach aircraft that could carry 100 to 200 passengers 6,000 nm at Mach 1.6 with an acceptable boom, and at Mach 2.0 without sonic boom restrictions. The acceptable sonic boom metric for all three aircraft was a noise that measured less than 70 PLdB.[40]

In December 2007, the White House released the first *National Plan for Aeronautics Research and Development.* It implemented Executive Order 13419 of December 20, 2006, and an accompanying *National Aeronautics Research and Development Policy* released with the Executive order. Among various future aircraft capabilities, "economically viable aircraft capable of supersonic speeds over land (with an acceptable sonic boom impact) are also envisioned."[41] Apparently, however, this was not an explicitly stated objective. In addressing R&D for the near term (less than 5 years), midterm (5 to 10 years), and far-term (more than 10 years), the interagency plan listed specific goals for improving supersonic cruise efficiency, reducing high-altitude emissions, and lowering supersonic jet noise, but as regards sonic boom, it called for reductions only "as regards military aircraft."[42] This, however, did not preclude NASA's Supersonics Project from actively continuing research for the eventual design of low-boom civilian aircraft.

One of the Supersonics Project's major technical challenges was to accurately model the propagation of sonic booms all the way from an aircraft to the ground incorporating all relevant physical phenomena and all flight conditions. These included realistic atmospheric conditions, especially turbulence, during which "the resultant variability in ground signatures is profound and important to the quantification of boom impact."[43] Another challenging goal was to model the effects of acoustic vibrations on structures and the people inside (an issue for which military firing ranges and the use of explosives had been the focus of most recent research). Developing these models would require continued advances in CFD capabilities, wind tunnel improvements, exploitation of existing databases, and additional flight tests.[44] The ARMD solicited proposals on meeting the goals of the Fundamental Aeronautics Program from both industry and academia.[45] Meanwhile, an extensive in-house study by experts at the Langley and Ames Centers on the potential of CFD tools (which to date had been mainly developed for aero-dynamic efficiency purposes) to make better sonic boom simulations would find reasons for optimism. "Given the encouraging nature of the preliminary results...it is reasonable to expect the expeditious development of an efficient sonic boom prediction methodology that will eventually become compatible with a [shaped boom] optimization environment."[46]

During this period, NASA's aeronautics budget continued to contract with requested funding dropping from $511.7 million for FY 2008 to $446 million for FY 2009. The bulk of this 13-percent reduction was for Aerospace Systems, down $26 million, and Fundamental Aeronautics, down $34 million.[47]

Meanwhile, the Fundamental Aeronautics Program actively continued to establish collaborative projects with the private sector, academia, and other Federal agencies. Many of these were through partnerships established in response to NASA Research Announcements (NRAs) under its Research Opportunities in Aviation (ROA) program.[48] In support of this outreach, the ARMD hosted 600 attendees at the second meeting on Fundamental Aeronautics in Atlanta from October 7 to October 9, 2008, timed to pre-cede the 61st anniversary of Chuck Yeager's historic flight. Jaiwon Shin, associate administrator for aeronautics since January, emphasized how the program was benefiting from innovative precompetitive research with approximately 100 industry and academic partners working on 219 studies and projects. The meeting included more than a dozen reports on sonic boom experimentation and modeling, most of them by experts from the Langley and Ames Centers.[49]

The Supersonics Project's many activities included continued research on ways to assess human responses to sonic booms.[50] Based on multiple stud-ies that had long cited the more bothersome effects of booms experienced indoors, Langley began in the summer of 2008 to build one of the most sophisticated sonic boom simulation systems yet. Completed in 2009, it consisted of a carefully constructed 12-foot-by-14-foot room with sound and pressure systems that would replicate all the noises and vibrations caused by various levels and types of sonic booms.[51] Such studies would be vital if most concepts for supersonic business jets were ever to be realized.

During the same month as the second Fundamental Aeronautics con-ference in Atlanta, NASA awarded advanced study contracts for the N+2 and N+3 quiet supersonic airplane concepts, each worth about $2 mil-lion, to teams led by Boeing and Lockheed Martin. It also began working with Japan's Aerospace Exploration Agency (JAXA) on supersonic research, including sonic boom modeling.[52] Although not yet resurrecting any firm plans for a new low-boom supersonic research airplane, NASA supported an application to the Air Force by its research partner Gulfstream that reserved the designation X-54A for when this might be done in the future.[53] Reflecting Gulfstream's progress toward a low-boom design, the company also trademarked "The Whisper" as a name for a future SSBJ. Despite its progress on controlling the sonic boom, Gulfstream's management did not believe there would be a business case for proceeding with an SSBJ until FAA and ICAO regulations were relaxed. Some of the other aircraft manufacturers, as well as Aerion and SAI, continued to work on their SSBJ concepts during 2008, but none were yet willing or able to invest the funding necessary to move beyond research and design activities.[54]

One of the most unusual and challenging concept explorations completed at this time (not part of the Supersonics Project) was a DARPA program, nicknamed Switchblade, to determine the feasibility of a supersonic oblique flying wing. Although the potential advantages of such a configuration were primarily military—the ability to loiter at slow speeds but fly efficiently at supersonic speeds by changing its angle—an oblique wing might also propagate a weaker-than-normal sonic boom carpet, projecting its strongest signature off to one side of its flightpath.[55] In March 2006, DARPA awarded $10.3 million to Northrop Grumman Integrated Systems for a Phase 1 preliminary design review, which the company hoped would lead to an experimental unpiloted technology demonstrator during Phase 2.[56] Controlling a tailless flying wing, with its engine pods kept pointing straight ahead as the rest of the airplane swiveled above, would have been one of an oblique flying wing's most difficult challenges.[57] Having to absorb cuts to its FY 2009 budget, DARPA decided not to continue Switchblade beyond the first phase, which by October 2008 had included more than 1,000 subsonic and supersonic wind tunnel runs.[58]

Shortly after the October 2008 Fundamental Aeronautics conference, the FAA—citing continued inquiries from the aircraft manufacturers and designers—slightly updated its policy on certification standards for supersonic aircraft noise. Although still putting off any changes to the supersonic prohibition pending future research and public participation, the FAA clarified that future supersonic aircraft flying at subsonic speeds would have to meet the same noise restrictions (Stage 4) as subsonic aircraft.[59]

Unfortunately for the near-term prospects of civilian supersonic flight, the autumn of 2008 also brought the near collapse of the American financial system, leading into a global recession followed by years of economic and fiscal problems in the United States and Europe. These developments negatively affected many industries not the least being air carriers and aircraft manufacturers. The impact on those recently thriving companies making business jets was aggravated by a populist and political backlash at

American corporate executives, some of them subsidized temporarily by the Federal Government, for their continued use of company jets. Lamenting this unsought negative publicity, *Aviation Week & Space Technology* examined the plight of the small-jet manufacturers in a story with this descriptive subheading: "As if the economy were not enough, business aviation becomes a scapegoat for executive excess."[60] Ironically, one early consequence of the recession was $150 million in stimulus funding added to NASA's FY 2009 aeronautics budget as an element in the Obama administration's American Recovery and Reinvestment Act.[61]

NASA's Fundamental Aeronautics Program kept its primary focus on the more distant future. Almost 600 people attended its third conference, held again in Atlanta from September 29 to October 1, 2009. The Supersonics Project had by now awarded more than $43 million in contracts to 68 commercial and educational partners for a wide variety of research projects. In the next year, the Supersonics Project planned to focus on several areas: designing simultaneously for both low boom and low drag; doing more flight testing and modeling of boom impacts on structures and people, especially for developing modeling capabilities; working with the FAA, ICAO, and other organizations on a roadmap for sonic boom acceptability; and continuing to explore approaches for large-scale sonic boom testing.[62]

As regards the latter possibility, Boeing—under a Supersonics Project contract—studied low-boom modifications for one of NASA's F-16XL aircraft (being kept in storage at Dryden) as a possible way to obtain a reduced-boom demonstrator. This relatively low-cost idea had been one of the options being considered during NASA's short-lived Sonic Boom Mitigation Project in 2005. In the case of the F-16XL, the modifications proposed by Boeing included an extended nose glove (reminiscent of the SSBD), lateral chines that blend into the wings (as with the SR-71), a sharpened V-shaped front canopy (like those of the F-106 and SR-71), an expanded nozzle for its jet engine (similar to those of F-15B number 837 described below), and a dorsal extension (called a "stinger") to lengthen the rear of the airplane. Although such add-ons would not offer the low-drag characteristics also desired in a demonstrator, Boeing felt that its "initial design studies have been encouraging with respect to shock mitigation of the forebody, canopy, inlet, wing leading edge, and aft lift/volume distribution features."[63]

Additional design work refined this concept with more extensive modifications, including a large horizontal stabilizer (shown in figure 9-3) to achieve the desired results.[64] The study did much to advance the

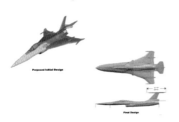

Figure 9-3. Proposed low-boom modifications to an F-16XL. (NASA)

F-15B number 836 flying with F-15B number 837 in January 2009. (NASA)

application of CFD and geometry shape optimization for a low-boom design, which in this case was achieved at a slightly higher undertrack noise level than desired. NASA deemed the extent of the required modifications (of which the H-tail was one option) to be too complex, expensive, and taxing on the flight control system of the F-16XL to pursue beyond the initial study.[65]

Measuring Tail Shocks: The LaNCETS Tests

Meanwhile, NASA already had another unique aircraft available to perform an innovative sonic boom experiment under the aegis of the Supersonics Project. NASA Dryden named this experiment the Lift and Nozzle Change Effects on Tail Shocks (LaNCETS). Both the SSBD and Quiet Spike experiments had involved only the shock waves generated from the front of an aircraft. Yet those from the rear of an aircraft as well as its jet-engine exhaust plumes also contribute to sonic booms—especially the strong recompression spike of the typical N-wave signature. These rear shocks had long been more difficult to analyze and control than those from the front of an aircraft. NASA initiated the LaNCETS experiment to address this issue.[66]

NASA Dryden had just the airplane with which to do this: the F-15B number 837. Originally built in 1973 as the Air Force's first preproduction TF-15A two-seat trainer (soon redesignated F-15B), it had been extensively modified for various experiments over its long lifespan. These included the Short Takeoff and Landing (STOL) Maneuvering Technology Demonstration, the High-Stability Engine Control project, the Advanced Control for Integrated Vehicles Experiment (ACTIVE), and the Intelligent Flight Control Systems (IFCS). This unique F-15 Eagle had the following special features: digital fly-by-wire controls, canards on the forebody (that could be used for adjusting longitudinal lift distribution), and thrust-vectoring variable area-ratio nozzles on its twin jet engines to change the pitch and yaw of the exhaust flow (that could also be used to constrict and expand plumes).[67] Researchers planned to use these capabilities for validating computational tools developed at Langley, Ames, and Dryden to predict the interactions between shocks from the tail and exhaust under various lift and plume conditions.

Tim Moes, one of the Supersonics Project's associate managers, was the LaNCETS project manager at Dryden. Jim Smolka, who had flown most of F-15B number 837's previous missions at Dryden, was its test pilot. He and Nils Larson in F-15B number 836 conducted Phase 1 of the test program with three missions from June 17 to June 19, 2008. They gathered high-quality baseline measurements with 29 probes, all at 40,000 feet and speeds of Mach 1.2, 1.4, and 1.6.[68] Figure 9-4 shows the shock wave pattern measured by one of these probes in relation to the tested F-15's modified airframe.[69]

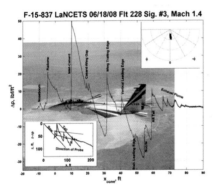

Figure 9-4. Shock wave signature of the highly modified F-15B number 837. (NASA)

Phase 2 of the project began on November 24, 2008. The LaNCETS team flew nine flight tests by December 11 before being interrupted by a freak snowstorm during the third week of December and then having to break for the holiday season.[70] The LaNCETS team completed the project with flight tests on January 12, 15, and 30, 2009. In all, Jim Smolka and flight engineer Mike Thomson flew a total of 13 missions in F-15B number 837, 11 of which included in-flight shock wave measurements by number 836 at distances of 100 feet to 500 feet. Nils Larson piloted the probing flights with Jason Cudnik or Carrie Rhoades in the back seat. The aircrews tested the effects of positive and negative canard trim at Mach 1.2, Mach 1.4, and Mach 1.6 as well as thrust vectoring at Mach 1.2 and Mach 1.4. They also gathered

Test pilot James Smolka, flight engineers Mike Thomson and Jason Cudnik, and test pilot Nils Larson in front of F-15Bs used for LaNCETS project. (NASA)

supersonic data on plume effects with different nozzle areas and exit-pressure ratios. Once again, Dryden's sophisticated GPS equipment recorded the exact locations of the two aircraft for each of the datasets.[71] On January 30, 2009, with Jim Smolka at the controls one more time, number 837 made its 251st NASA flight, the last before a well-earned retirement among the other historic aircraft on display at Dryden.[72]

In addition to its own researchers at the Langley, Ames, and Dryden Centers, NASA also made the large amount of data collected available to industry and academia as part of the Supersonics Project. For the first time, analysts and engineers would be able to use actual flight-test results to validate and improve CFD models on tail shocks and exhaust plumes—taking another important step toward the design of a truly low-boom supersonic airplane.[73]

Simulated, Softened, and Super Booms

Although the LaNCETS project was the most prominent of NASA's sonic boom tests during the last several years of the 21st century's first decade, Dryden continued adapting its Low Boom/No Boom aerial technique for testing the effects of reduced booms on people and buildings as well as for gathering data and sound recordings that could be used with sonic boom simulators. In June 2006, researchers installed 288 various accelerometers and microphones all over

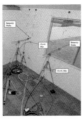

Figure 9-5. Some of the sensors in one room of the house used for sonic boom testing at Edwards AFB in 2006. (NASA)

and around a ranch-style house built at Edwards AFB in about 1960 and slated for demolition. As indicated by figure 9-5, the instrumentation installed inside the house was very extensive (as was that outdoors). During 6 days between June 13 and June 22, Dryden F/A-18Bs conducted 19 flight tests (with two aircraft flying on each mission) to pro-duce a total of 98 low-amplitude (0.05 psf to 0.8 psf) and 14 louder (0.84 psf to 1.8 psf) sonic booms. The flights resulted in the collection of a vast amount of detailed structural data as well as recordings for sonic boom simulations.[74] The researchers also recruited 77 volunteer listeners (divided into groups of about 20 per day) for a human-response survey. With two F/A-18s in the air at once, they were able to deliver a sonic boom every 3 minutes to help the volunteers compare one to another. By contrast to similar surveys in the past, immediately after listening to the sonic booms, the volunteers tended to give comparable annoyance ratings whether they were seated in the living room or outdoors in the backyard. When filling out questionnaires at the end of each day's tests, however, 63 percent of the volunteers concluded that the booms experienced inside the house, which was in rather poor repair, were more annoying.[75]

NASA and other participating researchers learned how sonic booms of varying intensity affected a more substantial home in a similar experiment, during July 2007, given the nickname House Variable Intensity Boom Effect on Structures (House VIBES). Acoustics specialists from Langley installed 112 sensors (again, a mix of accelerometers and microphones) inside the unoccu-pied half of a modern (late 1990s) duplex house. Other sensors were placed on and around the house and up a nearby 35-foot tower. These measured pressures and vibrations from 12 normal-intensity N-shaped booms (up to 2.2 psf) created by F/A-18s in steady and level flight at Mach 1.25 and 32,000 feet as well as 31 shaped booms (registering 0.08 psf to 0.7 psf) from F/A-18s using the Low Boom/No Boom flight profile (illustrated in figure 9-6 with a photograph showing one of their contrails).[76]

The quieter booms were similar to those that would be expected from an acceptable supersonic business jet. The specially instrumented F/A-18B number 852, with the RQDS air-data transmission system, performed six of the flights and an F/A-18A made one. As during the SSBE, an instrumented L-23 sailplane from the Air Force Test Pilot School recorded shock waves at pre-cise locations in the path of the focused booms above the surface boundary layer to account for atmospheric effects. The data from the indoor sensors confirmed considerably lower vibrations and noise levels in the modern house than had been the case with the older house. At the same time, data gathered by the outdoor sensors added significantly to NASA's variable-intensity sonic boom database. This would help to

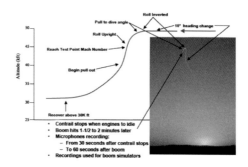

Figure 9-6. Flight profile used to deliver reduced sonic booms. (NASA)

program and validate sonic boom propagation codes for years to come, including more advanced three-dimensional versions of PCBoom.[77]

In 2009, NASA Dryden began a series of analogous tests in an experiment called Sonic Booms on Big Structures (SonicBOBS), the purpose of which was clearly indicated by its name. Sponsored by NASA's Supersonics Project, participants included Gulfstream Aerospace, Penn State University, and the Air Force Flight Test Center. The first phase, conducted on June 11, consisted of flights to calibrate a variety of sensors installed in and around the Air Force Flight Test Museum, which offered a large volume of interior space. As in previous testing, two F/A-18s flew both straight and level flights and looping Low Boom/No Boom profiles to create nine normal and nine quiet sonic booms.[78]

These profiles were repeated with similar flights on September 9 and September 12 for Phase I of SonicBOBS. NASA Dryden, NASA Langley, and Gulfstream provided a variety of sensors, including special microphones placed inside mannequin heads to better mimic what a person would hear. As a follow-on to House VIBES, the first day's experiment gathered data from an unoccupied residence in the base housing area. The second day focused on measurements at the Base Consolidated Support Facility, the Environmental Services office, and the Flight Test Museum.[79] Dryden later made the data recorded by its microphones and pressure sensors available to other researchers on a DVD. (This database also included a coincidental recording of the sonic boom from Discovery on its approach to Edwards AFB for the 48th and last Space Shuttle landing there.)[80]

Of special significance during the flight tests on September 9, the differences between normal and quiet sonic booms were experienced by a large delegation from the ICAO's Committee on Aviation Environmental Protection, including its Aircraft Noise Working Group and Supersonic Task Group. Peter Coen, who was there for the tests, thought "the visit was immensely successful.... From the perspective of sonic boom research and the prospect of potentially establishing a noise-based rule for supersonic overland flight, the visit was a major milestone."[81]

The Center resumed Sonic BOBS testing on October 14 and October 16, 2010, with the first day's flights being conducted on the 63rd anniversary of the XS-1 breaking the sound barrier. This time, the sensors were installed only in and near the base's consolidated support building. Although analysis of the data would require some time, what was being learned from the all these tests and other research led Peter Coen to observe, "I'm more convinced our big issue will be disturbances in large structures rather than houses."[82]

NASA Dryden continued to perform a variety of tests and experiments related to sonic booms into the century's second decade. In January 2011, the Center was involved in an unusual sonic boom research project: the Sonic Boom Resistant Earthquake Warning System (SonicBREWS). In anticipation of the installation of advanced seismic monitors called QuakeGuard, made by Seismic Warning Systems, Inc., in the Lancaster-Palmdale area, NASA Dryden installed some of the monitors in its main office building. This would help determine whether or not these ultrasensitive monitors would be able to filter out vibrations from the frequent sonic booms produced at Edwards AFB.[83]

The technology for in-flight measurements of shock waves also continued to advance. In February 2011, for example, NASA Dryden began flight testing two sophisticated prototype probe devices attached to the centerline pylon of an F-15B. Designed by Eagle Aeronautics and built by Triumph Aerospace Systems as part of the NASA Supersonics Project, one of the probes was conical for mounting on the nose of the trailing aircraft and the other was wedge-shaped for mounting on the generating aircraft. Combined with high-response transducers that could nearly instantaneously measure shock waves from the probed aircraft, the new devices could also monitor flight conditions such as Mach number, angle of attack, sideslip angle, temperature, and pressure without the lag time and other discrepancies encountered during existing probing missions.[84]

In a twist from the low-boom testing of recent years, in May 2011, Dryden began a major new flight test that could create louder than normal booms. Sponsored by the Supersonics Project, it was named the Superboom Caustic Analysis and Measurement Program (SCAMP). Its main purpose was to examine the critical transition from subsonic to supersonic speeds, which could create a focused boom two to five times louder than when cruising. A recently completed study of maneuver effects on a generic low-boom SSBJ using the latest computational and modeling techniques "found that the focus boom can be minimized by initiating transition and higher altitude and increasing climb angle. Acceleration rates have been found to have little influence."[85] More still needed to be learned about how to predict focused boom signatures and their locations and then design ways that quiet supersonic aircraft of the future could avoid them.

The SCAMP testing provided empirical data on this phenomenon. Researchers strung out an array of 81 microphones to record 70 localized sonic booms from 13 flights by an accelerating F/A-18B flown by Nils Larsen in the remote Black Mountain supersonic corridor north of Boron. As in other recent flight tests at Edwards, additional measurements were made above the ground—this time by a motorized glider flying at 4,000 feet to 10,000 feet and a 35-foot blimp tethered at 3,500 feet. The project involved a large team that included members from NASA Langley, Wyle Labs, Eagle, Boeing, Northrop Grumman, Gulfstream, Cessna, Penn State, Central Washington University, MetroLaser Inc., and Seismic Warning Systems. In some ways, the SCAMP was reminiscent of the SSBD and SSBE projects—and it certainly benefited from lessons learned then. According to Thomas Jones, Dryden's SCAMP manager, "It was operationally complex, given the number of team members, and logistically complicated, given the remote location of the microphone array, the unpaved roads leading to the site, and the communications between all the players, assets, and the control room at Dryden.... However, given the challenges, the SCAMP team worked together to gather one of the most interesting sets of supersonic flight research data...in some time."[86] Future analysis of this data could assist in the development of CFD codes for helping develop flight profiles and refine low-boom aircraft designs for mitigating transition focused booms.[87]

It was clear that, even without any immediate prospects for development of a new low-boom test bed, NASA's Supersonics Project was sustaining research on the acceptability of sonic booms. For example, in the fall of 2011, Dryden hosted a project dubbed Waveforms and Sonic Boom Perception and Response (WSPR) in conjunction with Langley, Wyle Labs, Gulfstream, Penn State, Fidell Associates, and Tetra Tech. Dryden's Larry Cliatt was the principal investigator. The WSPR project's primary purpose was developing data-collection methods and test protocols for future public perception studies in other communities where (unlike at Edwards) the residents were not used to hearing sonic booms. With data being recorded by 13 sensors in the base housing area, Dryden's F/A-18s flew 22 specified profiles from November 4 to November 18, generating 82 reduced and 5 normal sonic booms ranging from 0.08 psf up to 1.4 psf. Using a standard questionnaire, more than 100 volunteer residents reported their responses upon hearing these sonic booms on paper forms, at a Web site, or from smartphones using a special "app" supplied by Dryden.[88]

Keeping Focused on the Future

Although the manufacturers of large passenger aircraft were doing well in the first years of the 21st century's second decade, the business jet market remained mired in economic doldrums.[89] Even so, some companies still had supersonic aspirations, albeit at a very slow pace. The HISAC alliance in Europe was no longer active, but a few of its original members such as Sukhoi and Dassault continued some research on potential designs and discussions on joint developmental arrangements.[90] The Japan Aerospace Exploration Agency, a collaborator with NASA in sonic boom acceptability research, was also working on its own supersonic transport concept and an unpiloted sonic boom demonstrator.[91] In the United States, Gulfstream remained one of NASA's most active partners in the Supersonics Project while keeping its SSBJ design on the drawing board. As explained by Gulfstream's president, Joe Lombardo, "We believe we have the technology to limit the boom to a non-discernable level, but you will still have to change regulations that prohibit supersonic flight over land."[92] Boeing, although it had no announced plans for an SSBJ, continued to be interested in technology breakthroughs that might pave the way for a future SST.[93] Aerion, with its SSBJ designed for transonic cruise over land, was continuing research and seeking a manufacturing partner, but the company also began promoting its expertise in subsonic laminar flow technology.[94]

NASA held its fourth Fundamental Aeronautics meeting in Cleveland from March 15 to March 17, 2011.[95] Over the past 5 years, the Supersonics Project had supported a wide range of analysis capabilities and technologies. In addition to sonic boom research, these included advances in the following areas: aerodynamic design tools for more efficient supersonic cruise, powerful yet fuel-efficient engine technologies that addressed high-altitude emissions while making less noise around airports, advanced lightweight materials for innovative airframe construction, improved multidiscipline system-level design techniques, and (if the sonic boom issue could be solved) the integration of future supersonic airplanes into the FAA's next-generation (NextGen) air traffic control system.[96] As regards findings about how to achieve a truly quiet sonic boom signature, Peter Coen disclosed one preliminary result when interviewed for an article published the same month as the conference. "We're coming to the conclusion that the best thing to do is really try not to get that completely smooth pressure rise on the back end, but break the shock on the back end into several pieces so you can get the attenuation without the coalescing."[97] NASA and its Supersonics Project partners were also making significant progress in designing high-speed propulsion systems for both lower boom and quieter engines.[98]

As mentioned above, the Supersonics Project's sonic boom research had been a beneficiary of the portion of supplemental FY 2009 funds from the American Recovery and Reinvestment Act that went to Fundamental Aeronautics. These funds helped support sonic boom design validations by teams led by Boeing and Lockheed Martin, additional wind tunnel testing, focused boom research, and a new community-response pilot project. A number of previous projects recently had been completed, including the N+2 system-level study for a small supersonic airliner by 2020–2025. Boeing, in partnership with Pratt &

Figure 9-7. Some of the concepts for the Supersonics Project's N+1 SSBJ, N+2 small supersonic airliner, and N+3 supersonic transport. (NASA)

Whitney, Rolls-Royce, and Georgia Tech, had come up with two promising if still-conceptual configurations: a 100-passenger design optimized for cruise efficiency and a more extensively studied 30-passenger design optimized for a quieter sonic boom (65 PLdB to 70 PLdB).[99]

The system-level studies by Boeing and Lockheed Martin for the N+3 supersonic transport were perhaps the most futuristic of all the research sponsored by the Supersonics Project. By the targeted timeframe of 2030 to 2035 (three decades after the retirement of the Concorde), the companies thought both the necessary technologies and the air travel market could be ready for a new and economically supportable SST. Building upon their work during the QSP program and progress since then, both companies came up with interesting Mach 1.8 design concepts featuring long streamlined airframes with special sonic boom shaping, advanced materials, laminar flow, variable-cycle engines, and bleedless inlets. Boeing designed its more conventional configuration to carry 120 passengers. It included small canards, swept wings in the rear, a large upright V-tail, and two top-mounted engines.[100] Lockheed Martin's concept for carrying 100 passengers had larger canards and an inverted V-tail joined to swept wings, with four engines mounted below.[101] Although the shaped sonic booms from both were predicted to be at least 30 dB quieter than those from the similarly sized Concorde, neither would be as quiet as called for in NASA's N+3 goals.[102]

Size still mattered when it came to sonic boom signatures, but by 2012, continuing advances in CFD began showing great promise for designing N+2-size supersonic airliners able to carry 30 to 80 passengers while generating a sonic boom level of only 70 PLdB.[103] A briefing by the Supersonics Program manager highlighted this achievement as follows:

> Breakthrough Knowledge Advancement: Methodologies for the development of aircraft with shaped sonic boom signatures, particularly in the aft end of the vehicle where complex interaction between lift and volume effects takes place, have been applied to integrated systems level designs and validated through wind tunnel testing. Low boom targets for N+2 configurations have been met; methods are applicable to N+1 and N+3 vehicles as well.[104]

Although not yet budgeted, NASA had begun planning to have a low-boom experimental vehicle (LBEV) that would be available for flight testing by the end of FY 2018.[105] The successful wind tunnel validations of the latest design concepts increased interest in moving ahead with this aircraft, which, as mentioned previously, would be designated the X-54.[106] A study by the National Research Council that was completed in early 2012 strongly recommended that NASA resume focusing much of its diminished aeronautics budget on the development of X-planes, such as the sonic boom demonstrator.[107] Meanwhile, no matter how many advances some of the Nation's best aeronautical engineers were making on designing supersonic aircraft with quieter booms, the most immediate issues remained public acceptance and related national and international aviation regulations.

Here too, there had been some signs of progress, if still very tentative. The Federal Aviation Administration, with the support of NASA and other members of the Partnership for Air Transportation Noise and Emissions Reduction, held its first formal public forum on supersonic noise restrictions on October 24, 2008, in Chicago during a symposium on aircraft noise at O'Hare International Airport. Subsequently, the FAA held "Public Meetings on Advanced Technologies and Supersonics" in each of the next 3 years: on March 1, 2009, in Palm Springs, CA, along with an annual University of California symposium on aviation noise and air quality issues; on April 21, 2010, as part of an Acoustical Society of America conference in Baltimore; and on July 14, 2011, at DOT headquarters in Washington, DC.[108] These meetings, attended by professionals and anyone else who was interested in the issues, included presentations by experts from NASA, Penn State (representing the PARTNER Center), and four prospective supersonic manufacturers: Gulfstream, Aerion, Lockheed Martin, and Boeing. In addition to briefings on the recent progress in sonic boom research and discussions on the need for quality data with reliable evidence of acceptable sonic boom mitigation levels to justify amending the rules to allow the development of any future supersonic aircraft, the meetings featured realistic demonstrations of various types and amplitudes of sonic booms in Gulfstream's latest mobile audio booth, the Supersonic Acoustics Signature Simulator II (SASSII).[109]

At the meeting in July 2011, Lourdes Maurice of the FAA emphasized how the FAA, NASA, and the ICAO had initiated the development of a roadmap for researching public response to sonic booms. Her presentation defined key steps needed to assure a firm technical basis and noise standards for determining sonic boom acceptability.[110] Aerion and Gulfstream representatives gave details on their progress in researching and designing a quiet SSBJ, much of it in partnership with NASA, with the sonic boom "whisper" expected from Gulfstream's design demonstrated in the SASSII parked outside.[111] The NASA presentation on advancements being made in sonic boom research showed how recent designs could exponentially lower the perceived sound of sonic booms (figure 9-8), hopefully making them acceptable to the public.[112]

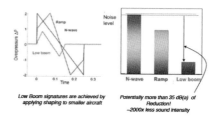

Low Boom signatures are achieved by applying shaping to smaller aircraft

Potentially more than 35 dB(a) of Reduction! ~2000x less sound intensity

Figure 9-8. Effects of sonic boom shaping and pressure rise on sound levels. (NASA)

Until a truly low-boom experimental airplane became available to give live demonstrations of the progress that designers had been making for the past decade, gaining this acceptance from the public would be difficult. In a recent article, Peter Coen summed up the impasse. "It's a real challenge for us moving forward to clearly identify and explain to the public that the sonic boom we're talking about now is completely different from what has ever been heard in the past."[113]

An environment of anemic economic activity and unstable financial markets continued to threaten much of the aerospace industry, including the business jet sector.[114] And with the Federal Government facing a looming fiscal crisis in a period of political deadlock in Washington, NASA's future budgets remained under threat.[115] Nevertheless, understandably cautious aircraft companies and a financially constrained NASA kept pressing on toward the ultimate goal of supersonic civilian flight. It was clear that the more than six decades of discoveries about sonic booms and the lessons learned on how to control them had begun to pay real dividends. Meanwhile, supersonic specialists in NASA and the private sector remained patiently committed to finding a way to follow up on Northrop Grumman's Shaped Sonic Boom Demonstrator with an airplane that could demonstrate the quietness of a carefully shaped and mitigated sonic boom.

As documented by this book, past expectations for a civilian supersonic airplane with an acceptable sonic boom to fly over populated areas have repeatedly run up against scientific, technical, economic, and political hurdles too high to overcome. That is why such an airplane has yet to fly almost three quarters of a century into the jet age. Yet the knowledge gained and lessons learned from each attempt attest to the value of persistence in pursuing both basic and applied research. Recent progress in controlling sonic booms builds upon the meticulous research, careful testing, and inventive experimentation by NASA and its partners in Government, industry, and academia over more

The F-5 Shaped Sonic Boom Demonstrator on display at the Valiant Air Command Museum in Titusville, FL. (Photo courtesy of the museum)

than six decades; the data and documentation preserved through NASA's scientific and technical information program; and the special facilities and test resources maintained and operated by NASA's research Centers. This book has emphasized one of the most important milestones on this long journey: Northrop Grumman's F-5 Shaped Sonic Boom Demonstrator. The success of the DARPA-sponsored Shaped Sonic Boom Demonstration and NASA's follow-on Shaped Sonic Boom Experiment exemplifies how a truly cooperative effort among Government agencies, private corporations, and academic institutions can produce noteworthy results in a short time at a reasonable cost.

Since the dawn of civilization, conquering the twin tyrannies of time and distance has been a powerful human aspiration, one that served as a catalyst for many technological innovations. It seems reasonable to assume that this need for speed may eventually break down the barriers in the way of practical supersonic transportation to include solving the problem of the sonic boom. If that time finally does come, a worn out former fighter plane, with the front of its fuselage modified to resemble a long pelican's beak, will have helped lead the way.

Endnotes

1. For a compendium of articles, photos, and videos in commemoration of the Concorde's last flights, see the British Broadcasting Corporation's (BBC's) "In Depth Farewell to Concorde," last updated August 15, 2007, accessed September 21, 2011, *http://news.bbc.co.uk/1/hi/in_depth/uk/2003/concorde_retirement/default.stm.*

2. Stuart F. Brown and Edward H. Phillips, "Mine's Faster than Yours," *Fortune* (June 28, 2004), accessed ca. September 1, 2011, *http://money.cnn.com/magazines/fortune/fortune_archive/2004/06/28/374394/index.htm.* For other examples, see Graham Warwick, "Quiet Progress: Aircraft Designers Believe They Can Take the Loud Boom Out of Supersonic Travel," *Flight International* (October 20, 2004): 32–33; Edward H. Phillips, "Boom Could Doom: Debate Over Hybrid SSBJ Versus Pure Supersonic is Heating Up," *Aviation Week* (June 13, 2005): 84–85; Francis Fiorino, "Lowering the Boom," *Aviation Week* (November 7, 2005): 72; "Supersonic Private Jets in Development," *Business Travel News Online* (October 26, 2006); John Wiley, "The Super-Slow Emergence of Supersonic," *Business and Commercial Aviation* (September 1, 2007): 48–50; Edward H. Phillips, "Shock Wave: Flying Faster Than Sound is The Holy Grail of Business Aviation," *Aviation Week* (October 8, 2007): 50–51.

3. Mark Huber, "Mach 1 for Millionaires," *Air & Space Magazine* (March–April 2006), accessed February 20, 2009, *http://www.airspacemag.com/flight-today/millionaire.html.*

4. David Collogan, "Manufacturers, NASA Working on Bizjet Sonic Boom Project," *The Weekly of Business Aviation*, July 18, 2005, 21.

5. Brenda M. Sullivan, "Research on Subjective Response to Simulated Sonic Booms at NASA Langley Research Center," paper presented at International Sonic Boom Forum, State College, PA, July 21–22, 2005.

6. Brenda M. Sullivan, Patricia Davis, Kathleen Hodgdon, Joseph A. Salamone, and Anthony Pilon, "Realism Assessment of Sonic Boom Simulators," CASI document no. 200800022677, January 2008.

7. Graham Warwick, "Quiet Progress: Aircraft Designers Believe They Can Take the Loud Boom Out...but Can They Convince Regulators?" *Flight International* (October 12, 2004): 32; Frances Fiorino, "Lowering the 'Boom'," *Aviation Week* (November 7, 2005): 72.

8. "Fast Money: Concepts Unveiled at…NBAA," *Flight International* (October 19, 2004): 3; David Collogan, "Two Groups Vie To Develop Supersonic Business Jets," *Aviation Week* (October 18, 2004), accessed November 18, 2008, *http://www.aviationweek. com/aw/generic/story_generic.jsp?channel=businessweekly&id=news/ SUPER10184.xml.*

9. Eric Hagerman, "All Sonic, No Boom," *Popular Science* (March 2007), accessed July 8, 2011, *http://www.popsci.com/military-aviation-space/article/2007-03/all-sonic-no-boom*; Huber, "Mach 1 for Millionaires"; Supersonic Business Jet (Web site), "Quiet Supersonic Transport (QSST)," accessed October 10, 2011, *http://www. supersonic-business-jet.com/prototypes/quiet_supersonic_transport.php.* SAI's own Web site apparently went offline before mid-2011.

10. Edward H. Phillips, "Revised Design: Aerion Aims for Improvements in Aerodynamics and Cabin Comfort," *Aviation Week* (November 7, 2005): 70; Huber, "Mach 1 for Millionaires"; "Aerion Refines Business Case for Supersonic Jet," *Flight International* (August 1, 2006): 19; Aerion Corporation, accessed October 22, 2011, *http://aerioncorp.com/*; Supersonic Business Jet (Web site), "Aerion SBJ," accessed October 22, 2011, *http://www.supersonic-business-jet.com/prototypes/quiet_supersonic_transport.php.*

11. See Kenneth J. Plotkin, Jason R. Matisheck, and Richard R. Tracy, "Sonic Boom Cutoff Across the United States," AIAA paper no. 2008-3033, 14th AIAA/CEAS Aeroacoustics Conference, Vancouver, B.C., May 5–7, 2008.

12. Peter Coen to Lawrence Benson, "Re: Chapter 9 of SSBD Book," e-mail, November 8, 2011.

13. Jefferson Morris, "NASA Budget Boosts Space Exploration, Cuts Aeronautics," *Aviation Week* (February 8, 2005), accessed ca. October 15, 2011, *http://www.aviationweek.com/aw =news/ NASABUDGET02085.xml* (site discontinued); Graham Warwick, "NASA Narrows R&D Agenda," *Flight International* (February 15, 2005): 28.

14. NASA Budget Request Summaries, FYs 2004–2007, accessed October 25, 2011, *http://www.nasa.gov/news/budget/index.html.* Congressional appropriations usually added funding to these requests.

15. Coen to Benson, November 8, 2011.

16. Frank Morring, "Rudderless: NASA Aeronautics Chief Hopes New Administrator Will Push for Clear Policy To Guide Facility Closings," *Aviation Week* (March 7, 2005): 38.

17. Collogan, "Manufacturers, NASA Working on Bizjet Sonic Boom Project," 21.
18. Ellen H. Thompson (NASA HQ), Gary Creech (Dryden), and Kathy Barnstorff (Langley), "NASA Funds Studies for Quieter Supersonic Boom," NASA news release 05-176, July 8, 2005; Biography, Robert E. Meyer Jr., accessed October 22, 2011, *http://www.nasa.gov/centers/dryden/news/Biographies/Management/meyer.html*; "Northrop Grumman to Help NASA Define Requirements for Quiet Sonic Boom Research Aircraft," Northrop Grumman news release, July 12, 2005.
19. Collogan, "Manufacturers, NASA Working on Bizjet Sonic Boom Project," 21.
20. Ibid. For Gulfstream's analysis of SSBJ prospects at the time, see Preston A. Henne, "Case for a Small Supersonic Civil Aircraft," *Journal of Aircraft* 42, no. 3 (May–June 2005): 765–774.
21. T.A. Heppenheimer, "The Boom Stops Here," *Air & Space Magazine* (October–November 2005), accessed January 3, 2009, *http://www.airspacemag.com/flight-today/boom.html*.
22. David Mould and Dean Acosta, "NASA Names New Associate Administrator," NASA news release 05-348, October 24, 2005; Michael A. Dornheim, "Will Low Boom Fly? NASA Cutbacks Delay Flight Test of Shaped Demonstrator…," *Aviation Week* (November 7, 2005): 68–69. There were no NASA news releases on the cancellation of the Sonic Boom Mitigation Project.
23. Gulfstream Aerospace Corporation, supersonic aircraft with spike for controlling and reducing sonic boom. US Patent 6,698,684, issued March 2, 2004, *http://www.patentstorm.us/patents/6698684/description.html*.
24. Jay Levine, "Lowering the Boom," *X-Press* (July 29, 2005), accessed January 4, 2009, *http://www.nasa.gov/centers/dryden/news/X-Press/stories/2005/072905*.
25. Edward A. Haering, James W. Smolka, James E. Murray, and Kenneth J. Plotkin, "Flight Demonstration of Low Overpressure N-Wave Sonic Booms and Evanescent Waves," 17th International Symposium on Nonlinear Acoustics, International Sonic Boom Forum, State College, PA, July 21–22, 2005.
26. Ibid.; Jay Levine, "Lowering the Boom."
27. US Patent 6,698,684 (previously cited); Graham Warwick, "Gulfstream Design Spikes Sonic Boom," *Flight International* (March 23, 2004): 32; Donald C. Howe et al., "Development of the Gulfstream Quiet Spike for Sonic Boom Minimization," AIAA

paper no. 2008-124, presented at 46th Aerospace Sciences Meeting, Reno, NV, January 7–10, 2008.

28. "Gulfstream Aerospace to Test Patented Spike for Controlling and Reducing Sonic Boom on NASA F-15," Gulfstream news release, July 17, 2006; Natalie D. Spivey et al., "Quiet Spike Build-up Ground Vibration Testing Approach," NASA TN 2007-214625 (November 2007); James W. Smolka, Robert A. Cowart, et al., "Flight Testing of the Gulfstream Quiet Spike on a NASA F-15B," paper presented to the Society of Experimental Test Pilots, Anaheim, CA, September 27, 2007, 1–24 (unpaginated copy provided in December 2008 to the author, who added page numbers), figure 9-1 extracted from 12; Stephen B. Cumming, Mark S. Smith, and Michael A. Frederick, "Aerodynamic Effects of a 24-foot Multi-segmented Telescoping Nose Boom on an F-15B Airplane," NASA TM 2008-214634 (April 2008).

29. The most critical flight safety concern was having to land with the spike fully extended.

30. "Gulfstream's Initial Flight Tests Validate Structural Soundness of Sonic Boom Mitigator," Gulfstream news release, October 16, 2006; "Gulfstream Quiet Spike Goes Supersonic," Gulfstream news release, October 23, 2006; Stephen B. Cummings, Mark S. Smith, and Michael A. Frederick, "Aerodynamic Effects of a 24-Foot Multi-segmented Telescoping Nose Boom on an F-15B Airplane," NASA TM 2008-214634 (April 2008); Jim Smolka, Leslie Molzahn, and Robbie Cowart, "Flight Testing of the Gulfstream Quiet Spike on a NASA F-15B," PowerPoint presentation, printed out for the author on December 11, 2008, while at Dryden.

31. Smolka et al., "Flight Testing of the Quiet Spike," 33-37, figure 9-2 extracted from 36.

32. As quoted by J.B. Wilson in "Quiet Spike: Softening the Sonic Boom," *Aerospace America* 45, no. 10 (October 2007): 39.

33. Edward A. Haering et al., "Preliminary Results from the Quiet Spike Flight Test," PowerPoint presentation, NASA Fundamental Aeronautics Program meeting, New Orleans, October 30–November 1, 2007.

34. Smolka et al, "Flight Testing of the Gulfstream Quiet Spike," 38.

35. "Aeronautics/Propulsion Laureate," *Aviation Week* (March 17, 2008): 40; "Gulfstream appoints Gerard Schkolnik as Director, Supersonic Technology Programs," Gulfstream news release, March 15, 2006; "Obituary: Gerard Schkolnik," *Aviation Week* (April 21, 2008): 22;

David Collogan, "Head of SST Research for Gulfstream killed in Crash," *Weekly of Business Aviation* 86, no. 16 (April 21, 2008): 177.

36. Michael Braukus and Doc Mirleson, "NASA Restructures Aeronautics Research," NASA news release 06-008, January 12, 2006; Lisa Porter, "Reshaping NASA's Aeronautics Program," PowerPoint presentation, January 12, 2006, 3, accessed February 2, 2009, *http://www.nasa.gov/home/hqnews/2006/jan/HQ_06008_ARMD_Restructuring.html*.

37. Peter Coen, Mary Jo Long-Davis, and Louis Povinelli, "Fundamental Aeronautics Program Supersonics Project, Reference Document," May 26, 2006, accessed January 5, 2009, *http://www.aeronautics.nasa.gov/fap/documents.html*.

38. For example, several articles in the October 8, 2007, issue of *Aviation Week*, with "Supersonics: Boom or Bust?" as its cover title; John Wiley, "The Super-Slow Emergence of Supersonic, Sixty Years After Glamorous Glennis Made History," *Business and Commercial Aviation* 101, no. 3 (September 2007): 48–50.

39. Lisa J. Porter, "NASA's Aeronautics Program," PowerPoint presentation, Fundamental Aeronautics Annual Meeting, New Orleans, LA, October 30, 2007, accessed January 5, 2009, *http://www.aeronautics.nasa.gov/fap/PowerPoints/ARMD&FA_Intro.pdf*.

40. Peter Coen, Lou Povinelli, and Kaz Civinskas, "Supersonics Program Overview," October 30, 2007, accessed January 5, 2009, *http://www.aeronautics.nasa.gov/fap/PowerPoints/SUP-Coen-final.pdf*.

41. John H. Marburger, *National Plan for Aeronautics Research and Development and Related Infrastructure* (Washington, DC: National Science and Technology Council, December 2007), accessed February 2, 2009, *http://www.whitehouse.gov/sites/default/files/microsites/ostp/aero-natplan-2007.pdf*, 16, 53.

42. Ibid., 53; Kristin Scuderi, "President Bush Approves National Plan for Aeronautics Research and Development and Related Infrastructure," White House Office of Science and Technology news release, December 21, 2007.

43. Coen et al., "Supersonics Project, Reference Document," 36.

44. Ibid., 19–22 and 42–49; NASA/ARMD, "Fundamental Aeronautics Reference Materials," PowerPoint presentation, 45th annual AIAA Aerospace Sciences Meeting, Reno, NV, January 2007, *http://www.aeronautics.nasa.gov/fap/documents.html*.

45. J.D. Harrington, "NASA Announces Aeronautic Research Opportunities," NASA news release 07-18, January 30, 2007.

46. J.H. Casper et al., "Assessment of Near-Field Sonic Boom Simulation Tools," AIAA paper no. 2008-6592, 26th Applied Aerodynamics Conference, Honolulu, HI, August 18–21, 2008.

47. NASA Budget Request Summaries, FY 2008 and FY 2009, accessed October 29, 2011, *http://www.nasa.gov/news/budget/index.html*; Daniel Morgan and Carl E. Behrens, "National Aeronautics and Space Administration: Overview, FY2009 Budget, and Issues for Congress," Congressional Research Service, February 26, 2008.

48. NASA/ARMD, "Fundamental Aeronautics Collaborative Research," December 2007, accessed January 5, 2009, *http://www.aeronautics. nasa.gov/fap/partners.html*; Jefferson Morris, "Quiet, Please: With More Emphasis on Partnering, NASA Continues Pursuit of Quieter Aircraft," *Aviation Week* (June 25, 2007): 57–58.

49. "Fundamental Aeronautics Program Annual Meeting—October 7–9, 2008," accessed January 5, 2009, *http://www.aeronautics.nasa. gov/fap/meeting/agenda.html*; Jaiwon Shin, "Opening Remarks, Fundamental Aeronautics Program 2008 Annual Meeting," PowerPoint presentation, October 7, 2008; Peter Coen, Lou Povinelli, and Kas Civinskas, "Supersonics Program Overview," PowerPoint presentation, same venue, October 7, 2008; "Marking 61 Years of Supersonic Curiosity," accessed January 5, 2009, *http:// www.nasa.gov/topics/aeronautics/features/x1_supersonics_prt.htm*.

50. Brenda M. Sullivan, "Sonic Boom Modeling Technical Challenges," CASI document no. 200700363733, October 31, 2007.

51. Brenda M. Sullivan, "Design of an Indoor Sonic Boom Simulator at NASA Langley Research Center," paper presented at Noise-Con 2008, Baltimore, MD, July 12–14, 2008, and "Research at NASA on Human Response to Sonic Booms," paper presented at 5th International Conference on Flow Dynamics, Sendai, Japan, November 17–20, 2008; Coen et al., "Supersonics Project Overview," October 7, 2008.

52. Beth Dickey, "NASA Awards Future Aircraft Research Contracts," contract release C08-060, October 6, 2008; Graham Warwick, "Forward Pitch," *Aviation Week* (October 20, 2008): 22–23; Beth Dickey, "NASA and JAXA To Conduct Joint Research on Sonic Boom Modeling," NASA news release 09-117, May 8, 2008.

53. "X-54A Designation Issued as Placeholder for Future Boom Research Aircraft," *Aerospace Daily* (July 21, 2008): 1.

54. John Croft, "Missing Ingredients Fail To Dent Enthusiasm," *Flight International* (May 13, 2008), accessed ca. October 1, 2011, *http://business.highbeam.com/411058/article-1G1-179313476/*;

Robert Wall, "Gulfstream Sees Need To Demonstrate Low-Noise Supersonic Flight before 2013," *Aviation Week* (May 28, 2007): 34; Graham Warwick, "Making Waves," *Aviation Week* (June 30, 2008): 44; "Supersonics Face Funding Barrier," *Flight International* (October 14, 2008), accessed ca. October 1, 2011, *http://business. highbeam.com/411058/article-1G1-186885117/.*

55. For example: Ilan Kroo and Alex VanDerVelden, "The Sonic Boom of an Oblique Flying Wing," AIAA paper no. 90-4002 (October 1990), and "Sonic Boom of the Oblique Flying Wing," *Journal of Aircraft* 31, no. 1 (January–February 1994): 19–25; Christopher A. Lee, "Design and Testing of Low Sonic Boom Configurations and an Oblique All-Wing Supersonic Transport," NASA CR 197744 (February 1995).

56. Jim Hart, "Northrop Grumman Selected...to Design First-Ever Supersonic 'Oblique Flying Wing' Aircraft," Northrop Grumman news release, March 23, 2006.

57. Graham Warwick, "Flying Sideways," *Flight International* (April 18, 2006), accessed August 22, 2011, *http://www.flightglobal.com/news/ articles/flying-sideways-206034/.*

58. Graham Warwick, "DARPA Kills Oblique Flying Wing," *Aviation Week* (October 1, 2008), accessed October 10, 2011, *http://www. aviationweek.com/aw/generic/story.jsp?id=news/OBLI10018c.xml.*

59. "FAA Updates Policy on SST Noise Certification," *The Weekly of Business Aviation* 87, no. 17 (October 27, 2008): 195.

60. Graham Warwick, "Open Season," *Aviation Week* (March 2, 2009): 20–21.

61. Peter Coen, "Fundamental Aeronautics Program Supersonics Project," PowerPoint presentation, NASA Fundamental Aeronautics Technical Conference, Cleveland, OH, March 15–17, 2011, accessed October 23, 2011, *http://www.aeronautics.nasa.gov/pdf/ supersonics.pdf,* slide no. 21.

62. Peter Coen, "Fundamental Aeronautics Program Supersonics Project," NASA Fundamental Aeronautics Program 2009 Annual Meeting, September 29, 2009, accessed October 22, 2011, *http:// www.aeronautics.nasa.gov/fap/SUP-Atlanta-2009-v2.pdf,* slide nos. 10, 34–36.

63. Graham Warwick, "Beyond the N-Wave: Modifying NASA's Arrow-Wing F-16XL Could Help Pave the Way for Low-Boom Supersonic Transports," *Aviation Week* (March 23, 2009): 52.

64. Coen, "Supersonics Project," September 29, 2009, slide no. 38 (source for figure 9-3).

65. Coen to Benson, November 8, 2011.

66. "Supersonics Project Reference Document," 43. The acronym "LaNCETS" was devised by Ed Haering: Interview by Lawrence Benson, Dryden Flight Research Center, December 12, 2008.

67. Dryden Flight Research Center, "F-15B #837," accessed February 15, 2009, *http://www.nasa.gov/centers/dryden/ aircraft/F-15B-837/ index.html*.

68. Larry Cliatt et al., "Overview of the LaNCETS Flight Experiment and CFD Analysis," PowerPoint presentation, Fundamental Aeronautics Annual Meeting, Atlanta, October 2, 2008.

69. Tim Moes, "Objectives and Flight Results of the Lift and Nozzle Change Effects on Tail Shock (LaNCETS) Project," PowerPoint presentation, International Test & Evaluation Association, Antelope Valley Chapter, February 24, 2009, with figure 9-4 copied from slide no. 23.

70. Single sorties were flown on November 24 and 25 and December 2, 3, 9, and 11. Three sorties were flown on December 4. Timothy Moes to Lawrence Benson, "Re: More Details on LaNCETS," e-mail, March 11, 2009.

71. Guy Norris, "Sonic Solutions: NASA Uses Unique F-15B To Complete Design Tools for Quiet Supersonic Aircraft," *Aviation Week* (January 5, 2009): 53; Gray Creech and Beth Dickey, "Lancets Flights Probe Supersonic Shockwaves," Dryden news release 09-04, January 22, 2009; Moes, "Flight Results of LaNCETS Project," February 24, 2009.

72. "NASA NF-15B Research Aircraft," fact sheet, as of March 11, 2009, accessed December 14, 2009, *http://www.nasa.gov/centers/ dryden/news/FactSheets/FS-048-DFRC_prt.htm*.

73. For an early analysis, see Trong T. Bui, "CFD Analysis of Nozzle Jet Plume Effects on Sonic Boom Signature," AIAA paper no. 2009-1054, 47th Aerospace Sciences Meeting, Orlando, FL, January 5–8, 2009.

74. Jacob Klos and R.D. Bruel, "Vibro-Acoustical Response of Buildings Due to Sonic Boom Exposure: June 2006 Field Test," NASA TM 2007-214900 (September 2007), with figure 9-5 extracted from 309 and 310; Denise M. Miller and Victor W. Sparrow, "Assessing Sonic Boom Responses to Changes in Listening Environment, Signature Type, and Testing Methodology," *JASA* 127, issue 3 (2010): 1898.

75. Brenda Sullivan et al., "Human Response to Low-Intensity Sonic Booms Heard Indoors and Outdoors," NASA TM 2010-216685 (April 2010).

76. Figure 9-6 copied from Edward A. Haering et al., "Initial Results from the Variable Intensity Sonic Boom Propagation Database," PowerPoint presentation, May 7, 2008, slide no. 4.

77. Gary Creech, "Sonic Boom Tests Scheduled," Dryden news release 07-38, July 5, 2007; Guy Norris, "Sonic Spike," *Aviation Week* (October 8, 2007): 52; Jacob Klos, "Vibro-Acoustic Response of Buildings Due to Sonic Boom Exposure: July 2007 Field Test," NASA TM 2008-215349 (September 2008); Edward A. Haering et al., "Initial Results from the Variable Intensity Sonic Boom Propagation Database," AIAA paper no. 2008-3034, presented at the 14th AIAA/CEAS Aeroacoustics Conference, Vancouver, May 5–7, 2008. Accompanied by a PowerPoint briefing with the same title.

78. Gray Creech, "Supersonic Diving: Quieting the Boom," Dryden Web page feature, June 18, 2009, accessed January 22, 2011, *http://www.nasa.gov/centers/dryden/Features/sonicbobs.html*.

79. Alan Brown, "NASA To Conduct Sonic Boom Research Over Edwards Next Week," Dryden news release 09-54, September 4, 2009.

80. Edward A. Haering and Sarah Renee Arnac, "Sonic Booms on Big Structures (SonicBOBS) Phase I Database," DFRC-2020, CASI document no. 20100024301, March 2010; "Space Shuttle: End of Mission Landings," accessed October 29, 2011, *http://www.nasa.gov/mission_pages/shuttle/launch/eomland.html*.

81. Gray Creech, "ICAO Team Witnesses Sonic Boom Research Flights," Dryden Web page feature, September 29, 2009, accessed October 21, 2011, *http://www.nasa.gov/centers/dryden/Features/icao_sonic_booms.html*; Gray Creech, Dryden news release 09-75, "2009—Another Year of Accomplishments at NASA Dryden," December 21, 2009.

82. "SonicBOBS: NASA Researching Reducing Intensity of Sonic Booms," Dryden Web page feature, October 7, 2010, accessed January 22, 2011, *http://www.nasa.gov/centers/dryden/Features/sonic_bobs_tests.html*; John Croft, "NASA Furthers Sonic Boom Irritation Studies," *Flight International* (November 24, 2010), accessed October 22, 2011, *http://www.flightglobal.com/news/articles/350137/*.

83. Gray Creech, "SonicBREWS Brewing Up an Earthquake Warning System," January 11, 2011, accessed October 22, 2011, *http://www.nasa.gov/centers/dryden/Features/SonicBREWS.html*.

84. "NASA Dryden Flies New Supersonic Shockwave Probes," *Aerotech News and Review* (February 25, 2011): 1, 11; Domenic Maglieri to

Lawrence Benson, "Comments on Chapter 9," e-mail, October 31, 2011.

85. Domenic J. Maglieri, Percy J. Bobbitt, Steven J. Massey, Kenneth J. Plotkin, Osama A. Kandil, and Xudong Zheng, "Focused and Steady-State Characteristics of Shaped Sonic Boom Signatures: Prediction and Analysis," NASA CR 2011-217156 (June 2011), 12.

86. Gray Creech, "Getting Loud Now: Enjoy Peace and Quiet Then," May 27, 2011, accessed July 20, 2011, *http://www.nasa.gov/centers/dryden/Features/scamp.html*.

87. Domenic Maglieri to Lawrence Benson, "Comments on Chap. 9," e-mail, October 31, 2011. Preliminary data indicated that the overpressure of the most powerful of the super booms was probably 11 psf: Kenneth Plotkin to Lawrence Benson, "Fw: AVTIP and SCAMP," e-mail, October 30, 2011.

88. "Sonic Boom Research Study Recruitment Begins," Dryden Web page feature, June 23, 2011, accessed July 26, 2011, *http://www.nasa.gov/centers/dryden/Features/sonic_boom_recruit_orig.html*; "More Edwards Residents Needed for Sonic Boom Study," Dryden Web page feature, August 4, 2011, accessed August 5, 2011, *http://www.nasa.gov/centers/dryden/Features/sonic_boom_recruit.html*; Gray Creech, "NASA Quiet Sonic Boom Research Effort Ends with a Whisper," December 1, 2011, Dryden Web page feature, *http://www.nasa.gov/centers/dryden/Features/WSPR_research_complete.html*.

89. Deliveries had dropped from 1,139 in 2008 to 732 in 2010 with as few as 600 projected for 2011. Joseph C. Anselmo and William Garvey, "Prolonged Pain: Honeywell's Bizjet Forecast Corks Any Premature Celebration," *Aviation Week* (October 10, 2011): 54–55.

90. Graham Warwick, "Sonic Overture: European Researchers Test the Waters on International Supersonic Collaboration," *Aviation Week* (July 13, 2009): 53–54; Bob Coppinger, "A Distant Boom," *Flight International* (April 27, 2010), accessed ca. September 15, 2011, *http://www.flightglobal.com/news/articles/ebace-a-distant-boom-340971/*; Robert Wall and Guy Norris, "Filling A Gap: Interest Rises in High-Speed Premium Air Travel," *Aviation Week* (June 27, 2011): 38.

91. Giko Sehata and Yomiuri Shimbun, "Can SSTs Be Made Quieter?: JAXA Advancing in Development of High-Speed Commercial Aircraft," *Daily Yomiuri* (Tokyo), May 9, 2010, 3; "Quiet Booms," *Aviation Week* (May 23, 2011): 13.

92. "Turbulent Times: Face to Face—Joe Lombardo," *Aviation Week* (October 18, 2010): 61–62.

93. Warwick, "Sonic Overture," 54; Graham Warwick, "Key Developments," *Aviation Week* (January 25, 2010), 104.

94. Aerion, "Order Book at $4 Billion as Aerion Nears Decision on Manufacturing Partner," news releases, June 15, 2009; "Aerion Taps SHS to Promote Development of New Supersonic Business Jet," March 15, 2010; "Aerion Diversifies to Meet Demand for Subsonic Natural Laminar Flow," May 16, 2011, *http://aerioncorp.com/media#press*. During July and August 2010, one of NASA Dryden's F-15Bs carried a scale model of Aerion's latest supersonic laminar flow wing design under its centerline on five flights at speeds up to Mach 2. Aeron, "Aerion Announces Details of Recent Supersonic Flight Tests," news release, October 18, 2011; Guy Norris, "Aerion Ambitions," *Aviation Week* (October 18, 2010): 42; William Garvey, "Looking for a Home: A Fast Promise After a Slow Start," *Aviation Week* (April 9, 2012): 16. By 2012, Aerion estimated that up to $3 billion would be needed to develop and produce its SSBJ.

95. NASA, "2011 Fundamental Aeronautics Technical Conference Recap," Cleveland, OH, March 15–17, 2011, accessed October 23, 2011, *http://www.aeronautics.nasa.gov/fap/meeting_recap_2011.html*. Just under 400 people were in attendance.

96. Peter Coen, "Fundamental Aeronautics Program Supersonics Project," 2011 Technical Conference, Cleveland, OH, March 15–17, 2011, accessed October 23, 2011, *http://www.aeronautics.nasa.gov/pdf/supersonics.pdf*.

97. Jim Banke, "Quieter Flight: A Balancing Act," *Aerospace America* (March 2011): 42–43.

98. Guy Norris and Graham Warwick, "Sound Barrier: Propulsion Integration Holds the Key to Low-Noise, Low-Boom Supersonic Transports," *Aviation Week* (June 4/11, 2012): 50–53.

99. Coen, "Supersonics Project," March 15, 2011; Harry R. Wedge et al., "N+2 Supersonic Concept Development and System Integration," NASA CR 2010-216842 (August 2010).

100. Robert H. Wedge et al. (incl. Kenneth Plotkin and Juliet Page), "N Plus 3 Advanced Concept Studies for Supersonic Commercial Transport Aircraft Entering Service in the 2030–2035 Period," NASA CR 2011-217084 (April 2011).

101. John Morgenstern, Nicole Norstrud, Marc Stelnack, and Craig Skoch, "Final Report for the Advanced Concept Studies for Supersonic Commercial Transports Entering Service in the 2030 to 2035 Period, N+3 Supersonic Program," NASA CR 2010-216796 (October 25, 2010).

102. Coen, "Supersonics Project," March 15, 2011, figure 9-7 extracted from slide no. 9; Graham Warwick, "Mach Work: Advances in Aerodynamics, Structures, and Engines Hold Promise for Quiet, Efficient Supersonic Transports," *Aviation Week* (May 17, 2010): 43.

103. By the spring of 2012, wind tunnel tests had verified the success of these improved CFD-aided designs. As explained by Peter Coen, "That was really a breakthrough for us. Not only that the tools worked, but that our tests show we could do even better in terms of reducing noise than we thought at the start of the effort." Jim Banke, "Sonic Boom Heads for a Thump," NASA Web Feature, May 8, 2012, accessed May 11, 2112, *http://www.nasa.gov/topics/aeronautics/features/sonic_boom_thump.html.*

104. Peter Coen, "Supersonics Project Overview," PowerPoint Presentation, Fundamental Aeronautics Program 2012 Technical Conference, Cleveland, OH, March 13–15, 2012, accessed April 15, 2012, *http://www.aeronautics.nasa.gov/fap/2012-PRESENTATIONS/SUP_2012_508.pdf,* slide no. 15.

105. Graham Warwick, "Raise the Limit: Research to Define Acceptable Boom Limit Paces Return of Supersonic Travel," *Aviation Week* (July 18/25, 2011): 57.

106. Graham Warwick, "Boom Time: Demonstrator Needed To Convince Regulators Shaped Booms Allow Quiet Supersonic Flight," *Aviation Week* (April 23/30, 2012): 54–55.

107. National Research Council, *Recapturing NASA's Aeronautics Flight Research Capabilities* (Washington DC: the National Academies Press, 2012), 32–36, accessed May 11, 2012, *http://www.nap.edu/catalog.php?record_id=13384.* As shown on page 9 of this report, NASA's recent aeronautics budgets had been $512 million in FY 2008, $500 million in FY 2009, $507 million in FY 2010, and $534 million in 2011.

108. FAA, "Supersonic Aircraft Noise," accessed October 15, 2011, *http://www.faa.gov/about/office_org/headquarters_offices/apl/noise_emissions/supersonic_aircraft_noise/.* This Web site has links to all the papers presented at the Civil Supersonic Aircraft Technical Workshop in 2003 and the FAA Public Meetings on Advanced Technologies and Supersonics in October 2009, March 2010, and July 2011.

109. Ibid.; FAA, "Civil Supersonic Aircraft Panel Discussion," *Federal Register* 76, no. 100 (May 24, 2011): 30231.

110. Lourdes Maurice, "Civil Supersonic Aircraft Advanced Noise Research," PowerPoint presentation at FAA Public Meeting on

Advanced Technologies and Supersonics, Washington, DC, July 14, 2011.

111. Richard Tracy, "Aerion Supersonic Business Jet," and Robbie Cowart, "Supersonic Technology Development," PowerPoint presentations at FAA Public Meeting on Advanced Technologies and Supersonics, Washington, DC, July 14, 2011.

112. Peter Coen, "Fixing the Sound Barrier: Three Generations of U.S. Research into Sonic Boom Reduction...and What it Means to the Future," PowerPoint presentation at FAA Public Meeting on Advanced Technologies and Supersonics, Washington, DC, July 14, 2011, figure 9-8 copied from slide no. 18. The signature identified as a "ramp" can be described more completely as a "symmetrical initial shock ramp" and the "low boom" as a "symmetrical ramp." Maglieri to Benson, October 31, 2011.

113. Banke, "Quieter Flight," 42.

114. Joseph C. Anselmo and William Garvey, "Prolonged Pain: Honeywell's Bizjet Forecast Corks any Premature Celebration," *Aviation Week* (October 10, 2011): 54–55.

115. In response to these budgetary pressures, the Fundamental Aeronautics Program merged the air-breathing portion of the Hypersonics Project with the Supersonics Project while research on Entry, Landing, and Descent (EDL) vehicles was transferred to the Space Technology Program. See Jaiwon Shin, "NASA Aeronautics Fundamental Aeronautics Program Technical Conference," PowerPoint presentation, 2012 Fundamental Aeronautics Technical Conference, Cleveland, OH, March 13–15, 2012, accessed April 15, *http://www.aeronautics.nasa.gov/fap/meeting_recap_2012.html*; Graham Warwick, "Cut To the Bone: Aeronautics Research Funding Decline Makes It Harder To Take Technologies to the Next Level," *Aviation Week* (February 27, 2012): 35.

A color-coded illustration of high-pressure shock waves and lower pressure expansions based on imagery that was generated by the Computational Fluid Dynamics (CFD) program used by its SSBD team in designing the F-5's modifications. A photo of the modified aircraft in flight has been inserted to complete the illustration. (Illustration: NGC, Photo: NASA)

Key SSBD and SSBE Personnel

SSBD Management Team

Richard Wlezien	DARPA	QSP Program Manager (2000–2003)
Steven Walker	DARPA	QSP Program Manager (2003–2004)
Lisa Veitch	IDA	QSP Deputy Manager
Charles Boccadoro	NGC	NGC QSP Program Manager
Peter Coen	NASA Langley	DARPA Technical Agent
Mark Gustafson	AFRL	DARPA Technical Agent
Joseph Pawlowski	NGC	SSBD Project Manager
Edward Haering	NASA Dryden	Principal Investigator
David Richwine	NASA Dryden	F-15B Manager

SSBD Working Group

Richard Wlezien	DARPA
Steven Walker	DARPA
Peter Coen	NASA Langley
Mark Gustafson	DARPA/AFRL
Lisa Veitch	IDA
Sue Morris	CENTRA*
Joe Pawlowski	NGC
Rich Biegner	NGC
Dave Graham	NGC
Steve Madison	NGC
Roy Martin	NGC
Keith Meredith	NGC
Ed Haering	NASA Dryden
Dave Richwine	NASA Dryden
David McCurdy	NASA Langley
David Hilliard	NASA Langley
Kevin Shepherd	NASA Langley

Domenic Maglieri	Eagle Aeronautics
Percy Bobbitt	Eagle Aeronautics
Ken Plotkin	Wyle Laboratories
Juliet Page	Wyle Laboratories
Todd Magee	Boeing
Peter Hartwich	Boeing
Tom Porter	Gulfstream
Joe Salamone	Gulfstream
Rob Wolz	Gulfstream
Tony Pilon	Lockheed Martin
John Morgenstern	Lockheed Martin
Joe Vadyak	Lockheed Martin
Sam Bruner	Raytheon Aircraft
Cathy Downen	Raytheon Aircraft
Kurt Schueler	Raytheon Aircraft
James Poncer	AFRL

*CENTRA Technology (a DARPA support contractor)

SSBE Program Management Team

Peter Coen	NASA Langley	SSBE Program Manager
Keith Numbers	AFRL	AVTIP Program Manager
Adam Harder	AFRL	AVTIP Contract Administrator
Charles Boccadoro	NGC	NGC QSP Program Manager
Joseph Pawlowski	NGC	NGC SSBE Project Manager
Edward Haering	NASA Dryden	Principal Investigator
Steven Corda	NASA Dryden	F-15B Manager

SSBE Working Group

Peter Coen	NASA Langley
Joe Pawlowski	NGC
Dave Graham	NGC
Steve Madison	NGC
Roy Martin	NGC
Keith Meredith	NGC
Dave Schein	NGC
Ed Haering	NASA Dryden

Gerard Schkolnik	NASA Dryden
Dave McCurdy	NASA Langley
Brenda Sullivan	NASA Langley
David Hilliard	NASA Langley
Kevin Shepherd	NASA Langley
Domenic Maglieri	Eagle Aeronautics
Percy Bobbitt	Eagle Aeronautics
Ken Plotkin	Wyle Laboratories
Juliet Page	Wyle Laboratories
Todd Magee	Boeing
Tom Porter	Gulfstream
Joe Salamone	Gulfstream
Rob Wolz	Gulfstream
Victor Lyman	Lockheed Martin
Tony Pilon	Lockheed Martin
John Morgenstern	Lockheed Martin
Joe Vadyak	Lockheed Martin
Sam Bruner	Raytheon Aircraft
Cathy Downen	Raytheon Aircraft
Adam Harder	AFRL
Keith Numbers	AFRL

SSBD Test Pilots

Roy Martin	Northrop Grumman	F-5 SSBD
Darryl "Spike" Long	NAWS China Lake	F-5 SSBD
Dwight "Tricky" Dick	NAS Fallon	F-5E
Edgar "Sting" Higgins	NAS Fallon	F-5E (ISSM)
Mike Bryan	Boeing	T-38
Dana Purifoy	NASA Dryden	F-15B
Jim Smolka	NASA Dryden	F/A-18

SSBE Test Pilots

Roy Martin	Northrop Grumman	F-5 SSBD
Dwight "Tricky" Dick	NAS Fallon	F-5E
Dana Purifoy	NASA Dryden	F-15B
Jim Smolka	NASA Dryden	F/A-18
Richard Ewers	NASA Dryden	F/A-18

Mark "Forger" Stucky	USAF Test Pilot School	Blanik L-23
Robert "Critter" Malacrida	USAF Test Pilot School	Blanik L-23
Vince "Opus" Sei	USAF Test Pilot School	Blanik L-23
Gary Aldrich	USAF Test Pilot School	Pawnee Pa-25

Northrop Grumman Vehicle Design Team

Keith Applewhite	Team Lead; Loads
Terry Britt	Dynamics
Dale Brownlow	Modification Lead
John Corskery	Mass Properties
Habeeb Hasseem	Finite Element Model
Dick Hong	Mass Properties
Joseph Lavoie	Composite Structure
Andy Limon	Structures Lead
Steve McCleskey	Stress
Bob Parker	Stress
Eddy Pedroza	Avionics/Electrical
Larry Roussel	Configuration Mgt.
Mark Sherman	Fairing Structure
Mark Smith	Mfg. Engineering
Aziz Soltani	Configuration Mgt.
Ron Srenco	Materials & Processes
Jerry Stuart	Nose Structure
Jim Ueda	F-5 Design Manager
Chris Yasaki	Nose Structure
Tom York	Fairing Structure

NGC Flight Test Support: St. Augustine

Steve Madison	Test Conductor
Keith Applewhite	Loads & Dynamics
Eric Vartio	Stability & Control
Al Scholz	Instrumentation
Pat Foster	Instrumentation
John Nevadomsky	Operations Mgr.
Dan Nehring	Flt. Test Engineer
Darren McPhillips	Crew Chief
Jim Fallica	Flt. Test Support

W.D. Thorne	Flt. Test Support
Diane Barnes	Quality Assurance
Nate McKendrick	Flight Dispatch
Merv Burne	Instrumentation
John Garry	Instrumentation
Larry Stencel	Instrumentation

NGC Flight Test Support: Palmdale

Steve Madison	Test Conductor
Keith Applewhite	Loads & Dynamics
Eric Vartio	Stability & Control
Al Scholz	Instrumentation
Pat Foster	Instrumentation
Mike Foxgrover	Crew Chief
Larry Baldini	Test Support
Dennis Cruickshank	Quality Assurance
Jim Difenderfer	Test Support
W.D. Thorne	Test Support
Mike Ingalls	Test Support
Chuck Rider*	F-5E Plane Capt.

* Sikorsky employee, NAS Fallon

NASA Dryden Flight Test Support: Edwards AFB

F-15B Operations:

Tim Moes	Chief Engineer
Keith Krake	Instrumentation
Martin Trout	Backseat
Michael Thomson	Backseat
Christine Visco	Ops. Engineer
Mark Collard	Ops. Engineer
John Spooner	Instr. Technician.
Corry Rung	Instr. Technician.
Perry Silva	Crew Chief
Tim Cutler	Ground Crew
Roger Lynn	Ground Crew
Tom Wolfe	Ground Crew

Carl Booker	Inspector

Other Support:

Patricia Kinn	Acft. Scheduling
Nancy Wilcox	Acft. Scheduling
Tracy Ackeret	Range Control
Kathleen Howell	Control Room
Russell James	Control Room
Rich Rood	Frequency Mgt.
Carla Thomas	Photography
Jim Ross	Photography
Tony Landis	Photography
Thomas Tschida	Photography

SSBD Data Collection Team

Ed Haering	NASA Dryden
Jim Murray	NASA Dryden
David Berger	NASA Dryden
Ellen Klingbeil	NASA Dryden
James Parie	NASA Dryden
Norma Campos	NASA Dryden
Jack Ehernberger	NASA Dryden
Ed Teets	NASA Dryden
Chris Ashburn	NASA Dryden
Peter Coen	NASA Langley
Dave McCurdy	NASA Langley
Ken Plotkin	Wyle Labs
Tom Baxter	Wyle Labs
John Swain	Wyle Labs
Joe Salamone	Gulfstream
Todd Magee	Boeing
Dave Graham	NGC
Dave Schein	NGC
Jonathan King	NGC
Steve Komadina	NGC
John Mangus	NGC
Andrew Maskiell	NGC
Dustin Okada	NGC
Anne Bender	NGC

Robert Ganguin	NGC
Leslie Smith	NGC
Rich Wasson	NGC
Bryan Westra	NGC
Joan Yazejian	NGC
Alan Arslan	Lockheed Martin
Herb Kuntz	Lockheed Martin
John Morgenstern	Lockheed Martin
Tony Pilon	Lockheed Martin

SSBE Data Collection Team

Ed Haering	NASA Dryden
Jim Murray	NASA Dryden
Gregory Noffz	NASA Dryden
Chris Ashburn	NASA Dryden
Peter Coen	NASA Langley
Brenda Sullivan	NASA Langley
Ken Plotkin	Wyle Labs
Tom Baxter	Wyle Labs
John Swain	Wyle Labs
Domenic Maglieri	Eagle Aeronautics
Joe Salamone	Gulfstream
Dave Graham	NGC
Dave Schein	NGC
Greg Epke	NGC
Joe Pawlowski	NGC
Eric Adamson	Boeing

APPENDIX B
F-5 SSBD Flight Chronology

Flight Number	Date (and Takeoff Time if Flight Test)	Capsule Summary
St. Augustine and Jacksonville, FL *Piloted by Roy Martin, NGC*		
QSP-1	July 24, 2003	Functional Check Flight (FCF), St. Johns Airport to Cecil Field, with Boeing T-38 chase plane.
QSP-2	July 27, 2003	FCF/envelope expansion, with T-38 chase.
QSP-3	July 28, 2003	FCF/envelope expansion, with T-38 chase.
Cross-Country to California *Piloted by Roy Martin, NGC*		
Ferry 1 & 2	July 28, 2003	Cecil Field via Huntsville, AL, to Tinker AFB, OK, with T-38.
Ferry 3 & 4	July 29, 2003	Tinker AFB via Roswell, NM, to Palmdale, CA, with T-38.
Shaped Sonic Boom Demonstration *All flown from Air Force Plant 42, Palmdale, to Edwards AFB restricted areas*		
QSP-4	August 2, 2003	FCF/envelope expansion piloted by Martin, with NASA F/A-18 chase.
QSP-5	August 2, 2003	FCF/envelope expansion by Martin, with F/A-18 chase; ground data measurements practice.
QSP-6	August 4, 2003	FCF/envelope expansion by Cmdr. Darryl Long, USN, with F/A-18 chase; ground data measurements practice.
QSP-7	August 15, 2003	FCF/envelope expansion by Martin, with F/A-18 chase; ground data measurements practice.
QSP-8	August 25, 2003, 1030	First sonic boom flight test, flown by Long, with F/A-18 chase; practice NASA F-15B probe at Mach 1.38 and ground data measurements.
QSP-9	August 27, 2003, 0626	Flown by Martin with Navy F-5E over ground array at Mach 1.36 and 32,000 feet; historic first shaped sonic boom measurement.
QSP-10	August 27, 2003, 0905	Flown by Long with F-5E, sonic boom run at Mach 1.38 over ground array.

| QSP-11 | August 28, 2003, 0620 | Flown by Martin with F-5E over ground array. |
| QSP-12 | August 29, 2003, 0830 | Flown by Long with F-15B for in-flight probe at Mach 1.34; F-5E had departed for NAS Fallon after photos by F/A-18B chase. |

Shaped Sonic Boom Experiment
All piloted by Roy Martin from Air Force Plant 42, Palmdale, to Edwards AFB restricted areas
All measured by ground sensor array

QSP-13	January 12, 2004, 0940	FCF with F/A-18B chase and L-23 sailplane; included Mach 1.4 practice focus boom from 32,000 feet.
QSP-14	January 12, 2004, 1318	F/A-15B in-flight probe aborted; Mach 1.4 run at 32,000 feet; Air Force L-23 below.
QSP-15	January 13, 2004, 0656	Mach 1.4 run at 32,000 feet with F-5E; L-23 below.
QSP-16	January 13, 2004, 1000	Mach 1.4 run at 32,000 feet with F-5E; L-23 below; F/A-18B chase.
QSP-17	January 13, 2004, 1300	Mach 1.4 run at 32,000 feet with F-5E; L-23 below.
QSP-18	January 14, 2004, 0957	Mach 1.43 run at 32,000 feet with F-5E; L-23 below.
QSP-19	January 14, 2004, 1327	Mach 1.35 run with F-5E; L-23 below; F/A-18B chase
QSP-20	January 15, 2004, 0655	Mach 1.43 run at 32,000 feet with F-5E, L-23 below.
QSP-21	January 15, 2004, 0957	Mach 1.35 run at 32,000 feet with F-5E; L-23 below.
QSP-22	January 15, 2004, 1257	Mach 1.35 close formation run at 32,000 feet with F-5E before it continued on to NAS Fallon; L-23 below.
QSP-23	January 16, 2004, 1503	Solo run at Mach 1.375 and 32,000 feet; L-23 below.
QSP-24	January 17, 2004, 0703	Solo run at Mach 1.375 and 32,000 feet to create focus boom.
QSP-25	January 17, 2004, 0945	Solo run at Mach 1.375 and 36,000 feet.
QSP-26	January 17, 2004, 1138	Solo run at Mach 1.45 and 36,000 feet.
QSP-27	January 19, 2004, 0659	Solo run at Mach 1.375 and 32,000 feet to create focus boom; L-23 below.
QSP-28	January 19, 2004, 0954	Two solo runs: Mach 1.375 and 32,000 feet; Mach 1.33 and 31,000 feet; L-23 below.

QSP-29	January 19, 2004, 1159	Two solo runs: Mach 1.40 and 32,000 feet; Mach 1.31 and 32,000 feet; L-23 below.
QSP-30	January 21, 2004, 0702	Two F-15B probing runs: Mach 1.4 and 32,000 feet; Mach 1.35 and 32,000 feet.
QSP-31	January 22, 2004, 1124	Two F-15B probing runs: both at Mach 1.375 and 32,000 feet.
QSP-32	January 22, 2004, 1342	Two F-15B probing runs: Mach 1.4 and Mach 1.35, both at 32,000 feet.
QSP-33	January 22, 2004, 1534	Two F-15B probing runs: Mach 1.375 and Mach 1.4, both at 32,000 feet.

Cross-Country to Florida *Piloted by Roy Martin*		
Ferry 5 & 6	January 23, 2004	From Palmdale via Albuquerque to Tinker AFB; accompanied by F/A-18 from NAWS China Lake.
Ferry 7	January 24, 2004	From Tinker AFB to Birmingham; F/A-18 diverted to Memphis for maintenance.
Ferry 8	January 27, 2004	From Birmingham to St. Augustine after weather delay; accompanied by NGC Citation XL corporate jet.

Northrop Grumman SSBD/SSBE postflight reports; PowerPoint tables: "Shaped Sonic Boom Demo Flight Test Program," and "Shaped Sonic Boom Flight Test Summary," August 17, 2004; Telephonic interviews of Roy Martin by Lawrence Benson, May 31 and November 6, 2011.

F-5E SSBD Modifications and Specifications

Key F-5 SSBD Specifications:

length:	49.8 feet
height:	13.4 feet
wingspan:	26.7 feet
area:	186 square feet.
takeoff weight:	~15,000 lbs *(internal fuel only)*
maximum authorized speed with modifications:	Mach 1.45

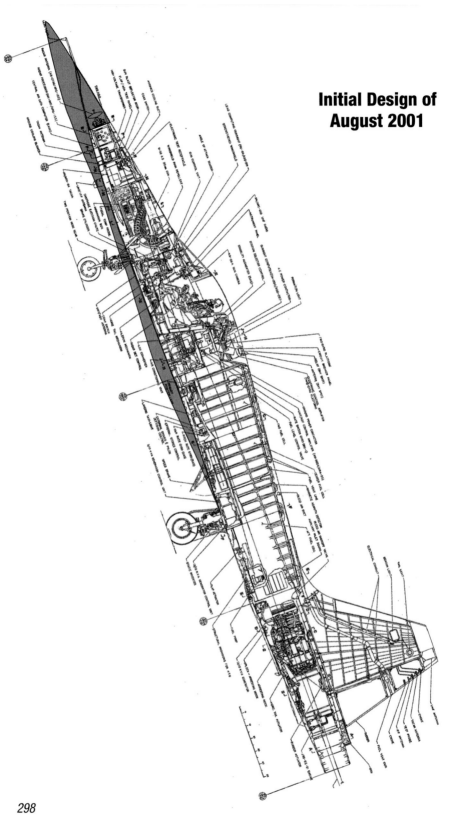

Initial Design of August 2001

**Nearly Final Design
of January 2002**

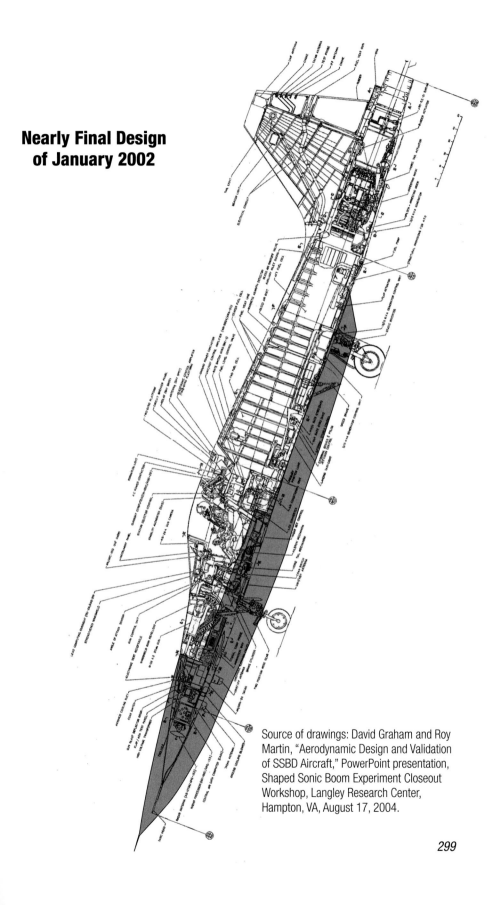

Source of drawings: David Graham and Roy
Martin, "Aerodynamic Design and Validation
of SSBD Aircraft," PowerPoint presentation,
Shaped Sonic Boom Experiment Closeout
Workshop, Langley Research Center,
Hampton, VA, August 17, 2004.

APPENDIX D
Key SSBD and SSBE Reports

Kenneth J. Plotkin, Juliet Page, David H. Graham, Joseph W. Pawlowski, David B. Schein, Peter G. Coen, David A. McGurdy, Edward A. Haering, James E. Murray, L.J. Ehernberger, Domenic J. Maglieri, Percy J. Bobbitt, Anthony Pilon, and Joe Salamone, "Ground Measurements of a Shaped Sonic Boom," AIAA paper no. 2004-2923, 10th AIAA-CEAS Aeroacoustics Conference, Manchester, England, May 10–12, 2004.

Joseph W. Pawlowski, David H. Graham, Charles H. Boccadoro, Peter G. Coen, and Domenic J. Maglieri, "Origins and Overview of the Shaped Sonic Boom Demonstration Program," AIAA paper no. 2005-5, 43rd Aerospace Sciences Meeting, Reno, NV, January 10–13, 2005.

Keith B. Meredith, John A. Dahlin, David H. Graham, Michael B. Malone, Edward A. Haering, Juliet A. Page, and Kenneth J. Plotkin, "Computational Fluid Dynamics Comparison and Flight Test Measurement of F-5E Off-Body Pressures," AIAA paper no. 2005-6, 43rd Aerospace Sciences Meeting, Reno, NV, January 10–13, 2005.

David H. Graham, John A. Dahlin, Juliet A. Page, Kenneth J. Plotkin, and Peter G. Coen, "Wind Tunnel Validation of Shaped Sonic Boom Demonstration Aircraft Design," AIAA paper no. 2005-7, 43rd Aerospace Sciences Meeting, Reno, NV, January 10-13, 2005.

David H. Graham, John A. Dahlin, Keith B. Meredith, and Jay L. Vadnais, "Aerodynamic Design of Shaped Sonic Boom Demonstration Aircraft," AIAA paper no. 2005-8, 43rd Aerospace Sciences Meeting, Reno, NV, January 10–13, 2005.

Edward A. Haering, James E. Murray, Dana D. Purifoy, David H. Graham, Keith B. Meredith, Christopher E. Ashburn, and Lt. Col. Mark Stucky, "Airborne Shaped Sonic Boom Demonstration Pressure Measurements with Computational Fluid Dynamics," AIAA paper no. 2005-9, 43rd Aerospace Sciences Meeting, Reno, NV, January 10–13, 2005.

Kenneth J. Plotkin, Edward A. Haering, James E. Murray, Domenic J. Maglieri, Joseph Salamone, Brenda M. Sullivan, and David Schein, "Ground Data Collection of Shaped Sonic Boom Experiment Aircraft Pressure Signatures," AIAA paper no. 2005-10, 43rd Aerospace Sciences Meeting, Reno, NV, January 10–13, 2005.

Kenneth J. Plotkin, Roy Martin, Domenic J. Maglieri, Edward A. Haering, and James E. Murray, "Pushover Focus Booms from the Shaped Sonic Boom Demonstrator," AIAA paper no. 2005-11, 43rd Aerospace Sciences Meeting, Reno, NV, January 10–13, 2005.

John M. Morgenstern, Alan Arslan, Victor Lyman, and Joseph Vadyak, "F-5 Shaped Sonic Boom Persistence of Boom Shaping Reduction through the Atmosphere," AIAA paper no. 2005-12, 43rd Aerospace Sciences Meeting, Reno, NV, January 10–13 2005.

Kenneth J. Plotkin, Domenic Maglieri, and Brenda M. Sullivan, "Measured Effects of Turbulence on the Loudness and Waveforms of Conventional and Shaped Minimized Sonic Booms," AIAA paper no. 2005-2949, 11th AIAA/CEAS Aeroacoustics Conference, Monterey, CA, May 23–25, 2005.

Selected Bibliography

Books, Monographs, and Conference Proceedings

Air Force Flight Test Center History Office. *Ad Inexplorata: The Evolution of Flight Testing at Edwards Air Force Base*. Edwards AFB: AFFTC, 1996.

Aircraft Engine Noise and Sonic Boom. Conference Proceedings (CP) no. 42. Presented at the North Atlantic Treaty Organization (NATO) Advisory Group for Aerospace Research and Development (AGARD), Neuilly sur Seine, France, 1969.

Anderegg, C.R. *Sierra Hotel: Flying Air Force Fighters in the Decade after Vietnam*. Washington, DC: Air Force History and Museums Program, 2001.

Angelucci, Enzo. *The Rand McNally Encyclopedia of Military Aircraft, 1914–1980*. New York: The Military Press, 1983.

Baals, Donald D., and William R. Corliss, *Wind Tunnels of NASA*. NASA SP-440. Washington, DC: NASA, 1981.

Baize, Daniel G., ed. *1995 NASA High-Speed Research Program Sonic Boom Workshop Held at Langley Research Center, Hampton, Virginia, September 12–13, 1995* 1. NASA CP-3335. Hampton, VA: Langley Research Center, July 1996. *Configuration, Design, Analysis, and Testing* 2. NASA CP-1999-209520. Hampton, VA: Langley Research Center, December 1999.

Brandt, Steven A., John J. Bertin, Randall J. Stiles, and Ray Whitford. *Introduction to Aeronautics: A Design Perspective*. Reston, VA: AIAA, 2004.

Chambers, Joseph R. *Innovation in Flight; Research of the NASA Langley Research Center on Revolutionary Concepts for Aeronautics*. NASA SP-2005-4539. Washington, DC: NASA, 2005.

Conway, Erik M. *High-Speed Dreams: NASA and the Technopolitics of Supersonic Transportation, 1945–1999.* Baltimore: Johns Hopkins University Press, 2005.

Darden, Christine M., ed. *High-Speed Research: Sonic Boom; Proceedings of a Conference Held at Langley Research Center, Hampton, Virginia, February 25–27, 1992.* 2 vols. NASA CP-3172 and CP-3173. Washington, DC: NASA, 1992.

Darden, Christine, Albert R. George, Wallace D. Hayes, Allan D. Pierce, and Clemans Powell. *Status of Sonic Boom Methodology and Understanding; Proceedings of a Workshop Sponsored by the National Aeronautics and Space Administration and held at NASA Langley Research Center, Hampton, Virginia, January 19–20, 1988.* NASA CP-3027. Washington, DC: NASA, 1989.

Darling, David. *The Complete Book of Spaceflight: From Apollo 1 to Zero Gravity.* Hoboken, NJ: John Wiley and Sons, 2003.

Dick, Steven J., ed. *NASA's First 50 Years: Historical Perspectives; NASA 50th Anniversary Proceedings.* NASA SP-2010-4704. Washington, DC: NASA, 2010.

Driver, Cornelius (conference chairman). *Proceedings of the SCAR Conference Held at Langley Research Center, Hampton, Virginia, November 9–12, 1976.* Parts 1 and 2. NASA CP-001. Hampton, VA: Langley Research Center, 1976.

Driver, Cornelius, and Hal T. Barber (conference cochairmen). *Supersonic Cruise Research '79: Proceedings of a Conference Held at Langley Research Center, Hampton, VA, November 13–16, 1979.* Parts 1 and 2. NASA CP-2108. Hampton, VA: Langley Research Center, 1980.

Edwards, Thomas A., ed. *High-Speed Research: Sonic Boom; Proceedings of a Conference Held at Ames Research Center, Moffett Field, California, May 12–14, 1993* 1. NASA CP-10132. Moffett Field, CA: Ames Research Center, 1994.

Gorn, Michael H. *Expanding the Envelope: Flight Research at NACA and NASA.* Lexington, KY: University of Kentucky Press, 2001.

Hallion, Richard P. *Supersonic Flight: Breaking the Sound Barrier and Beyond—The Story of the Bell X-1 and Douglas D-558.* New York: The Macmillan Co. in association with the Smithsonian Institution National Air and Space Museum, 1972.

Hallion, Richard P., and Michael H. Gorn. *On the Frontier: Experimental Flight at NASA Dryden.* Washington, DC: Smithsonian Institution Press, 2003.

Hallion, Richard P., ed. *NASA's Contributions to Aeronautics.* 2 vols. NASA SP-2010-570. Washington, DC: NASA, 2010.

Horwitch, Mel. *Clipped Wings: The American SST Conflict.* Cambridge, MA: MIT Press, 1982.

Huebner, Lawrence D., Scott C. Asbury, John E. Lamar, Robert E. McKinley, Robert C. Scott, William J. Small, and Abel O. Torres (compilers). *Transportation Beyond 2000: Technologies Needed for Engineering Design, Proceedings of a Workshop Held at Hampton, Virginia, September 26–28, 1995.* 2 parts. NASA CP-10184. February 1996.

James and Associates, ed. *YF-12 Experiments Symposium: A Conference Held at Dryden Flight Research Center, Edwards, California, September 13–15, 1978.* NASA CP-2054. 1978.

Jenkins, Dennis R., and Tony R. Landis, *Valkyrie: North America's Mach 3 Superbomber.* North Branch, MN: Specialty Press, 2004.

Johnsen, Frederick A. *Northrop F-5/F-20/T-38.* Vol. 44, Warbird Tech Series. North Branch, MN: Specialty Press, 2006.

Kempel, Robert W. *The Conquest of the Sound Barrier.* Beirut: HPM Publications, 2007.

Kent, Richard J. *Safe, Separated, and Soaring: A History of Civil Aviation Policy, 1961–1972.* Washington, DC: FAA, 1980.

Knaack, Marcelle S. *Post–World War II Bombers, 1945–1973* 2, *Encyclopedia of U.S. Air Force Aircraft and Missile Systems.* Washington, DC: GPO for Office of Air Force History, 1988.

Langley Research Center. *Aerodynamic Analysis Requiring Advanced Computers.* Parts 1 and 2 [conference proceedings]. NASA SP-347. Washington, DC: National Technical Information Service, 1975.

Marburger, John H., signatory. *National Plan for Aeronautics Research and Development and Related Infrastructure.* Washington, DC: National Science and Technology Council, December 2007. *http://www.whitehouse.gov/sites/default/files/microsites/ostp/aero-natplan-2007.pdf.*

McCurdy, David A., ed. *High-Speed Research: 1994 Sonic Boom Workshop: Proceedings of a Workshop Held in Hampton, Virginia, June 1–3, 1994* 1, *Atmospheric Propagation and Acceptability Studies.* NASA CP-3279. Hampton, VA: Langley Research Center, October 1994. 2, *Configuration, Design, Analysis, and Testing.* NASA CP-1999-209699. Hampton, VA: Langley Research Center, December 1999.

McLean, F. Edward. *Supersonic Cruise Technology.* NASA SP-472. Washington, DC: NASA, 1985.

McLucas, John L., Kenneth J. Alnwick, and Lawrence R. Benson. *Reflections of a Technocrat: Managing Defense, Air, and Space Programs During the Cold War.* Montgomery, AL: Air University Press, 2006.

Merlin, Peter W. *From Archangel to Senior Crown: Design and Development of the Blackbird.* Reston, VA: AIAA, 2008.

National Research Council. *Commercial Supersonic Technology: The Way Ahead.* Washington, DC: National Academies Press, 2001.

National Research Council. *Recapturing NASA's Aeronautics Flight Research Capabilities.* Washington, DC: National Academies Press, 2012.

National Research Council. *U.S. Supersonic Aircraft: Assessing NASA's High-Speed Research Program.* Washington, DC: National Academies Press, 1997.

Neufeld, Michael J. *Von Braun: Dreamer of Space, Engineer of War.* New York: Alfred A. Knopf, 2007.

Rochester, Stuart I. *Takeoff at Mid-Century: Federal Civil Aviation Policy in the Eisenhower Years, 1953–1961.* Washington, DC: FAA, 1976.

Rumerman, Judy A. *NASA Historical Data Book Volume VI: NASA Space Applications, Aeronautics, Space Research and Technology, Tracking and Data Acquisition/Space Operations, Commercial Programs, and Resources, 1979–1988.* NASA SP-2000-4012. Washington, DC: NASA, 2000.

Schwartz, Ira R., ed. *Second Conference on Sonic Boom Research: Proceedings of a Conference Held at the National Aeronautics and Space Administration, Washington, DC, May 9–10, 1968.* NASA SP-180. Washington, DC: NASA, 1968.

Schwartz, Ira R., ed. *Third Conference on Sonic Boom Research: Proceedings of a Conference Held at the National Aeronautics and Space Administration, Washington, DC, October 29–30, 1970.* NASA SP-255. Washington, DC: NASA, 1971.

Seebass, A. Richard, ed. *Sonic Boom Research: Proceedings of a Conference Held at the National Aeronautics and Space Administration, Washington, DC, April 12, 1967.* NASA SP-147. Washington, DC: NASA, 1967.

Stillwell, Wendell H. *X-15 Research Results.* NASA SP-60. Washington, DC: NASA, 1965.

Wallace, Lane E. *Flights of Discovery: Sixty Years of Flight Research at Dryden Flight Research Center.* NASA SP-2006-4318. Washington, DC: NASA, 2006.

Ward, Bob. *Dr. Space: The Life of Wernher von Braun.* Annapolis, MD: Naval Institute, 2005.

Whitehead, Allen H., ed. *First Annual High-Speed Research Workshop; Proceedings of a Workshop Sponsored by the National Aeronautics and Space Administration, Washington, DC, and Held in Williamsburg, Virginia, May 14–16, 1991.* 4 parts. NASA CP-10087. Hampton, VA: Langley Research Center, 1992.

Wolfe, Tom. *The Right Stuff.* New York: Bantam Books, 1980.

Yeager, Chuck, and Leo Janis. *Yeager.* New York: Bantam Books, 1985.

Reports, Papers, Presentations, and Articles[1]

Aboulafia, Richard. "The Business Case for Higher Speed." *Aerospace America Online*. Accessed July 8, 2011, *http://www.aiaa.org/aerospace/ Article.cfm?issuetocid=ArchiveIssueID=16*.

Alford, William J., and Cornelius Driver. "Recent Supersonic Transport Research." *Astronautics & Aeronautics* 2, no. 9 (September 1964): 26–37.

Anderson, John D. "NASA and the Evolution of Computational Fluid Dynamics." In Hallion, *NASA's Contributions to Aeronautics* 1, 431–434.

Anderson, Robert E. "First Annual HSR Program Workshop: Headquarters Perspective." In Whitehead, *1991 HSR Workshop*, 3–24.

Andrews, William H. "Summary of Preliminary Data Derived from the XB-70 Airplanes." NASA TM X-1240 (June 1966).

Anselmo, Joseph C., and William Garvey. "Prolonged Pain: Honeywell's Bizjet Forecast Corks Any Premature Celebration." *Aviation Week*, October 10, 2011, 54–55.

Bahm, Catherine M., and Edward A. Haering. "Ground-Recorded Sonic Boom Signatures of F/A-18 Aircraft in Formation Flight." In Baize, *1995 Sonic Boom Workshop* 1, 220–243.

Baize, Donald G., Peter G. Coen, Chris S. Domack, James A. Fenbert, Karl A. Geiselhart, Marcus O. McElroy, Kathy E. Needleman, and Lori P. Ozororski. "A Performance Assessment of Eight Low-Boom High-Speed Civil Transport Concepts." In McCurdy, *1994 Sonic Boom Workshop: Configuration Design, Analysis, and Testing*, 149–170.

Banke, Jim. "Quieter Flight: A Balancing Act." *Aerospace America*, March 2011, 38–43.

Barger, Raymond L. "Sonic-Boom Wave-Front Shapes and Curvatures Associated with Maneuvering Flight." NASA TP-1611 (December 1979).

---

1. See notes for additional magazine and newspaper articles, news releases, and similar sources.

Barger, Raymond L., W.D. Beasley, and J.D. Brooks. "Laboratory Investigation of Diffraction and Reflection of Sonic Booms by Buildings." NASA TN D-5830 (June 1970).

Beasly, W.D., J.D. Brooks, and R.L. Barger. "A Laboratory Investigation of N-Wave Focusing." NASA TN D-5306 (July 1969).

Beissner, F.L. "Application of Near-Term Technology to a Mach 2.0 Variable Sweep Wing Supersonic Cruise Executive Jet." NASA CR-172321 (March 1984).

Beissner, F.L. "Effects of Emerging Technology on a Convertible, Business/Interceptor, Supersonic Cruise Jet." NASA CR-178097 (May 1986).

Beissner, F.L., W.A. Lovell, A. Warner Robins, and E.E. Swanson. "Effects of Advanced Technology and a Fuel-Efficient Engine on a Supersonic-Cruise Executive Jet with a Small Cabin." NASA CR-172190 (August 1983).

Bishop, Dwight E. "Noise and Sonic Boom Impact Technology: PCBOOM Computer Program for Sonic Boom Research Technical Report" 1. HSD-TR-88-014 (October 1988).

Bloom, A.J., G. Kost, J. Prouix, and R.L. Sharpe, "Response of Structures to Sonic Booms Produced by XB-70, B-58, and F-104 Aircraft…at Edwards Air Force Base, Final Report," NSBEO 2-67 (October 1967).

Bobbitt, Percy J. "Application of Computational Fluid Dynamics and Laminar Flow Technology for Improved Performance and Sonic Boom Reduction." In Darden, *1992 Sonic Boom Workshop* 2, 137–144.

Bobbitt, Percy J., and Domenic Maglieri. "Dr. Antonio Ferri's Contribution to Supersonic Transport Sonic-Boom Technology." *Journal of Spacecraft and Rockets* 40, no. 4 (July–August 2003): 459–466.

Boeing Commercial Airplanes. "High-Speed Civil Transport Study; Final Report." NASA CR-4234 (September 1989).

Borsky, Paul M. "Community Reactions to Sonic Booms in the Oklahoma City Area" 2, "Data on Community Reactions and Interpretations." USAF Aerospace Medical Research Laboratory (August 1965). *http://*

www.norc.org/NR/rdonlyres/255A2AA2-B953-4305-9AD0-B8ABC-
C824FA9/0/NORCRpt_101B.pdf.

Brain, Marshall. "How Stereolithography 3-D Layering Works." Accessed July 30, 2011, *http://computer.howstuffworks.com/stereolith.htm.*

Bruner, H.S. "SSBJ—A Technological Challenge." *ICAO Journal* 46, no. 8 (August 1991): 9–13.

Bui, Trong T. "CFD Analysis of Nozzle Jet Plume Effects on Sonic Boom Signature." AIAA paper no. 2009-1054 (January 2009).

Busemann, Adolf. "Sonic Boom Reduction." In Seebass, *Sonic Boom Research: Proceedings of a Conference,* 79–82.

Carlson, Harry W. "An Investigation of Some Aspects of the Sonic Boom by Means of Wind-Tunnel Measurements of Pressures about Several Bodies at a Mach Number of 2.01." NASA TN D-161 (December 1959).

Carlson, Harry W. "Configuration Effects on Sonic Boom." In NASA Langley, "Proceedings of NASA Conference on SST Feasibility." 381–398 (December 1963).

Carlson, Harry W. "Correlation of Sonic-Boom Theory with Wind Tunnel and Flight Measurements." NASA TR-R-213 (December 1964).

Carlson, Harry W. "Experimental and Analytical Research on Sonic Boom Generation at NASA." In Seebass, *Sonic Boom Research: Proceedings of a Conference,* 9–23.

Carlson, Harry W. "Simplified Sonic-Boom Prediction." NASA TP-1122 (March 1978).

Carlson, Harry W. "Some Notes on the Status of Sonic Boom Prediction and Minimization Research." In Schwartz, *Third Conference on Sonic Boom Research,* 395–400.

Carlson, Harry W. "The Lower Bound of Attainable Sonic-Boom Overpressure and Design Methods of Approaching This Limit." NASA TN D-1494 (October 1962).

Carlson, Harry W. "Wind Tunnel Measurements of the Sonic-Boom Characteristics of a Supersonic Bomber Model and a Correlation with Flight-Test Ground Measurements." NASA TM X-700 (July 1962).

Casper, J.H., S.E. Cliff, D.A. Durston, M.S. McMullen, J.E. Melton, M.A. Park, and S.D. Thomas. "Assessment of Near-Field Sonic Boom Simulation Tools." AIAA paper no. 2008-6592 (August 2008).

Cheung, Samsun, and Thomas A. Edwards. "Supersonic Airplane Design Optimization Method for Aerodynamic Performance and Low Sonic Boom." In Darden, *1992 Sonic Boom Workshop* 2, 31–44.

Cheung, Samsun. "Supersonic Civil Airplane Study and Design: Performance and Sonic Boom." NASA CR-197745 (January 1995).

Clark, Buhr, and Nexen (company name). "Studies of Sonic Boom Damage." NASA CR-227 (May 1965).

Clark, Evert. "Reduced Sonic Boom Foreseen for New High-Speed Airliner." *New York Times*, January 14, 1965, 7, 12.

Cleveland, Robin O., David T. Blackstock, and Mark F. Hamilton. "Effect of Stratification and Geometrical Spreading on Sonic Boom Rise Time." In McCurdy, *1994 Sonic Boom Workshop: Configuration, Design, Analysis, and Testing*, 19–38.

Cliatt, Larry J., Trong Bui, and Edward A. Haering. "Overview of the LaNCETS Flight Experiment and CFD Analysis." PowerPoint presentation. Fundamental Aeronautics Annual Meeting, Atlanta, GA, October 2, 2008.

Cliff, Susan E. "Computational/Experimental Analysis of Three Low Sonic Boom Configurations with Design Modifications." In Darden, *1992 Sonic Boom Workshop* 2, 89–118.

Cliff, Susan E., Timothy J. Baker, and Raymond M. Hicks. "Design and Computational/Experimental Analysis of Low Sonic Boom Configurations." In McCurdy, *1994 Sonic Boom Workshop: Configuration Design, Analysis, and Testing*, 33–57.

Coen, Peter G. "Development of a Computer Technique for Prediction of Transport Aircraft Flight Profile Sonic Boom Signatures." NASA CR-188117 (March 1991).

Coen, Peter G. "Fixing the Sound Barrier: Three Generations of U.S. Research into Sonic Boom Reduction...and What It Means to the Future." PowerPoint presentation. FAA Public Meeting on Advanced Technologies and Supersonics, Washington, DC, July 14, 2011.

Coen, Peter G. "Fundamental Aeronautics Program Supersonics Project." PowerPoint presentation. Fundamental Aeronautics Technical Conference, Cleveland, OH, March 15–17, 2011.

Coen, Peter G. "Fundamental Aeronautics Program Supersonics Project." PowerPoint presentation. Fundamental Aeronautics Technical Conference, Cleveland, OH, March 13–15, 2012.

Coen, Peter G. "Supersonic Vehicles Technology: Sonic Boom Technology Development and Demonstration." PowerPoint presentation. FAA Civil Supersonic Aircraft Technical Workshop, Arlington, VA, November 13, 2003.

Coen, Peter G. "Supersonics Program Overview." PowerPoint presentation. NASA Fundamental Aeronautics Annual Meeting, New Orleans, LA, October 30, 2007.

Coen, Peter G. (for Kevin P. Shepherd). "Human Response to Sonic Booms." PowerPoint presentation, FAA Civil Supersonic Aircraft Technical Workshop, Arlington, VA, November 13, 2003.

Coen, Peter G., and Roy Martin. "Fixing the Sound Barrier," PowerPoint presentation. Experimental Aircraft Assoc. AirVenture, Oshkosh, WI, July 2004.

Coen, Peter G., David H. Graham, Domenic J. Maglieri, and Joseph W. Pawlowski. "QSP Shaped Sonic Boom Demo." PowerPoint presentation. 43rd AIAA Aerospace Sciences Meeting, Reno, NV, January 10, 2005.

Coen, Peter G., David H. Graham, Domenic J. Maglieri, and Joseph W. Pawlowski. "Origins and Overview of the Demonstration Program."

PowerPoint presentation. 43rd AIAA Aerospace Sciences Meeting, Reno, NV, January 10, 2005.

Coen, Peter G., Lou Povinelli, and Kas Civinskas, "Supersonics Program Overview." PowerPoint presentation. NASA Fundamental Aeronautics Program Annual Meeting, Atlanta, GA, October 7–9, 2008.

Coen, Peter G., Mary Jo Long-Davis, and Louis Povinelli. "Fundamental Aeronautics Program Supersonics Project, Reference Document." May 26, 2006. Accessed March 15, 2009, *http://www.aeronautics.nasa.gov/fap/documents/html.*

Cole, M.J., and M.B. Freeman. "Analysis of Multiple Scattering of Shock Waves by a Turbulent Atmosphere." In Schwartz, *Third Conference on Sonic Boom Research*, 67–74.

Collogan, David. "Manufacturers, NASA, Working on BizJet Sonic Boom Project." *Aviation Week*, July 18, 2005. *http://www.aviationweek.com/aw/generic/story_generic.jsp?channel=businessweekly&id=news/SONB07185.xml&headline=Manufacturers,%20NASA%20Working%20On%20Bizjet%20Sonic%20Boom%20Project.*

Collogan, David. "Two Groups Vie To Develop Supersonic Business Jets," *Aviation Week*, October 18, 2004. *http://www.aviationweek.com/aw/generic/story_generic.jsp?channel=businessweekly&id=news/SUPER10184.xml.*

Covault, Craig. "NASA Advances Supersonic Technology." *Aviation Week* (January 10, 1977): 16–18.

Cowart, Robbie. "Supersonic Technology Development." PowerPoint presentation. FAA Public Meeting on Advanced Technologies and Supersonics, Washington, DC, July 14, 2011.

Creech, Gray. "Getting Loud Now: Enjoy Peace and Quiet Then." Dryden Flight Research Center, May 27, 2011. *http://www.nasa.gov/centers/dryden/Features/scamp.html.*

Creech, Gray. "ICAO Team Witnesses Sonic Boom Research Flights." Dryden Flight Research Center, September 29, 2009. *http://www.nasa.gov/centers/dryden/Features/icao_sonic_booms.html.*

Creech, Gray. "Supersonic Diving: Quieting the Boom." Dryden Flight Research Center, June 18, 2009. *http://www.nasa.gov/centers/dryden/Features/sonicbobs.html.*

Croft, John. "Engineering through the Sound Barrier." *Aerospace America* 42, no. 9 (September 2004): 24–31.

Cumming, Stephen B., Michael A. Frederick, and Mark S. Smith. "Aerodynamic Effects of a 24-foot Multi-segmented Telescoping Nose Boom on an F-15B Airplane." NASA TM-2008-214634 (April 2008). Also published as AIAA paper no. 2007-6638 (August 2007).

Cunningham, Aimee. "Sonic Booms and Human Ears: How Much Can the Public Tolerate?" *Popular Science*, July 30, 2004. *www.popsci.com/military-aviation-space/article/2004-07/sonic-booms-and-human-ears.*

Dahlke, Hugo E., G.T. Kantarges, T.E. Siddon, and J.J. Van Houten. "The Shock-Expansion Tube and Its Application as a Sonic Boom Simulator." NASA CR-1055 (June 1968).

Darden, Christine M. "Affordable/Acceptable Supersonic Flight: Is It Near?" 40th Aircraft Symposium, Japan Society for Aeronautical and Space Sciences (JSASS), Yokohama (October 2002).

Darden, Christine M. "Charts for Determining Potential Minimum Sonic-Boom Overpressures for Supersonic Cruise Aircraft." NASA TP-1820 (May 1981).

Darden, Christine M. "Minimization of Sonic-Boom Parameters in Real and Isothermal Atmospheres." NASA TN D-7842 (March 1975).

Darden, Christine M. "Progress in Sonic-Boom Understanding: Lessons Learned and Next Steps." In McCurdy, *1994 Sonic Boom Workshop, Configuration, Design, and Testing*, 269–292.

Darden, Christine M. "Sonic Boom Minimization with Nose-Bluntness Relaxation." NASA TP-1348 (January 1979).

Darden, Christine M. "Sonic Boom Theory—Its Status in Prediction and Minimization." AIAA paper no. 76-1 (January 1976).

Darden, Christine M., and Robert J. Mack. "Current Research in Sonic-Boom Minimization." In Driver, *Proceedings of the SCAR Conference* (1976), pt. 1, 525–541.

Darden, Christine M., Robert J. Mack, Kathy E. Needleman, Daniel G. Baize, Peter G. Coen, Raymond L. Barger, N. Duane Melson, Mary S. Adams, Elwood W. Shields, and Marvin E. McGraw. "Design and Analysis of Low Boom Concepts at Langley Research Center." In Whitehead, *1991 HSR Workshop*, pt. 2, 673–700.

DeMeis, Richard. "Sukhoi and Gulfstream Go Supersonic." *Aerospace America* 28, no. 4 (April 1990): 40–42.

Dornheim, Michael A. "Will Low Boom Fly? NASA Cutbacks Delay Flight Test of Shaped Demonstrator, But Work on It Continues." *Aviation Week* (November 7, 2005): 68–69.

Douglas Aircraft Company. "1989 High-Speed Civil Transport Studies." NASA CR-4375 (May 1991).

Downing, J. Micah. "Lateral Spread of Sonic Boom Measurements from U.S. Air Force Boomfile Flight Tests." In Darden, *1992 Sonic Boom Workshop* 1, 117–136.

Downing, J. Micah, Noel Zemat, Chris Moss, Daniel Wolski, E. Chung Sukhway, Kenneth Plotkin, and Domenic Maglieri. "Measurement of Controlled Focused Sonic Booms from Maneuvering Aircraft." *Journal of the Acoustical Society of America* 104, no. 1 (July 1998): 112–121.

Dumond, J.W.M., E.R. Cohen, W.K.H. Panfsky, and E. Deeds. "A Determination of the Wave Forms and Laws of Propagation and Dissipation of Ballistic Shock Waves," *Journal of the Acoustical Society of America* 18, no. 1 (January 1946): 97–118.

Edge, Philip M., and Harvey H. Hubbard. "Review of Sonic Boom Simulation Devices and Techniques," *Journal of the Acoustical Society of America* 51, no. 2, pt. 3 (February 1972): 724–728.

Edge, Philip M., and William H. Mayes. "Description of Langley Low-Frequency Noise Facility and Study of Human Response to Noise Frequencies Below 50 cps." NASA TN D-3204 (January 1966).

Edwards, Thomas A., Raymond Hicks, Samson Cheung, Susan Cliff-Hovey, Mike Madison, and Joel Mendoza. "Sonic Boom Prediction and Minimization Using Computational Fluid Dynamics." In Whitehead, *1991 HSR Workshop*, pt. 2, 721–738.

FAA, "Civil Aircraft Sonic Boom," *Federal Register* 35, no. 4 (April 16, 1970): 6189–6190.

Ferri, Antonio, and Ahmed Ismail. "Analysis of Configurations." In Schwartz, *Second Conference on Sonic Boom Research*, 73–88.

Ferri, Antonio, Huai-Chu Wang, and Hans Sorensen. "Experimental Verification of Low Sonic Boom Configuration." NASA CR-2070 (June 1973).

Ferri, Antonio. "Airplane Configurations for Low Sonic Boom." In Schwartz, *Third Conference on Sonic Boom Research*, 255–276.

Fields, James M. "Reactions of Residents to Long-Term Sonic Boom Noise Environments," NASA CR-201704 (June 1997).

Fields, James M., Carey Moulton, Robert M. Baumgartner, and Jeff Thomas. "Residents' Reactions to Long-Term Sonic Boom Exposure: Preliminary Results." In McCurdy, *1994 Sonic Boom Workshop: Atmospheric Propagation and Acceptability*, 209–217.

Findley, Donald S. "Vibration Responses of Test Structure no. 2 during the Edwards AFB Phase of the National Sonic Boom Program," NASA TM-X-72704 (June 1975).

Findley, Donald S., Vera Huckel, and Herbert R. Henderson. "Vibration Responses of Test Structure no. 1 During the Edwards AFB Phase of the National Sonic Boom Program," NASA TM-X-72706 (June 1975).

Flynn, Edward O. "Lowering the Boom on Supersonic Flight Noise." *Aerospace America* 40, no. 2 (February 2002): 20–21.

Francombe, R. "LTM—A Flexible Processing Technology for Polymer Composite Structures." Paper presented at NATO Research and Technology Organization (RTO) Applied Vehicle Technology (AVT) specialists meeting on low-cost composite structures, Loen, Norway,

May 7–11, 2001. Accessed August 28, 2011. *ftp.rta.nato.int/public//PubFullText/RTO/...///MP-069(II)-(SM1)-12.pdf.*

Friedman, M.P., E.J. Kane, and A. Sigalla. "Effects of Atmosphere and Aircraft Motion on the Location and Intensity of a Sonic Boom." *AIAA Journal* 1, no. 6 (June 1963): 1327–1335.

Fryer, B.A., "Publications in Acoustics and Noise Control from the NASA Langley Research Center During 1940–1976." NASA TM X-7402 (July 1977).

Gardner, John H., and Peter H. Rogers. "Thermospheric Propagation of Sonic Booms from the Concorde Supersonic Transport." Naval Research Laboratory Memo Report 3904 (February 1979).

George, Albert R. "Lower Bounds for Sonic Booms in the Midfield." *AIAA Journal* 7, no. 8 (August 1969): 1542–1545.

George, Albert R. "The Effects of Atmospheric Inhomogeneities on Sonic Boom." In Schwartz, *Third Conference on Sonic Boom Research*, 33–58.

George, Albert R. "The Possibilities for Reducing Sonic Booms by Lateral Redistribution." In Seebass, *Sonic Boom Research: Proceedings of a Conference*, 83–93.

George, Albert R., and R. Seebass. "Sonic Boom Minimization Including Both Front and Rear Shocks." *AIAA Journal* 9, no. 10 (October 1971): 2091–2093.

George, Albert R., and R. Seebass. "Sonic-Boom Minimization," *Journal of the Acoustical Society of America* 51, no. 2, pt. 3 (February 1972): 686–694.

Gill, Peter M. "Nonlinear Acoustic Behavior at a Caustic." Ph.D. thesis. Cornell University, June 1974.

Graham, David H. "Shaped Sonic Boom Demonstrator Program." PowerPoint presentation. FAA Civil Supersonic Aircraft Technical Workshop, Arlington, VA, November 13, 2003.

Graham, David H., and Roy Martin. "Aerodynamic Design and Validation of SSBD Aircraft." Northrop Grumman Corp. PowerPoint presentation. Shaped Sonic Boom Experiment Closeout Workshop, NASA Langley Research Center, Hampton, VA, August 17, 2004.

Graham, David H., John A. Dahlin, Juliet A. Page, Kenneth J. Plotkin, and Peter G. Coen. "Wind Tunnel Validation of Shaped Sonic Boom Demonstration Aircraft Design." AIAA paper no. 2005-7 (January 2005).

Graham, David H., John A. Dahlin, Keith B. Meredith, and Jay L. Vadnais, "Aerodynamic Design of Shaped Sonic Boom Demonstration Aircraft." AIAA paper no. 2005-8 (January 2005).

Green, Karen S., and Terrill W. Putnam. "Measurements of Sonic Booms Generated by an Airplane Flying at Mach 3.5 and 4.8." NASA TM X-3126 (October 1974).

Greene, Randall, and Richard Seebass. "A Corporate Supersonic Transport." In Huebner et al., *Transportation Beyond 2000: Technologies Needed for Engineering Design, Proceedings of a Workshop Held in Hampton, Virginia, September 26–28, 1995*. Pt. 1. NASA CP-10184 (February 1996), 491–508.

Guiraud, J.P. "Focalization in Short Non-Linear Waves," NASA Technical Translation F-12,442 (September 1969).

Haering, Edward A., and James E. Murray. "Shaped Sonic Boom Demonstration/Experiment Airborne Data: SSBD Final Review." PowerPoint presentation. Shaped Sonic Boom Experiment Closeout Workshop. NASA Langley Research Center, Hampton, VA, August 17, 2004.

Haering, Edward A., James E. Murray, Dana D. Purifoy, David H. Graham, Keith B. Meredith, Christopher E. Ashburn, and Lt. Col. Mark Stucky. "Airborne Shaped Sonic Boom Demonstration Pressure Measurements with Computational Fluid Dynamics." AIAA paper no. 2005-9 (January 2005).

Haering, Edward A., James W. Smolka, James E. Murray, and Kenneth J. Plotkin. "Flight Demonstration of Low Overpressure N-Wave

Sonic Booms and Evanescent Waves." Presented at 17th International Symposium on Nonlinear Acoustics. International Sonic Boom Forum, State College, PA, July 21–22, 2005.

Haering, Edward A., Jr., and James E. Murray. "Shaped Sonic Boom Demonstration/Experiment Airborne Data SSBD Final Review." PowerPoint presentation. NASA Langley Research Center, Hampton, VA, August 17, 2004.

Haering, Edward A., L.J. Ehernberger, and Stephen A. Whitmore. "Preliminary Airborne Experiments for the SR-71 Sonic Boom Propagation Experiment." In Baize, *1995 Sonic Boom Workshop* 1, 176–198.

Haering, Edward A., Larry J. Cliatt, Thomas J. Bunce, Thomas B. Gabrielson, Victor W. Sparrow, and Lance L. Locey. "Initial Results from the Variable Intensity Sonic Boom Propagation Database." AIAA paper no. 2008-3034 (May 2008).

Haering, Edward A., Stephen A. Whitmore, and L.J. Ehernberger. "Measurement of the Basic SR-71 Airplane Near-Field Signature." In McCurdy, *1994 Sonic Boom Workshop: Configuration, Design, Analysis, and Testing*, 171–197.

Hagerman, Eric. "All Sonic, No Boom." *Popular Science*, March 2007. *http://www.popsci.com/military-aviation-space/article/2007-03/all-sonic-no-boom.*

Haglund, George T. "HSCT Designs for Reduced Sonic Boom." AIAA paper no. 91-3103 (September 1991).

Haglund, George T., and Edward J. Kane. "Flight Test Measurements and Analysis of Sonic Boom Phenomena Near the Shock Wave Extremity." NASA CR-2167 (February 1973).

Haglund, George T., and Steven S. Ogg. "Two HSCT Mach 1.7 Low Sonic Boom Designs." In Darden, *1992 Sonic Boom Workshop* 2, 65–88.

Hailey, Albion B. "AF Expert Dodges Efforts To Detail 'Sonic Boom' Loss." *Washington Post*, August 25, 1960, A15.

Hallion, Richard P. "Richard Whitcomb's Triple Play." *Air Force Magazine* 93, no. 2 (February, 2010): 68–72.

Hartwich, Peter M., Billy A. Burroughs, James S. Herzberg, and Curtiss D. Wiler. "Design Development Strategies and Technology Integration for Supersonic Aircraft of Low Perceived Sonic Boom." AIAA paper no. 2003-0556 (January 2003).

Hayes, Wallace, and Rudolph C. Haefeli. "The ARAP Sonic Boom Computer Program." In Schwartz, *Second Conference on Sonic Boom Research*, 151–158.

Hayes, Wallace, Rudolph C. Haefeli, and H.E. Kulsrud. "Sonic Boom Propagation in a Stratified Atmosphere with Computer Program." NASA CR-1299 (April 1969).

Hayes, Wallace. "Brief Review of the Basic Theory," *Journal of the Acoustical Society of America* 39, no. 5, pt. 2 (November 1966): 3–6.

Henne, Preston A. "Case for a Small Supersonic Civil Aircraft." *Journal of Aircraft* 42, no. 3 (May–June 2005): 765–774.

Heppenheimer, T.A. "The Boom Stops Here." *Air & Space Magazine*, October–November 2005. *http://www.airspacemag.com/flight-today/boom.html*.

Herbert, Adam J. "Long-Range Strike in a Hurry." *Air Force Magazine* 87, no. 11 (November 2004): 27–31.

Hicks, Raymond M., and Joel P. Mendoza. "Prediction of Sonic Boom Characteristics from Experimental Near Field Results." NASA TM-X-1477 (November 1967).

Higgins, Thomas H. "Sonic Boom Research and Design Considerations in the Development of a Commercial Supersonic Transport." *Journal of the Acoustical Society of America* 39, no. 5, pt. 2 (November 1966): 526–531.

Hilton, David A., D.J. Maglieri, and N.J. McLeod. "Summary of Variations of Sonic Boom Signatures Resulting from Atmospheric Effects." NASA TM-X-59633 (February 1967).

Hilton, David A., D.J. Maglieri, and R. Steiner. "Sonic-Boom Exposures During FAA Community Response Studies over a 6-Month Period in the Oklahoma City Area." NASA-TN-D-2539 (December 1964).

Hilton, David A., Harvey. H. Hubbard, Vera Huckel, and Domenic J. Maglieri. "Ground Measurements of Sonic-Boom Pressures for the Altitude Range of 10,000 to 75,000 Feet." NASA TR R-198 (July 1964).

Hilton, David A., Vera Huckel, and Domenic J. Maglieri. "Sonic Boom Measurements During Bomber Training Operations in the Chicago Area." NASA TN D-3655 (October 1966).

Hoffman, Sherwood. "Bibliography of Supersonic Cruise Aircraft Research (SCAR)" [1972-77]. NASA RP-1003 (November 1977).

Hoffman, Sherwood. "Bibliography of Supersonic Cruise Research (SCR) Program from 1977 to Mid-1980." NASA RP-1063 (December 1980).

Holloway, P.F., G.A. Wilhold, J.H. Jones, F. Garcia, and R.M. Hicks. "Shuttle Sonic Boom—Technology and Predictions." AIAA paper no. 73-1039 (October 1973).

Howe, Donald C., Frank Simmons, and Donald Freund. "Development of the Gulfstream Quiet Spike for Sonic Boom Minimization." AIAA paper no. 2008-124 (January 2008).

Hubbard, Harvey H., and Domenic J. Maglieri. "Factors Affecting Community Acceptance of the Sonic Boom." In NASA Langley, "Proceedings of NASA Conference on SST Feasibility." December 1963, 399–412.

Hubbard, Harvey H., and Domenic J. Maglieri. "Sonic Boom Signature Data from Cruciform Microphone Array Experiments During the 1966–67 EAFB National Sonic Boom Evaluation Program." NASA TN D-6823 (May 1972).

Hubbard, Harvey H., Domenic J. Maglieri, and David G. Stephens. "Sonic Boom Research—Selected Bibliography with Annotation." NASA TM 87685 (September 1986).

Huber, Mark. "Mach 1 for Millionaires." *Air & Space Magazine*, March–April 2006. *http://www.airspacemag.com/flight-today/millionaire.html.*

Jones, L.B. "Lower Bounds for Sonic Bang in the Far Field." *Aeronautical Quarterly* 18, pt. 1 (February 1967): 1–21.

Jones, L.B. "Lower Bounds for Sonic Bangs." *Journal of the Royal Aeronautical Society* 65, no. 606 (June 1961): 433–436.

Jones, L.B. "Lower Bounds for the Pressure Jump of the Bow Shock of a Supersonic Transport." *Aeronautical Quarterly* 21 (February 1970): 1–17.

Jones, Stacy V. "Sonic Boom Researchers Use Simulator," *New York Times*, May 10, 1969, 37, 41.

Jordan, Gareth H. "Flight Measurements of Sonic Booms and Effects of Shock Waves on Aircraft." *Society of Experimental Test Pilots Quarterly Review* 5, no. 1 (1961): 117–131.

Kamran Fouladi. "CFD Predictions of Sonic-Boom Characteristics for Unmodified and Modified SR-71 Configurations." In McCurdy, *1994 Sonic Boom Workshop: Configuration, Design, Analysis, and Testing,* 219–235.

Kane, Edward J. "A Study to Determine the Feasibility of a Low Sonic Boom Supersonic Transport." AIAA paper no. 73-1035 (October 1973).

Kane, Edward J. "Some Effects of the Atmosphere on Sonic Boom." In Seebass, *Sonic Boom Research: Proceedings of a Conference,* 49–64

Keefe, Fred, and Grover Amen. "Boom." *The New Yorker*, May 16, 1962, 33–34.

Kerr, Richard. "East Coast Mystery Booms: Mystery Gone but Booms Linger On." *Science* 203, no. 4337 (January 19, 1979): 256.

Kinney, Jeremy. "NASA and the Evolution of the Wind Tunnel." In Hallion, *NASA's Contributions to Aeronautics* 2, 310–365.

Klinger, R.T. "YF-12A Flight Test Sonic Boom Measurements." Lockheed Advanced Development Projects Report SP-815 (June 1965).

Klos, Jacob, and R.D. Bruel. "Vibro-Acoustical Response of Buildings Due to Sonic Boom Exposure: June 2006 Field Test." NASA TM 2007-214900 (September 2007).

Klos, Jacob. "Vibro-Acoustic Response of Buildings Due to Sonic Boom Exposure: July 2007 Field Test." NASA TM-2008-215349 (September 2008).

Komadina, Steve. "Quiet Supersonic Platform (QSP)." PowerPoint presentation. FAA Civil Supersonic Aircraft Technical Workshop, Arlington, VA, November 13, 2003.

Kroo, Ilan, and Alex VanDerVelden. "The Sonic Boom of an Oblique Flying Wing." AIAA paper no. 90-4002 (October 1990).

Kroo, Ilan. "Sonic Boom of the Oblique Flying Wing." *Journal of Aircraft* 31, no. 1 (January–February 1994): 19–25.

Lee, Christopher A. "Design and Testing of Low Sonic Boom Configurations and an Oblique All-Wing Supersonic Transport." NASA CR-197744 (February 1995).

Lee, Robert E., and J. Micah Downing. "Boom Event Analyzer Recorder: the USAF Unmanned Sonic Boom Monitor." AIAA paper no. 93-4431 (October 1993).

Levine, Jay. "Lowering the Boom." *X-Press*, July 29, 2005. *http://www.nasa.gov/centers/dryden/news/X-Press/stories/2005/072905.*

Lina, Lindsay J., and Domenic J. Maglieri. "Ground Measurements of Airplane Shock-Wave Noise at Mach Numbers to 2.0 and at Altitudes to 60,000 Feet." NASA TN-D-235 (March 1960).

Lipkens, Bart, and David T. Blackstock. "Model Experiments to Study the Effects of Turbulence on Risetime and Waveform of N Waves." In Darden, *1992 Sonic Boom Workshop* 1, 97–108.

Lomax, Harvard. "Preliminary Investigation of Flow Field Analysis on Digital Computers with Graphic Display." In Schwartz, *Second Conference on Sonic Boom Research*, 67–72.

Lundberg, B.K.O. "Aviation Safety and the SST." *Astronautics and Aeronautics* 3, no. 1 (January 1966): 25–30.

Lung, J.L., B. Tiegerman, N.J. Yu, and A.R. Seebass. "Advances in Sonic Boom Theory." In *Aerodynamic Analyses Requiring Advanced Computers*, pt. 2, 1033–1047.

Lux, David, L.J. Ehernberger, Timothy R. Moes, and Edward A. Haering. "Low-Boom SR-71 Modified Signature Demonstration Program." In McCurdy, *1994 Sonic Boom Workshop: Configuration, Design, Analysis and Testing*, 237–248.

MacDonald, G.J., S.M. Flatte, R.L. Garwin, and F.W. Perkins. "Jason 1978 Sonic Boom Report." SRI International, Arlington, VA. JSR-78-09 (November 1978).

Mack, Robert J. "A Whitham-Theory Sonic-Boom Analysis of the TU-144 Aircraft at Mach Number of 2.2." In McCurdy, *1995 Sonic Boom Workshop* 2, 1–16.

Mack, Robert J. "Wind-Tunnel Overpressure Signatures from a Low-Boom HSCT Concept with Aft-Mounted Engines." In McCurdy, *1994 Sonic Boom Workshop: Configuration Design, Analysis, and Testing*, 59–70.

Mack, Robert J., and Christine M. Darden. "Limitations on Wind-Tunnel Pressure Signature Extrapolation." In Darden, *1992 Sonic Boom Workshop*, 2, 201–220.

Mack, Robert J., and Christine M. Darden. "Some Effects of Applying Sonic Boom Minimization to Supersonic Cruise Aircraft Design." AIAA paper no. 79-0652 (March 1979).

Mack, Robert J., and Christine M. Darden. "Wind-Tunnel Investigation of the Validity of a Sonic-Boom-Minimization Concept." NASA TP-1421 (October 1979).

Mack, Robert J., and Kathy E. Needleman. "A Methodology for Designing Aircraft to Low Sonic Boom Constraints." NASA TM-4246 (February 1996).

Maglieri, Domenic J. "A Brief Review of the National Aero-Space Plane Sonic Booms Final Report." USAF Aeronautical Systems Center. TR-94-9344 (December 1992).

Maglieri, Domenic J. "Compilation and Review of Supersonic Business Jet Studies from 1963 through 1995," NASA CR-2011-217144 (May 2011).

Maglieri, Domenic J. "Sonic Boom Research: Some Effects of Airplane Operations and the Atmosphere on Sonic Boom Signatures." NASA SP-147 (1967).

Maglieri, Domenic J., and Donald L. Lansing. "Sonic Booms from Aircraft in Maneuvers." NASA TN D-2370 (July 1964).

Maglieri, Domenic J., and Garland J. Morris. "Measurement of Response of Two Light Airplanes to Sonic Booms." NASA TN D-1941 (August 1963).

Maglieri, Domenic J., and Harry W. Carlson. "The Shock-Wave Noise Problem of Supersonic Aircraft in Steady Flight." NASA Memo 3-4-59L (April 1959).

Maglieri, Domenic J., and Harvey H. Hubbard. "Ground Measurements of the Shock-Wave Noise from Supersonic Bomber Airplanes in the Altitude Range from 30,000 to 50,000 Feet." NASA TN-D-880 (July 1961).

Maglieri, Domenic J., and Percy J. Bobbitt. "History of Sonic Boom Technology Including Minimization." Hampton, VA: Eagle Aeronautics, November 2001.

Maglieri, Domenic J., and V.S. Richie. "In-Flight Shock-Wave Measurements Above and Below a Bomber Airplane at Mach Numbers from 1.42 to 1.69." NASA TN-D-1968 (October 1963).

Maglieri, Domenic J., David A. Hilton, Vera Huckel, Herbert R. Henderson, and Norman J. McLeod. "Measurements of Sonic Boom Signatures from Flights at Cutoff Mach Number." In Schwartz, *Third Conference on Sonic Boom Research*, 243–254.

Maglieri, Domenic J., Harvey H. Hubbard, and Donald L. Lansing, "Ground Measurements of the Shock-Wave Noise from Airplanes in Level Flight at Mach Numbers to 1.4 and Altitudes to 45,000 Feet." NASA TN D-48 (September 1959).

Maglieri, Domenic J., J.O. Powers, and J.M. Sands. "Survey of United States Sonic Boom Overflight Experimentation." NASA TM-X-66339 (May 1969).

Maglieri, Domenic J., Percy J. Bobbitt, Steven J. Massey, Kenneth J. Plotkin, Osama A. Kandil, and Xudong Zheng. "Focused and Steady-State Characteristics of Shaped Sonic Boom Signatures: Prediction and Analysis." NASA CR-2011-217156 (June 2011).

Maglieri, Domenic J., Vera Huckel, and H.R. Henderson. "Sonic Boom Measurements for SR-71 Aircraft Operating at Mach Numbers to 3.0 and Altitudes to 24834 Meters." NASA TN D-6823 (September 1972).

Maglieri, Domenic J., Vera Huckel, and Tony L. Parrott. "Ground Measurements of Shock-Wave Pressure for Fighter Airplanes Flying at Very Low Altitudes." NASA TN D-3443 (July 1966).

Maglieri, Domenic J., Victor E. Sothcott, and Thomas N. Keefer. "A Summary of XB-70 Sonic Boom Signature Data, Final Report." NASA CR 189630 (April 1992).

Maglieri, Domenic J., Victor E. Sothcott, and Thomas N. Keefer. "Feasibility Study on Conducting Overflight Measurements of Shaped Sonic Boom Signatures Using the Firebee BQM-34E RPV." NASA CR-189715 (February 1993).

Maglieri, Domenic J., Victor E. Sothcott, Thomas N. Deffer, and Percy J. Bobbitt. "Overview of a Feasibility Study on Conducting Overflight Measurements of Shaped Sonic Boom Signatures Using RPV's." In Whitehead, *1991 HSR Workshop*, pt. 2, 787–807.

Maglieri, Domenic J., Victor E. Sothcroft, and John Hicks. "Influence of Vehicle Configurations and Flight Profile on X-30 Sonic Booms." AIAA paper no. 90-5224 (October 1990).

Malkin, Myron S., ed. "Environmental Impact Statement: Space Shuttle Program (Final)," Headquarters NASA (April 1978).

Marconi, Frank, Larry Yeager, and H. Harris Hamilton. "Computation of High-Speed Inviscid Flows About Real Configurations." In *Aerodynamic Analyses Requiring Advanced Computers*, pt. 2, 1411–1453.

Marconi, Frank, Manuel Salas, and Larry Yeager. "Development of Computer Code for Calculating the Steady Super/Hypersonic Inviscid Flow Around Real Configurations." NASA CR-2675 (April 1976).

Mascitti, Vincent R. "A Preliminary Study of the Performance and Characteristics of a Supersonic Executive Aircraft." NASA TM-74055 (September 1977).

Maurice, Lourdes. "Civil Supersonic Aircraft Advanced Noise Research." PowerPoint presentation. FAA Public Meeting on Advanced Technologies and Supersonics, Washington, DC, July 14, 2011.

McAnich, Gerry L. "Atmospheric Effects on Sonic Boom—A Program Review." In Darden, *1991 HSR Workshop*, pt. 1, 1201–1207.

McCurdy, David A. and Sherilyn A. Brown. "Subjective Response to Simulated Sonic Boom in Homes." In Blaize, *1995 Sonic Boom Workshop* 1, 278–297.

McCurdy, David A., Sherilyn A. Brown, and R. David Hilliard. "An In-Home Study of Subjective Response to Simulated Sonic Booms." In McCurdy, *1994 Sonic Boom Workshop: Atmospheric Propagation and Acceptability*, 193–207.

McLean, F. Edward, and Barrett L. Shrout. "A Wind Tunnel Study of Sonic-Boom Characteristics for Basic and Modified Models of a Supersonic Transport Configuration." NASA TM X-1236 (May 1966).

McLean, F. Edward, and Barrett L. Shrout. "Design Methods for Minimization of Sonic Boom Pressure-Field Disturbances." *Journal of the Acoustical Society of America* 39, no. 5, pt. 2 (November 1966): 519–525.

McLean, F. Edward, Raymond L. Barger, and Robert J. Mack. "Application of Sonic-Boom Minimization Concepts in Supersonic Transport Design." NASA TN-D-7218 (June 1973).

McLean, F. Edward, Robert J. Mack, and Odell A. Morris. "Sonic Boom Pressure-Field Estimation Techniques," *Journal of the Acoustical Society of America* 39, no. 5, pt. 2 (November 1966): 510–518.

McLean, F. Edward. "SCAR Program Overview," In *Proceedings of the SCAR Held at Langley Research Center, Hampton, Virginia, November 9–12, 1976*, pt. 1, 1–4.

McLean, F. Edward. "Some Nonasymptotic Effects of the Sonic Boom of Large Airplanes." NASA TN D-2877 (June 1965).

Meredith, Keith. "SSBE—Flight Test to CFD Comparison." PowerPoint presentation. Langley Research Center, Hampton, VA, August 17, 2004.

Meredith, Keith., John A. Dahlin, David H. Graham, Michael B. Malone, Edward A. Haering, Juliet A. Page, and Kenneth J. Plotkin. "Computational Fluid Dynamics Comparison and Flight Test Measurement of F-5E Off-Body Pressures." AIAA paper no. 2005-6 (January 2005).

Miller, David S., and Harry W. Carlson. "A Study of the Application of Heat or Force Fields to the Sonic-Boom-Minimization Problem." NASA TN D-5582 (December 1969).

Miller, Denise M., and Victor W. Sparrow. "Assessing Sonic Boom Responses to Changes in Listening Environment, Signature Type, and Testing Methodology." *Journal of the Acoustical Society of America* 127, issue 3 (2010): 1898.

Miranda, L.R., R.D. Elliott, and W.M. Baker. "A Generalized Vortex Lattice Method for Subsonic and Supersonic Flow Applications." NASA CR-2895 (December 1977).

Moes, Timothy R. "Objectives and Flight Results of the Lift and Nozzle Change Effects on Tail Shock (LaNCETS) Project." PowerPoint presentation. International Test & Evaluation Association, Antelope Valley Chapter, February 24, 2009.

Mola, Roger A. "This Is Only a Test." *Air & Space Magazine*, March–April 2006. *http://www.airspacemag.com/history-of-flight/this-is-only-a-test.html*.

Morgan, Daniel, and Carl E. Behrens. "National Aeronautics and Space Administration: Overview, FY2009 Budget, and Issues for Congress." Congressional Research Service, February 26, 2008.

Morgenstern, John M. "Low Sonic Boom Design and Performance of a Mach 2.4/1.8 Overland High-Speed Civil Transport." In Darden, *1992 Sonic Boom Workshop* 2, 55–63.

Morgenstern, John M., Alan Arslan, Victor Lyman, and Joseph Vadyak. "F-5 Shaped Sonic Boom Persistence of Boom Shaping Reduction Through the Atmosphere." AIAA paper no. 2005-12 (January 2005).

Morgenstern, John M., David D. Bruns, and Peter P. Camacho. "SR-71A Reduced Sonic Boom Modification Design." In McCurdy, *1994 Sonic Boom Workshop: Configuration, Design, Analysis, and Testing*, 199–217.

Morgenstern, John M., Nicole Norstrud, Marc Stelnack, and Craig Skoch. "Final Report for the Advanced Concept Studies for Supersonic Commercial Transports Entering Service in the 2030 to 2035 Period, N+3 Supersonic Program." NASA CR-2010-216796 (October 2010).

Morring, Frank. "Rudderless: NASA Aeronautics Chief Hopes New Administrator Will Push for Clear Policy To Guide Facility Closings." *Aviation Week*, March 7, 2005, 28–30.

Morris, Jefferson. "Quiet, Please: With More Emphasis on Partnering, NASA Continues Pursuit of Quieter Aircraft." *Aviation Week*, June 25, 2007, 57–58.

NASA, Langley Research Center. "Proceedings of NASA Conference on Supersonic-Transport Feasibility Studies and Supporting Research, September 17–19, 1963." NASA TM X-905 (December 1963).

Nixon, Charles W., and Harvey H. Hubbard. "Results of the USAF-NASA-FAA Flight Program To Study Community Response to Sonic Booms in the Greater St. Louis Area." NASA-TN-D-2705 (May 1965).

Nixon, Charles W., H.K. Hille, H.C. Somner, and E. Guild. "Sonic Booms Resulting from Extremely Low Altitude Supersonic Flight: Measurements and Observations on Houses, Livestock, and People." AMRL TR-68-52 (October 1968).

Norris, Guy, and Graham Warwick. "Sound Barrier," *Aviation Week*, June 4/11, 2012, 50–53.

Norris, Stephen R., Edward A. Haering, and James E. Murray. "Ground-Based Sensors for the SR-71 Sonic Boom Propagation Experiment." *1995 Sonic Boom Workshop* 1, 199–218.

Northrop Grumman Corporation. "Shaped Sonic Boom Demo Flight Test Program." PowerPoint presentation. NASA Langley Research Center, Hampton, VA, August 17, 2004.

Page, Juliet A., and K.J. Plotkin. "An Efficient Method for Incorporating Computational Fluid Dynamics into Sonic Boom Theory." AIAA paper no. 91-3275 (September 1991).

Pawlowski, Joseph W., David H. Graham, Charles H. Boccadoro, Peter G. Coen, and Domenic J. Maglieri. "Origins and Overview of the Shaped Sonic Boom Demonstration Program." AIAA paper no. 2005-5 (January 2005).

Phillips, Edward H. "Shock Wave: Flying Faster than Sound Is the Holy Grail of Business Aviation." *Aviation Week*, October 8, 2007, 50–51.

Pierce, Allan D., and Victor W. Sparrow. "Relaxation and Turbulence Effects on Sonic Boom Signatures." In Whitehead, *First Annual HSR Workshop*, pt. 1, 1211–1234.

Plattner, C.M. "XB-70A Flight Research: Phase 2 To Emphasize Operational Data." *Aviation Week*, June 13, 1966, 60–62.

Plotkin, Kenneth J. "Perturbations Behind Thickened Shock Waves." In Schwartz, *Third Conference on Sonic Boom Research*, 59–66.

Plotkin, Kenneth J. "State of the Art of Sonic Boom Modeling." *Journal of the Acoustical Society of America* 111, no. 1, pt. 3 (January 2002): 530–536.

Plotkin, Kenneth J. "The Effect of Turbulence and Molecular Relaxation on Sonic Boom Signatures." In Whitehead, *1991 HSR Workshop*, pt. 1, 1241–1261.

Plotkin, Kenneth J. "Theoretical Basis for Finite Difference Extrapolation of Sonic Boom Signatures." In Baize, *1995 Sonic Boom Workshop* 1, 54–67.

Plotkin, Kenneth J., and Domenic J. Maglieri. "Sonic Boom Research: History and Future." AIAA paper no. 2003-3575 (June 2003).

Plotkin, Kenneth J., and Fabio Grandi. "Computer Models for Sonic Boom Analysis: PCBoom4, CABoom, BooMap, CORBoom." Wyle Report WR-02-11 (June 2002).

Plotkin, Kenneth J., and J.M. Cantril. "Prediction of Sonic Boom at a Focus." AIAA paper no. 76–72 (January 1976).

Plotkin, Kenneth J., and Roy Martin. "NASA Shaped Sonic Boom Experiment: Pushover Focus Maneuver, Final Data Review." PowerPoint presentation. NASA Langley Research Center, Hampton, VA, August 17, 2004.

Plotkin, Kenneth J., C.L. Moulton, V.R. Desai, and M.J. Lucas. "Sonic Boom Environment Under a Supersonic Military Operating Area." *Journal of Aircraft* 29, no. 6 (November–December 1992): 1069–1072.

Plotkin, Kenneth J., Domenic J. Maglieri, and Brenda M. Sullivan. "Measured Effects of Turbulence on the Loudness and Waveforms of Conventional and Shaped Minimized Sonic Booms." AIAA paper no. 2005-2949 (May 2005).

Plotkin, Kenneth J., Edward A. Haering, James E. Murray, Domenic J. Maglieri, Joseph Salamone, Brenda M. Sullivan, and David Schein. "Ground Data Collection of Shaped Sonic Boom Experiment Aircraft Pressure Signatures." AIAA paper no. 2005-10 (January 2005).

Plotkin, Kenneth J., Jason R. Matisheck, and Richard R. Tracy. "Sonic Boom Cutoff Across the United States." AIAA paper no. 2008-3033 (May 2008).

Plotkin, Kenneth J., Juliet Page, and Michah Downing. "USAF Single-Event Sonic Boom Prediction Model: PCBoom3." In McCurdy, *1994 Sonic Boom Workshop: Atmospheric Propagation and Acceptability Studies*, 171–184.

Plotkin, Kenneth J., Juliet Page, David H. Graham, Joseph W. Pawlowski, David B. Schein, Peter G. Coen, David A. McGurdy, Edward A. Haering, James E. Murray, L.J. Ehernberger, Domenic J. Maglieri, Percy J. Bobbitt, Anthony Pilon, and Joe Salamone. "Ground Measurements of a Shaped Sonic Boom." AIAA paper no. 2004-2923 (May 2004).

Plotkin, Kenneth J., Roy Martin, Domenic J. Maglieri, Edward A. Haering, and James E. Murray. "Pushover Focus Booms from the Shaped Sonic Boom Demonstrator." AIAA paper no. 2005-11 (January 2005).

Poling, Hugh W. "Sonic Boom Propagation Codes Validated by Flight Test." NASA CR-201634 (October 1996).

Porter, Lisa J. "NASA's Aeronautics Program." PowerPoint presentation. Fundamental Aeronautics Annual Meeting, New Orleans, October 30, 2007. *http://www.aeronautics.nasa.gov/fap/PowerPoints/ARMD&FA_Intro.pdf*.

Raspet, Richard, Henry Bass, Lixin Yao, and Wenliang Wu. "Steady State Risetimes of Shock Waves in the Atmosphere." In Darden, *1992 Sonic Boom Workshop* 1, 109–116.

Roberds, Richard M. "Sonic Boom and the Supersonic Transport." *Air University Review* 22, no. 7 (July–August 1971): 25–33.

Runyan, Larry J., and Edward J. Kane. "Sonic Boom Literature Survey2, Capsule Summaries." Boeing Commercial Airplane Company for the FAA. DTIC AD-771274 (September 1973).

Sands, Johnny M. "Sonic Boom Research (1958–1968)." FAA DTIC AD 684806 (November 1968).

Schultz, James. "HSR Leaves Legacy of Spinoffs." *Aerospace America* 37, no. 9 (September 1999): 28–32.

Schwartz, Ira R. "Sonic Boom Simulation Facilities," In AGARD, *Aircraft Engine Noise and Sonic Boom*, paper 29.

Seebass, A. Richard, and A.R. George. "Sonic Boom Minimization Through Aircraft Design and Operation." AIAA paper no. 73-241 (January 1973).

Seebass, A. Richard, and A.R. George. "Design and Operation of Aircraft To Minimize their Sonic Boom." *Journal of Aircraft* 11, no. 9 (September 1974): 509–517.

Seebass, A. Richard, and Brian Argrow. "Sonic Boom Minimization Revisited." AIAA paper no. 98-2956 (November 1998).

Seebass, A. Richard. "Comments on Sonic Boom Research." In Schwartz, *Third Conference on Sonic Boom Research*, 411.

Seebass, A. Richard. "History and Economics of, and Prospects for, Commercial Supersonic Transport." Paper 1 in *NATO Research and Technology Organization, Fluid Dynamics Research on Supersonic Aircraft*. Proceedings of a course held in Rhode Saint-Genèse, Belgium, May 25–29, 1998. RTO-EN-4 (November 1998).

Seebass, A. Richard. "Nonlinear Acoustic Behavior at a Caustic." In Schwartz, *Third Conference on Sonic Boom Research*, 87–122.

Seebass, A. Richard. "Sonic Boom Minimization." Paper 6 in *NATO Research and Technology Organization, Fluid Dynamics Research on Supersonic Aircraft*. Proceedings of a course held in Rhode Saint-Genèse, Belgium, May 25–29, 1998. RTO-EN-4 (November 1998).

Shapely, Deborah. "East Coast Mystery Booms: A Scientific Suspense Tale." *Science* 199, no. 4336 (March 31, 1978): 1416–1417.

Shepherd, Kevin P. "Overview of NASA Human Response to Sonic Boom Program." In Whitehead, *1991 HSR Workshop*, pt. 3, 1287–1291.

Shepherd, Kevin P., Brenda M. Sullivan, Jack E. Leatherwood, and David A. McCurdy. "Sonic Boom Acceptability Studies." In Whitehead, *1991 HSR Workshop*, pt. 3, 1295–1311.

Shin, Jaiwon. "ARMD Overview." PowerPoint presentation, NASA Aeronautics Fundamental Aeronautics Program Technical Conference, Cleveland, OH, March 13, 2012.

Siclari, M.J. "Ground Extrapolation of Three-Dimensional Near-Field CFD Predictions for Several HSCT Configurations," In Darden, *1992 Sonic Boom Workshop* 2, 175–200.

Siclari, M.J. "Sonic Boom Predictions Using a Modified Euler Code." *1991 HSR Workshop*, pt. 2, 760–784.

Siclari, M.J. "The Analysis and Design of Sonic Boom Configurations Using CFD and Numerical Optimization Techniques." In McCurdy, *1994 Sonic Boom Workshop: Configuration Design, Analysis, and Testing,* 107–128.

Smith, Harriet J. "Experimental and Calculated Flow Fields Produced by Airplanes Flying at Supersonic Speeds." NASA TN-D-621 (November 1960).

Smolka, James W., Leslie Molzahn, and Robbie Cowart. "Flight Testing of the Gulfstream Quiet Spike on a NASA F-15B." PowerPoint presentation. December 11, 2008.

Smolka, James W., Robert A. Cowart, and nine others. "Flight Testing of the Gulfstream Quiet Spike on a NASA F-15B." Paper presented to the Society of Experimental Test Pilots, Anaheim, CA, September 27, 2007.

Spearman, M. Leroy. "The Evolution of the High-Speed Civil Transport." NASA TM-109089 (February 1994).

Spivey, Natalie D., Claudia Y. Herrera, Roger Turax, Chan-gi Pak, and Donald Freund. "Quiet Spike Build-up Ground Vibration Testing Approach." NASA TN 2007-214625 (November 2007).

Stanford Research Institute. "Sonic Boom Experiments at Edwards Air Force Base; Interim Report." National Sonic Boom Evaluation Office. DTIC AD 0655310 (July 1967).

Stuart, William G. "Northrop F-5 Case Study in Aircraft Design." AIAA (September 1978).

Sturgielski, R.T., L.E. Fugelso, L.B. Holmes, and W.J. Byrne. "The Development of a Sonic Boom Simulator with Detonable Gases." NASA CR-1844 (November 1971).

Sullivan, Brenda M. "Design of an Indoor Sonic Boom Simulator at NASA Langley Research Center." Paper presented at Noise-Con 2008, Baltimore, MD, July 12–14, 2008.

Sullivan, Brenda M. "Metrics for Human Response to Sonic Booms." PowerPoint presentation. FAA Civil Supersonic Aircraft Technical Workshop, Arlington, VA, November 13, 2003.

Sullivan, Brenda M. "Research at NASA on Human Response to Sonic Booms." Paper presented at 5th International Conference on Flow Dynamics, Sendai, Japan, November 17–20, 2008.

Sullivan, Brenda M. "Research on Subjective Response to Simulated Sonic Booms at NASA Langley Research Center." Paper presented at International Sonic Boom Forum, State College, PA, July 21–22, 2005.

Sullivan, Brenda M. "Sonic Boom Modeling Technical Challenges." CASI document no. 200700363733 (October 2007).

Sullivan, Brenda M. Jacob Klos, Ralph D. Buehrle, David A. McCurdy, and Edward A. Haering. "Human Response to Low-Intensity Sonic Booms Heard Indoors and Outdoors." NASA TM-2010-216685 (April 2010).

Sweetman, Bill. "Back to the Bomber." *Jane's International Defence Review* 37, no. 6 (June 2004): 54–59.

Sweetman, Bill. "Whooshhh!" *Popular Science*. Posted on July 30, 2004, accessed on November 13, 2011. *http//www.popsci.com/military-aviation-space/article/2004-07/whooshhh*.

Taylor, Albion D. "The TRAPS Sonic Boom Program." NOAA Technical Memorandum ERL ARL-87 (July 1980).

Taylor, J.P., and E.G.R, "A Brief Legal History of the Sonic Boom in America." In *Aircraft Engine Noise and Sonic Boom*. Conference paper no. 42. NATO Advisory Group for Aerospace Research and Development, Neuilly sur Seine, France (1969).

Thomas, Charles L. "Extrapolation of Sonic Boom Pressure Signatures by the Waveform Parameter Method." NASA TN D-6823 (June 1972).

Tirpak, John. "The Bomber Roadmap." *Air Force Magazine* 82, no. 6 (June 1999): 30–36.

Tomboulian, Roger. "Research and Development of a Sonic Boom Simulation Device." NASA CR-1378 (July 1969).

TRACOR, Inc. "Public Reactions to Sonic Booms." NASA CR-1665 (September 1970).

Tracy, Richard. "Aerion Supersonic Business Jet." PowerPoint presentation. FAA Public Meeting on Advanced Technologies and Supersonics, Washington, DC, July 14, 2011.

Tu, Eugene, Samson Cheung, and Thomas Edwards. "Sonic Boom Prediction Exercise: Experimental Comparisons." In McCurdy, *1994 Sonic Boom Workshop: Configuration Design, Analysis, and Testing*, 13–32.

Unger, George. "HSR Community Noise Reduction Technology Status Report." In Whitehead, *1991 HSR Workshop*, pt. 1, 259–283.

Wall, Robert. "New Technologies in Quest of Quiet Flight." *Aviation Week*, January 8, 2001, 61–62.

Wallace, Lane E. "The Whitcomb Area Rule: NACA Aerodynamics Research and Innovation." Chapter 5 in Pamela E. Mack, ed. *From Engineering Science to Big Science*. NASA SP-4219 (1998). *http://history.nasa.gov/ SP-4219.htm*.

Warwick, Graham, William Garvey, Joseph C. Anselmo, and Robert Wall. "Open Season: As if the Economy Were Not Enough, Business Aviation Becomes a Scapegoat for Executive Excess." *Aviation Week*, March 2, 2009, 20–22.

Warwick, Graham. "Cut to the Bone." *Aviation Week*, February 27, 2012, 33.

Warwick, Graham. "DARPA Kills Oblique Flying Wing." *Aviation Week*, October 1, 2008. *http://www.aviationweek.com/aw/generic/story. jsp?id=news/OBLI10018c.xml*.

Warwick, Graham. "Fight for Flight." *Aviation Week*, April 23/30, 2012, 52–54.

Warwick, Graham. "Forward Pitch." *Aviation Week*, October 20, 2008, 22–23.

Warwick, Graham. "Quiet Progress: Aircraft Designers Believe They Can Take the Loud Boom out of Supersonic Travel." *Flight International*, October 20, 2004, 32–33.

Warwick, Graham. "Sonic Dreams." *Flight International*, May 6, 2003, 34–37.

Warwick, Graham. "Sonic Overture: European Researchers Test the Waters on International Supersonic Collaboration." *Aviation Week*, July 13, 2009, 53–54.

Wedge, Harry R., and 13 coauthors. "N+2 Supersonic Concept Development and System Integration." NASA CR-2010-216842 (August 2010).

Wedge, Robert H. and 20 co-authors. "N Plus 3 Advanced Concept Studies for Supersonic Commercial Transport Aircraft Entering Service in the 2030–2035 Period." NASA CR-2011-217084 (April 2011).

Weinstein, Leonard M. "An Electronic Schlieren Camera for Aircraft Shock Wave Visualization." In Baize, *1995 Sonic Boom Workshop* 1, 244–258.

Weinstein, Leonard M. "An Optical Technique for Examining Aircraft Shock Wave Structures in Flight." In McCurdy, *1994 Sonic Boom Workshop, Atmospheric Propagation and Acceptability*, 1–18.

Wenk, Edward. "SST—Implications of a Political Decision." *Astronautics and Aeronautics* 9, no. 10 (October 1971): 40–49.

Whitham, G.B. "On the Propagation of Weak Shock Waves." *Journal of Fluid Dynamics* 1, no. 3 (September 1956): 290–318. *http://journals. cambridge.org/action/displayJournal?jid=JFM*.

Whitham, G.B. "The Flow Pattern of a Supersonic Projectile." *Communications on Pure and Applied Mathematics* 5, no. 3 (1952). *http:// www3.interscience.wiley.com/journal/113395160/issue*.

Wiley, John. "The Super-Slow Emergence of Supersonic, Sixty Years after Glamorous Glennis Made History." *Business & Commercial Aviation* 101, no. 3 (September 2007): 48–50.

Willshire, William L., and David Chestnut, eds. "Joint Acoustic Propagation Experiment (JAPE-91) Workshop." NASA CR 3231 (1993).

Willshire, William L., and David W. DeVilbiss. "Preliminary Results from the White Sands Missile Range Sonic Boom Propagation Experiment." In Darden, *1992 Sonic Boom Workshop* 1, 137–144.

Wilson, J.B. "Quiet Spike: Softening the Sonic Boom." *Aerospace America* 45, no. 10 (October 2007): 38–42.

Wlezien, Richard, and Lisa Veitch. "Quiet Supersonic Platform Program." AIAA paper no. 2002-0143 (January 2002).

Yao, Lixin, Henry E. Bass, Richard Raspert, and Walter E. McBride. "Statistical and Numerical Study of the Relation Between Weather and Sonic Boom." In Whitehead, *1991 HSR Workshop*, pt. 3, 1263–1284.

Web Sites

Aerion Corporation. Accessed October 12, 2011. *http://aerioncorp.com/*.

Aerographer/Meteorology. Table 1-6, U.S. Standard Atmosphere Heights and Temperatures. Accessed August 27, 2011. *http://www.tpub.com/content/aerographer/14269/css/14269_75.htm*.

Airpower: Missiles and Rockets in Warfare. Accessed January 15, 2009. *http://www.centennialofflight.gov/essay/Air_Power/Missiles /AP29.htm*.

BBC. In Depth Farewell to Concorde. Last updated August 15, 2007. Accessed September 21, 2011. *http://news.bbc.co.uk/1/hi/in_depth/uk/2003/concorde_retirement/*.

Call for Information on Supersonic Aircraft Noise. *Federal Register*, May 23, 20003. *http://www.federalregister.gov/articles/2003/05/1308/*.

Department of Defense Appropriations Act 2001, Public Law 106-259, 114
Stat. 656 (August 9, 2001). Accessed November 13, 2011. *http://www.
gpo.gov/fdsys/pkg/PLAW-106publ259/pdf.*

Department of Defense Fiscal Year (FY) 2004/FY 2005 Biennial Budget
Estimates: Research, Development, Test and Evaluation Defense-Wide
1. Defense Advanced Research Projects Agency (February 2003).
Accessed June 29, 2011. *http://www.darpa.mil/WorkArea/DownloadAsset.
aspx?id=1635.*

Department of Defense Fiscal Year 2005 Biennial Budget Estimates:
Research, Development, Test and Evaluation Defense-Wide 1. Defense
Advanced Research Projects Agency (February 2004). Accessed June 29,
2011. *http://www.darpa.mil/WorkArea/DownloadAsset.aspx?id=1634.*

Department of Defense FY 2001 Budget Estimates: Research, Development,
Test and Evaluation Defense-Wide 1. Defense Advanced Research
Projects Agency (February 2000). Accessed June 29, 2011. *http://www.
darpa.mil/WorkArea/DownloadAsset.aspx?id=1638.*

Department of Defense FY 2002 Amended Budget Submission: Research,
Development, Test and Evaluation Defense-Wide 1. Defense Advanced
Projects Agency (June 2001). Accessed June 29, 2011. *http://www.darpa.
mil/WorkArea/DownloadAsset.aspx?id=1637.*

Department of Defense FY 2003 Budget Submission: Research,
Development, Test and Evaluation Defense-Wide 1, Defense Advanced
Projects Agency (February 2002). Accessed June 29, 2011. *http://
comptroller.defense.gov/defbudget/fy2003/budget_justification/pdfs/03_
RDT_and_E/darpa_vol1.pdf.*

FAA Historical Chronology, 1926–1996. Accessed February 1, 2009. *http://
www.faa.gov/about/media/b-chron.pdf.*

FAA. Supersonic Aircraft Noise. Accessed October 25, 2011. *http://www.faa.
gov/about/office_org/headquarters_offices/apl/noise_emissions/supersonic_air-
craft_noise/.* Contains links to related documents, including "Public
Meetings on Advanced Technologies and Supersonics," held in 2009,
2010, and 2011.

Gulfstream Aerospace Corp. "The History of Gulfstream, 1958–2008." Accessed November 13, 2011. *http://www.gulfstream.com/history/*.

Jones, Robert T. "Adolf Busemann, 1901–1986." *Memorial Tributes* 3. National Academy of Engineering (1989), 62–67. Accessed July 3, 2011. *http://www.nap.edu/openbook.php?record_id=1384*.

Krause, Fred. "Naval Air Station Fallon Adversaries, Part One: VFC-13 Saints." Accessed July 21, 2011. *http://modelingmadness.com/scotts/features/krausevfc13g.htm*.

Marine Fighter Training Squadron 401. Accessed July 21, 2011. *http://www.yuma.usmc.mil/tenantcommands/vmft401.html*.

NASA Aeronautics Research Mission Directorate. 2011 Fundamental Aeronautics Technical Conference Recap. Accessed October 2, 2011. *http://www.aeronautics.nasa.gov/fap/meeting_recap_2011.html*. Contains links to 2011 presentations as well as to 2007, 2008, and 2009 conference recaps and related presentations.

NASA Aeronautics Research Mission Directorate. Fundamental Aeronautics Program Documents Library. Last updated July 25, 2011. Accessed October 30, 2011. *http://www.aeronautics.nasa.gov/fap/documents.html*.

NASA. Budget Request Summaries. FYs 2004–2009. Accessed September 25, 2011. *http://www.nasa.gov/news/budget/index.html*.

Naval Air Station Fallon. Accessed July 21, 2011. *http://www.cnic.navy.mil/Fallon/About/index.htm*.

Northrop F-5. *Wikipedia*. Accessed July 18, 2011. *http://en.wikipedia.org/wiki/Northrop_F-5*.

Northrop Grumman Corp. "F-5 Tiger." Accessed July 18, 2011. *http://www.as.northropgrumman.com/products/f5tiger/index.html*.

Pojman, Paul. "Ernst Mach." *The Stanford Encyclopedia of Philosophy* (*Summer 2011 Edition*), edited by Edward N. Zalta. Accessed November 5, 2011. *http://plato.stanford.edu/archives/sum2011/entries/ernst-mach/*.

Quiet Supersonic Platform (QSP) Systems Studies and Technology Integration. *Commerce Business Daily*, SOL RA 00-48. Posted on CBDNet, August 16, 2000. Accessed August 18, 2011. *http://www. fedmine.us/freedownload/CBD/CBD-2000/CBD-2000-18au00.html.*

Sonic Boom. USAF Fact Sheet. October 2005. *http://www.af.mil/factsheets/ fsID=184.*

Supersonic Business Jet. "Quiet Supersonic Transport (QSST)." Accessed October 10, 2011. *http://www.supersonic-business-jet.com/prototypes/quiet_ supersonic_transport.php.*

Time&Date.com. "Sunrise and Sunset/Moonrise and Moonset in Los Angeles." August 2003. Accessed August 30, 2011. *http://www.timeanddate.com/ worldclock/astronomy.html?n=137&month=8&year=2003.*

University of Colorado. Biography of A. Richard Seebass. Accessed April 27, 2011. *http://www.colorado.edu/aerospace/ARichardSeebass.html.*

Valiant Air Command Warbird Museum. Accessed October 29, 2011. *http:// www.vacwarbirds.org/.*

Weather Warehouse. "Jacksonville Cecil Field Airport, 7/24/03." Accessed August 11, 2011. *https://weather-warehouse.com/WxHubP/ WxSPM531874661_174.28.158.213/1_Jacksonville.*

Windows to the Universe. "Layers of the Earth's Atmosphere." Accessed June 15, 2011. *http://www.windows2universe.org/earth/Atmosphere/layers.html.*

Wlezien, Richard. Biography. Accessed April 30, 2011. *http://www. richwelzein.com/Personal/Wlezien.html.*

XB-70. NASA Dryden Fact Sheet. Accessed March 10, 2009. *http://www. nasa.gov/centers/dryden/new/FactSheets/FS-084-DFRC_prt.htm.*

Acronyms List

ACTIVE	Advanced Control for Integrated Vehicles Experiment
AEDC	Arnold Engineering Development Center
AFFTC	Air Force Flight Test Center
AFRL	Air Force Research Laboratory
AGARD	Advisory Group for Aerospace Research and Development
AIAA	American Institute of Aeronautics and Astronautics
AMRL	Aerospace Medical Research Laboratory
AOA	angle of attack
ARAP	Aeronautical Research Associates of Princeton
ARC3D	Ames Research Center 3-Dimensional (CFD code)
ARMD	Aeronautics Research Mission Directorate
ASA	Acoustical Society of America
AST	Advanced Supersonic Technology
AVTIP	Air Vehicles Technology Integration Program
B&K	Brül and Kjaer
BAC	British Aircraft Corporation
BADS	Boom Amplitude and Direction Sensor
BASS	Boom Amplitude and Shape Sensor
BBC	British Broadcasting Corporation
BEAR	Boom Event Analyzer Recorder
BLM	Bureau of Land Management
BREN	Bare Reactor Experiment Nevada
BUNO	Bureau Number
CAD	computer-aided design
CASA	*Construcciones Aeronauticas Sociedad Anonima*
CASI	Center for AeroSpace Information
CDR	critical design review
CFD	computational fluid dynamics
CFR	Code of Federal Regulations
CP	Conference Publication
CR	Contractor Report
CRADA	Cooperative Research and Development Agreement
DACT	Dissimilar Air Combat Tactics
DAQ	data acquisition

DARPA	Defense Advanced Research Projects Agency
DAT	digital audio tape
dB	decibel
DFRC	Dryden Flight Research Center
DGPS	Differential Global Positioning System
DOD	U.S. Department of Defense
DOT	U.S. Department of Transportation
DTIC	Defense Technical Information Center
EAA	Experimental Aircraft Association
ECS	environmental control system
EIS	environmental impact statement
F	Fahrenheit
FAA	Federal Aviation Agency, later Administration
FAR	Federal Aviation Regulation
FCC	Federal Communications Commission
FCF	functional check flight
FORTRAN	Formula Translation
FRC	Flight Research Center
FRR	flight readiness review
FY	fiscal year
GASL	General Applied Sciences Laboratories
GCNSfv	Generalized Compressible Navier-Stokes Finite Volume (CFD code)
GPO	Government Printing Office
GPS	Global Positioning System
HISAC	French acronym for High-Speed Aircraft Industrial Project
HLFC	hybrid laminar flow control
HPCCP	High Performance Computing and Communications Program
HSCT	High-Speed Civil Transport
HSFRS	High-Speed Flight Research Station
HSR	High-Speed Research
HUD	heads-up display
Hz	hertz
IBM	formerly International Business Machines
ICAO	International Civil Aviation Organization
IDR	interim design review
IFCS	Intelligent Flight Control Systems
IFF	identification friend or foe
IHONORS	In-Home Noise Generation/Response System

IPT	integrated product team
IRIG	Inter-Range Instrumentation Group
IRIS	Information Retrieval and Indexing System
ISSM	Inlet Spillage Shock Measurement
JAPE	Joint Acoustic Propagation Experiment
JASA	*Journal of the Acoustical Society of America*
JAXA	Japan Aerospace Exploration Agency
km	kilometers
kph	kilometers per hour
L/D	lift-to-drag ratio
LaNCETS	Lift and Nozzle Change Effects on Tail Shocks LBEV low-boom experimental vehicle
lb/ft^2	pounds per square foot
LEX	leading edge extension
MCAS	Marine Corps Air Station
MIL-STD	Military Standard
MIM3D-SB	Multigrid Implicit Marching in Three Dimensions for Sonic Booms
MIT	Massachusetts Institute of Technology
MLT	modified linear theory
NACA	National Advisory Committee for Aeronautics
NAS	Naval Air Station
NASP	National Aerospace Plane
NAVAIR	Naval Air Systems Command
NAWS	Naval Air Weapons Station
NBAA	National Business Aviation Association
NGC	Northrop Grumman Corporation
NGSA	Northrop Grumman [facility], St. Augustine
NOAA	National Oceanic and Atmospheric Administration
NORAD	North American Air Defense Command
NRA	NASA Research Announcement
NRL	Naval Research Laboratory
NSBEO	National Sonic Boom Evaluation Office
NSBIT	Noise and Sonic Boom Impact Technology
NTRS	NASA Technical Reports Server
NYU	New York University
OAST	Office of Aeronautics and Space Technology
OEM	original equipment manufacturer
OML	outer mold line
OST	Office of Science and Technology
PAC	Presidential Advisory Committee

PARTNER	Partnership for Air Transportation Noise and Emissions Reduction
PDR	preliminary design review
PDT	Pacific daylight time
PE	program element
PLdB	perceived-level decibel
PNdB	perceived noise decibel
psf	pounds per square foot
QSP	Quiet Supersonic Platform
QSST	Quiet Small Supersonic Transport
R&D	Research and Development
RAF	Royal Air Force
RFI	Request for Information
RFP	Request for Proposal
ROA	Research Opportunities in Aviation
RP	Reference Publication
RPV	remotely piloted vehicle
RQDS	Research Quick Data System
SABER	Small Airborne Boom Event Recorder
SAC	Strategic Air Command
SAI	Supersonic Aerospace International
SASS	Supersonic Acoustics Signature Simulator
SBDWG	Sonic Boom Demonstration Working Group
SCAMP	Superboom Caustic Analysis and Measurement Program
SCAR	Supersonic Cruise Aircraft Research
SCAT	Supersonic Commercial Air Transport
SCIA	Supersonic Cruise Industry Alliance
SCR	Supersonic Cruise Research
SETP	Society of Experimental Test Pilots
SonicBOBS	Sonic Booms on Big Structures
SonicBREWS	Sonic Boom Resistant Earthquake Warning System
SP	Special Publication
SPF/DB	superplastic forming and diffusion bonding
SRI	Stanford Research Institute
SSBD	Shaped Sonic Boom Demonstrator/Demonstration
SSBDWG	Shaped Sonic Boom Demonstration Working Group
SSBE	Shaped Sonic Boom Experiment
SSBJ	supersonic business jet
SST	Supersonic Transport (program)
STOL	short takeoff and landing
STS	Space Transportation System (Shuttle)

TACTS	Tactical Air Combat Training System
TEAC	formerly Tokyo Electro Acoustic Company
TEAM	Three-dimensional Euler/Navier-Stokes Aerodynamic Method
TM	Technical Memorandum
TM-X	formerly classified Technical Memorandum
TN	Technical Note
TP	Technical Paper
TPS	Test Pilot School
TR	Technical Report
TRAPS	Tracing Rays and Aging Pressure Signatures
TYCOM	Type Command (U.S. Navy)
UHF	ultrahigh frequency
UK	United Kingdom
UPS3D	Universal Parabolized Simplified Navier-Stokes Code
USAF	United States Air Force
VCE	variable cycle engine
VFC	Composite Fighter Squadron
VIBES	Variable Intensity Boom Effect on Structures
WSPR	Waveforms and Sonic Booms Perception and Response

About the Author

Lawrence R. Benson attended the University of Maryland at College Park, where he received a master's degree in 1967 specializing in military and diplomatic history. After serving in the U.S. Army with a tour in Vietnam, Mr. Benson became a civilian employee of the Air Force in 1971. During the next 30 years, he worked in a variety of administrative positions and as a historian at 10 locations in the United States, Turkey, and Germany. He has researched and written numerous official histories, monographs, articles, book reviews, and studies on a range of topics related to military operations, international relations, and aerospace technology. After retiring in 2000 as chief of the Air Force Historian's Pentagon Office, he coauthored *Reflections of a Technocrat: Managing Defense, Air, and Space Programs during the Cold War* with former Secretary of the Air Force John. L. McLucas. Most recently, Benson wrote "Softening the Sonic Boom: 50 Years of NASA Research" in *NASA's Contributions to Aeronautics*, a case study that laid the foundation for writing this book. He and his wife, Carolyn, live in Albuquerque, NM.

Index

Numbers in **bold** indicate pages with photos or figures

H

highest airspeed flight test, 224, 241n81

honors and tributes given to participants, 235

lessons learned, 267

near-field shock wave research, 232–33, **232**, **233**

preparations for, 216–21

program management team, 216–17, 288

pushover maneuver and focused booms, 217, 222, 223, 224, 231–32, **231**

QSP program and, 136

reports and technical publications, 300–301

success of, 272

test pilots, 289–90

timeline and program activities, **135**, 136

turbulence and sonic boom research, 217, 225, 229–31, **230**, 232

Working Group formation and membership, 216–17, 229, 288–89

Shepherd, Kevin, 105, 248, 287, 289

Sherman, Mark, 290

Shin, Jaiwon, 258

SHOCKN code, 108

shock waves

acoustic rays, 54, **55**, 56, 64n68, 71, 231, **231**

aircraft design to reduce, x–xi, 6, 9–12, 42–50, 69–72, **72**

air pressure and pressure changes, 3, **3**, 16, **16**

animals, effects on, 12, 76, 77, 107

atmospheric conditions and visibility of, 161–62

bow waves, 3–4, **3**, 16, **16**, 41, 45, 54, 72, 75, **75**, 79, 234–35

CFD and shock wave calculations, 96–98, **97**

characteristics of, 54, 64n68

compression lift and B-70 design, 10

diffraction of into shadow zones, 49

digital images of, 117n63

expansion of and energy dissemination, 54, **55**, 64n68

F-5 shock waves, **286**

far field, 40–41, 44, 71, 108, 234

flattop signature, 47, **47**, 79

in-flight measurement of, 4, 16, **16**, 22, 266

freezing signature through effects in atmosphere, 49

lowering bow and tail waves and sonic boom reduction, 54, 63–64n65

multiple waves, 16, 40

near field, 40–41, 44, 45–46, 49, 71, 103, 106–7, 155, 234–35, **234**

near-field shock wave experiments, 103, 155, 232–33, **232**, **233**

nonlinear shock wave behavior, 49

N-wave signature, 16, **16**, 23–24, **24**

N-wave signature and acceptable noise level, 53–54

N-wave signature and shock cone, 3–4, **3**, 6, 64n68

overpressure measurements, 13–19, 21

overpressure prediction, 57

overpressure reduction solutions, 42, 45–47, 48, 56–58, **57**, 70, 79, **79**

people, effects on, 12, 14, 76, 77

phantom body to eliminate, 57, 69, 76

photos of, 1

prediction of evolution of, 56

quiet sonic boom signature, conditions for, 268

research on, 12

schlieren imaging system photos of, **104**, 105

speed of sound, sonic booms, and, 1–2, 3–5, 27–28nn5–8, 28n10

structures, effects on, 12, 76, 77

successful flattened signature shape, 212, **212**, 213–14, 216, 234–35, **234**

tail waves, 3–4, **3**, 6, 16, **16**, 54, 79

temperature and, 1–2, 18, 27n2, 78